AF388995

Computational Intelligence for Modeling, Control, Optimization, Forecasting and Diagnostics in Photovoltaic Applications

Computational Intelligence for Modeling, Control, Optimization, Forecasting and Diagnostics in Photovoltaic Applications

Editors

Massimo Vitelli
Luigi Costanzo

MDPI • Basel • Beijing • Wuhan • Barcelona • Belgrade • Manchester • Tokyo • Cluj • Tianjin

Editors
Massimo Vitelli
Università degli Studi della Campania Luigi Vanvitelli
Italy

Luigi Costanzo
Università degli Studi della Campania Luigi Vanvitelli
Italy

Editorial Office
MDPI
St. Alban-Anlage 66
4052 Basel, Switzerland

This is a reprint of articles from the Special Issue published online in the open access journal *Energies* (ISSN 1996-1073) (available at: https://www.mdpi.com/journal/energies/special_issues/ Computational_Intelligence_for_Modeling_Control_Optimization_Forecasting_and_Diagnostics_in_ Photovoltaic_Applications).

For citation purposes, cite each article independently as indicated on the article page online and as indicated below:

LastName, A.A.; LastName, B.B.; LastName, C.C. Article Title. *Journal Name* **Year**, *Article Number*, Page Range.

ISBN 978-3-03943-200-4 (Hbk)
ISBN 978-3-03943-201-1 (PDF)

Contents

About the Editors

Massimo Vitelli was born in Caserta, Italy, in 1967. He received the laurea degree (cum laude) in electrical engineering from the University of Naples Federico II, Naples, Italy, in 1992. He is currently a Full Professor at the Department of Engineering, Università degli Studi della Campania "Luigi Vanvitelli," where he teaches electro technics and power electronics. He has been engaged in many scientific national and international projects. He has co-authored several national and international patents. His main research interests include maximum power point tracking techniques in photovoltaic applications, power electronics circuits for renewable energy sources, vibration energy harvesting systems, methods to harvest and store energy from any available source, and methods for analysis and design of switching converters. He is an Associate Editor of the *IEEE Transactions on Power Electronics*.

Luigi Costanzo was born in Villaricca (Napoli), Italy, in 1989. He received his master's degree (cum laude) in electronic engineering from the Second University of Naples, Naples, Italy, in 2014, and Ph.D. degree in industrial and information engineering from the Department of Industrial and Information Engineering, Università degli Studi della Campania "Luigi Vanvitelli," Caserta, Italy, in 2017. He is currently a Research Fellow at the Department of Engineering of Università degli Studi della Campania "Luigi Vanvitelli." He is the co-author of more than 30 Publications and 2 Patents. His main research interests include maximum power point tracking techniques in photovoltaic applications, vibration energy harvesting systems, power electronics circuits for renewable energy sources, methods for analysis and design of switching converters, methods to harvest and store energy from any available source.

Preface to "Computational Intelligence for Modeling, Control, Optimization, Forecasting and Diagnostics in Photovoltaic Applications"

In recent years, a growing number of scientific papers has appeared on computational intelligence (CI) applied to photovoltaic (PV) systems. CI can be profitably used to carry out the following tasks in PV applications: modeling, sizing and control of stand-alone and grid-connected PV systems, centralized maximum power point tracking (MPPT), distributed MPPT, PV arrays reconfiguration, storage sizing, control optimization, detection of mismatching operating conditions, fault diagnosis, maintenance programming, prediction and modeling of solar radiation, and output power plants forecast of PV systems. The aims of the Special Issue are as follows:

- Focus on the latest theoretical studies, numerical algorithms, scientific results, and applications of CI in PV systems;

- Bring together scientists adopting several approaches and working on the above topics;

- Promote and share as much as possible top-level research in the field of CI in PV systems.

Massimo Vitelli, Luigi Costanzo
Editors

Article

A Novel MPPT Technique for Single Stage Grid-Connected PV Systems: T4S

Luigi Costanzo * and Massimo Vitelli

Department of Engineering, Università degli Studi della Campania "Luigi Vanvitelli", 81031 Aversa (CE), Italy; massimo.vitelli@unicampania.it
* Correspondence: luigi.costanzo@unicampania.it; Tel.: +39-501-0212

Received: 21 October 2019; Accepted: 25 November 2019; Published: 26 November 2019

Abstract: In this paper, a novel maximum power point tracking (MPPT) technique, which has been named T4S (a technique based on the proper setting of the sign of the slope of the photovoltaic voltage reference signal), is presented and discussed. It is specifically designed with reference to a single-stage grid-connected PV system. Its performance is numerically compared with that of the well-known and widely used perturb and observe (P&O) MPPT technique. The results of the numerical simulations confirm the validity of the proposed MPPT technique which exhibited a slightly better performance, under stationary and also time-varying irradiance conditions. In addition, the T4S technique is characterized by the following features: it does not require explicit power detection or calculation and, moreover, it allows the tracking of the maximum average power injected into the grid rather than the tracking of the maximum instantaneous power extracted by the PV source.

Keywords: photovoltaic systems; maximum power point tracking; single stage grid connected systems

---

1. Introduction

Maximum power point tracking (MPPT) is one of the main functions carried out by power electronic interfaces that connect photovoltaic (PV) sources to loads and grid. Many MPPT techniques have been presented and discussed in the literature [1–5]. Among them, the perturb and observe (P&O) technique has emerged as one of the simplest, but nonetheless efficient, MPPT techniques and is widely accepted as a benchmark MPPT technique [6–8]. It can be adopted in both single- and double-stage grid-connected PV systems [9,10]. It works by perturbing the operating point of the PV source and observing the corresponding variation of the PV power in order to detect the direction of perturbation of the operating point (increase or decrease of the PV voltage) leading to the maximization of the PV extracted power [11,12]. In this paper, a novel MPPT technique is introduced and analyzed. It has been named T4S (a technique based on the proper setting of the sign of the slope of the photovoltaic voltage reference signal). It is designed with specific reference to a single-stage grid-connected PV system [13–15]. As shown in this paper, the tracking efficiency of the proposed T4S technique is more or less comparable to that of the P&O technique. However, the T4S technique is characterized by the following advantages with respect to the P&O technique: it does not require PV current detection and the power stage efficiency profile is automatically taken into account by tracking the maximum power injected into the grid rather than the maximum power extracted by the PV source.

The rest of the paper is organized as it follows: In Section 2 a typical single-stage grid-connected PV system is described together with the working of the P&O MPPT technique. In Section 3, the proposed T4S MPPT technique is discussed and analyzed in detail. The main parameters affecting its performance are identified and proper guidelines for their choice are provided. In Section 4, numerical results are reported and discussed. In particular, both constant and time-varying irradiance profiles are considered in the numerical simulations. In Section 5, the performance of the proposed T4S MPPT technique is compared with the P&O technique under both stationary and time-varying irradiance conditions. Finally, the conclusions are presented at the end of the paper.

2. Single-Stage Grid-Connected PV System with P&O Controller

A typical single-stage grid-connected PV system is shown in Figure 1 [13–15]. The DC/AC converter is an active full bridge with a control circuitry exhibiting the presence of two control loops, an inner current control loop, and an outer voltage control loop [16,17].

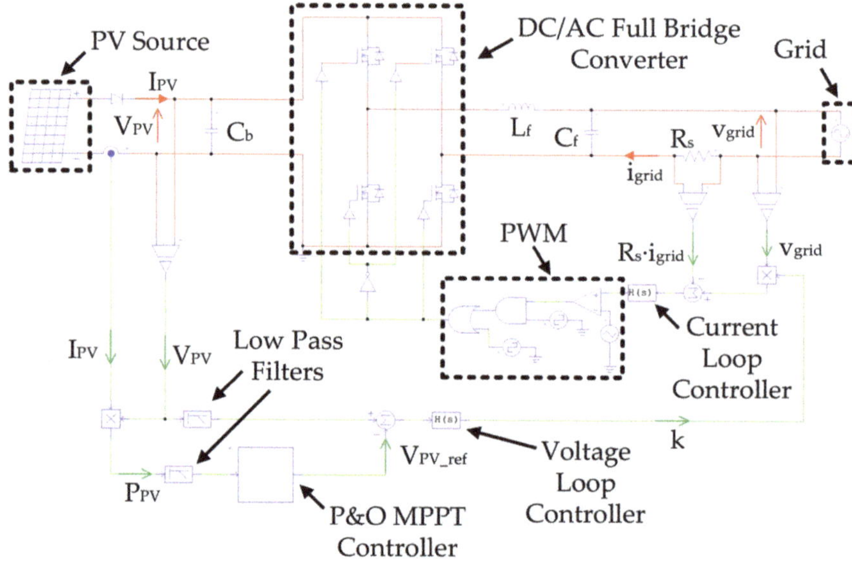

Figure 1. Single-stage grid-connected photovoltaic (PV) system equipped with the perturb and observe (P&O) maximum power point tracking (MPPT) technique.

A fast internal current control loop is aimed at injecting an AC current into the grid with a power factor ideally equal to one. The interested reader can find all the details concerning the proper design of such a control loop in [16]. In this paper, ideal working of the internal current control loop (zero error) is assumed, and therefore the following equation represents the starting point for the analysis:

$$R_s \cdot i_{grid}(t) = k(t) \cdot v_{grid}(t), \tag{1}$$

where R_s is the sensing resistance, $v_{grid}(t) = \sqrt{2} \cdot V_{grid} \cdot \sin(2\pi f \cdot t)$ is the grid voltage, and f is the line frequency. It is worth noting that k(t) is a slowly varying signal [16], hence from (1) it is possible to state that $i_{grid}(t) = \sqrt{2} \cdot I_{grid} \cdot \sin(2\pi f \cdot t)$, that is, $i_{grid}(t)$ and $v_{grid}(t)$ are in phase. The slowly varying signal, k, dictates the value of the average (over the period of the line frequency f) power, P_{grid}, injected into the grid, i.e., the higher the value of k, the higher the active power P_{grid}. In fact, it is

$$P_{grid} = I_{grid} \cdot V_{grid} = \frac{k}{R_s} \cdot V_{grid}^2. \tag{2}$$

A slow external voltage control loop is aimed at the proper regulation of the PV voltage, V_{PV}. The reference voltage, V_{PV_ref}, is provided, as shown in Figure 1, by the MPPT controller [18,19]. It is worth noting, as shown in Figure 1, that two low-pass filters appear. One for the PV voltage, V_{PV}, and the other for the PV power, P_{PV}. Filtering is necessary in order to avoid errors of the MPPT technique, which can be deceived by the presence of harmonics (at the double of the line frequency and at the frequencies of switching harmonics). In addition, the working of the voltage control loop can be negatively affected by such harmonics. Henceforth, for brevity, we refer to V_{PV} and P_{PV} even if we indeed refer to their filtered versions.

As stated in the Introduction, the P&O technique is one of the most exploited MPPT techniques since it is simple but nonetheless very efficient. The P&O algorithm operates by periodically changing, step-by-step, the reference voltage, V_{PV_ref}, and by measuring the corresponding PV power, $P_{PV} = V_{PV} \cdot I_{PV}$ [18]. After each perturbation of V_{PV_ref}, the P&O controller waits for the settling of the system steady-state and, subsequently, compares the new steady-state value of P_{PV} with the older one in order to drive the operating point towards the MPP. In particular, if after a perturbation of V_{PV_ref} the steady-state value of P_{PV} has increased (decreased), it means that the operating point has moved towards (away from) the MPP. Therefore, the next perturbation of V_{PV_ref} will have the same (opposite) sign as the previous perturbation, and so on. In summary, the control law that is implemented in the P&O MPPT controller of Figure 1 is as follows:

$$
\begin{aligned}
V_{PV\_ref}(t)\big|_{(r+1)\cdot T_{P\&O} < t < (r+2)\cdot T_{P\&O}} &= V_{PV\_ref}(t)\big|_{t = r\cdot T_{P\&O}} + \\
+\Big[V_{PV\_ref}(t)\big|_{t = r\cdot T_{P\&O}} - V_{PV\_ref}(t)\big|_{t = (r-1)\cdot T_{P\&O}} \Big] &\cdot \mathrm{sign}\Big\{ P_{PV}(t)\big|_{t = r\cdot T_{P\&O}} - P_{PV}(t)\big|_{t = (r-1)\cdot T_{P\&O}} \Big\},
\end{aligned}
\tag{3}
$$

where r is a growing integer index starting from 0 ($r = 0, 1, 2, \dots$). It is worth noting that in order to evaluate the PV extracted power, P_{PV}, appearing in (3), the PV current detection is necessary. Moreover, the objective of the P&O controller is the maximization of the power extracted by the PV source. Since the power stage efficiency $\eta_{power\_stage}(V_{PV}, P_{grid})$ [20–25] is not taken into account, the maximization of the PV power does not necessarily correspond to the maximization of the power $P_{grid} = P_{PV} \cdot \eta_{power\_stage}(V_{PV}, P_{grid})$ injected into the grid.

3. T4S MPPT Technique

As stated in the previous Section, k is a slowly varying signal [16] which dictates the value of the average power, P_{grid}, in particular, the higher the k signal, the higher the amplitude of $i_{grid}(t)$, and hence the higher the P_{grid}. On the basis of such a consideration, the objective of the proposed T4S MPPT technique is represented by the identification of the PV voltage reference, V_{PV_ref}, that leads to the maximization of the signal k. A single-stage grid-connected PV system equipped with the T4S MPPT controller is shown in Figure 2. From Figure 2, it is evident that for the T4S MPPT technique to function it needs only the sensing of the PV voltage, V_{PV}, and the signal k that is an already available internal signal. Even in this case, two low-pass filters appear and the same considerations, made with reference to Figure 1, apply for the quantities, k and V_{PV}. It is worth noting, different from the P&O technique, that no PV current detection is necessary.

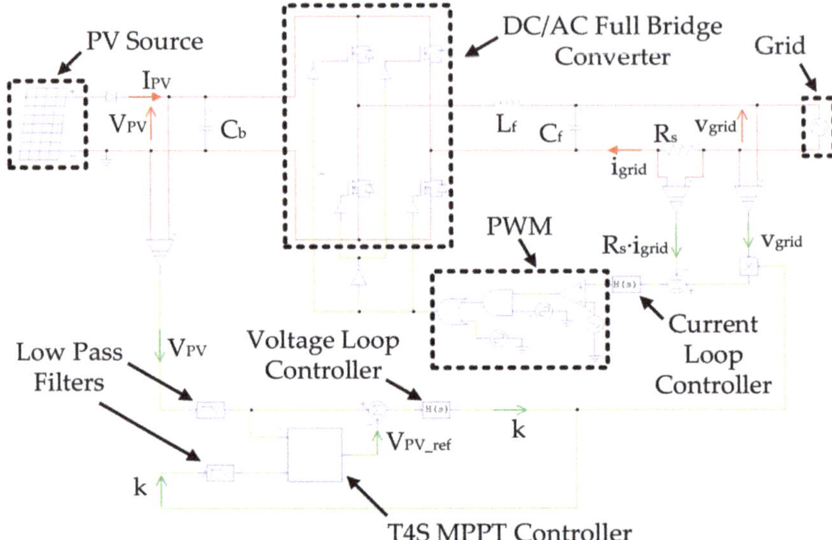

Figure 2. Single-stage grid-connected PV system equipped with the T4S MPPT technique.

The operating principle of the T4S technique is well illustrated by the following control law that is implemented in the T4S controller of Figure 2:

$$V_{PV\_ref}(t) = V_{ref\_0} + K_I \cdot \int_0^t \text{sign}\left\{ \frac{dk(\tau)}{d\tau} \cdot \frac{dV_{PV}(\tau)}{d\tau} \right\}\Bigg|_{\tau = r \cdot T_{SH}} d\tau \forall t \in [r \cdot T_{SH}; \ (r + 1) \cdot T_{SH}], \quad (4)$$

where K_I is a positive constant value and V_{ref_0} is the initial value of V_{PV_ref}. In particular, the PV voltage reference, V_{PV_ref}, provided by the T4S controller is the output of an integrator, as shown in Figure 3. Such an integrator processes a signal whose absolute value is the constant, K_I, and whose sign changes every T_{SH} second according to the sign of the product between the time derivatives of k and V_{PV}. Hence, V_{PV_ref} is a ramp function characterized by a slope equal to $\pm K_I$.

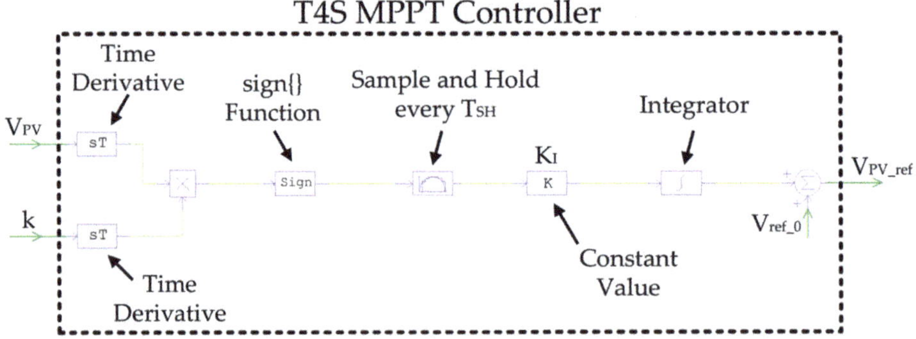

Figure 3. Block diagram of the proposed T4S MPPT technique.

In practice, the T4S MPPT technique is aimed at tracking the peak value of k with respect to V_{PV} since, as already evidenced above, the higher the value of the k signal, the higher the P_{grid}. Therefore, the output V_{PV_ref} of the T4S controller will have a positive slope ($+K_I$) in $[r \cdot T_{SH}^+, \ (r + 1) \cdot T_{SH}^-]$ if k and

V_{PV} are both increasing at $t = r \cdot T_{SH}^-$ (case A in Figure 4a, where the operating point is located at the left of the MPP and is moving towards the MPP) or if k and V_{PV} are both decreasing at $t = r \cdot T_{SH}^-$ (case B in Figure 4a, where the operating point is located at the left of the MPP and is moving away from the MPP). Instead, V_{PV_ref} will have a negative slope ($-K_I$) in $[r \cdot T_{SH}^+, (r + 1) \cdot T_{SH}^-]$ if k is increasing and V_{PV} is decreasing at $t = r \cdot T_{SH}^-$ (case C in Figure 4a where the operating point is located at the right of the MPP and it is moving towards the MPP) or if k is decreasing and V_{PV} is increasing (case D in Figure 4a where the operating point is located at the right of the MPP and it is moving away from the MPP). In practice, the working of the T4S technique is based on the proper setting, every T_{SH} second, of the sign of the slope of the photovoltaic voltage reference signal, V_{PV_ref}, from which the name was chosen for the proposed MPPT technique.

At steady-state, in the neighborhood of the MPP, the T4S technique leads to a V_{PV_ref} waveform characterized by a piecewise ramp profile with a period equal to $2 \cdot T_{SH}$ and a peak-peak amplitude equal to $K_I \cdot T_{SH}$ (Figure 4b). Hence, the steady-state performance of the T4S technique is strictly linked to the values assumed by the two parameters, T_{SH} and K_I. T_{SH} must be greater than the settling time of the filtered versions of the two input signals V_{PV} and k. Then, the proper value of K_I can be identified on the basis of the maximum allowed steady-state variation $\Delta V_{PV} = K_I \cdot T_{SH}$ of the PV operating point. It is worth noting that the higher the value of ΔV_{PV}, the higher the steady-state waste of available PV power.

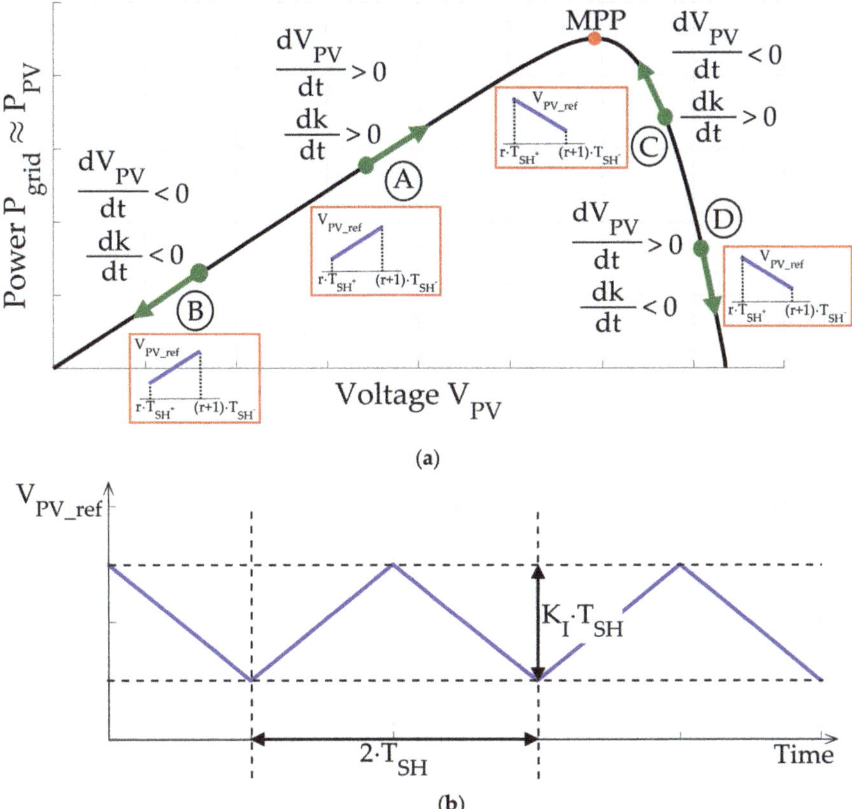

Figure 4. (a) The operation of the proposed T4S MPPT technique by assuming $\eta_{power\_stage}(V_{PV}, P_{grid})$ $\cong 1$ and (b) piecewise ramp waveform assumed by the photovoltaic voltage reference signal, V_{PV_ref}, at the steady-state in the maximum power point (MPP).

4. Numerical Results

In this section, numerical results concerning the working of a grid-connected single-stage PV system equipped with a T4S MPPT controller are reported and discussed. In particular, the system, as shown in Figure 2, has been implemented and tested in a PSIM environment [26]. The RMS grid voltage, V_{grid}, was set equal to 230 V and the line frequency, f, was set equal to 50 Hz. The considered PV source is characterized, in STC, by $P_{MPP\_STC} = 3058$ W and $V_{MPP\_STC} = 401.5$ V. With regard to the current control loop, its crossover frequency, f_{CI}, was set equal to 5 kHz, and the switching frequency, f_{SI}, of the PWM modulator has been set equal to 50 kHz. Moreover, $R_s = 0.1\ \Omega$, $C_b = 10$ mF, $L_f = 330\ \mu$H, and $C_f = 47\ \mu$F. With respect to the slow voltage control loop, its crossover frequency, f_{CV}, was set equal to 5 Hz and the low pass filters cut-off frequency, f_{LPF}, was set equal to 10 Hz. The maximum allowed steady-state variation, ΔV_{PV}, of the PV operating point was set equal to 10 V; such a value leads to a maximum steady-state PV power waste equal to about 0.5%.

The parameters of the T4S technique were identified as explained in the previous section. In particular, in [11] the interested reader can find a detailed analysis and discussion on the guidelines for the proper choice of perturbation periods in MPPT techniques with specific reference to the P&O technique. Basically, such perturbation periods must be greater than the system settling time in order to avoid errors in the MPPT process. In practice, the same considerations also hold for the proposed T4S MPPT technique. In particular, in the considered case, the estimated system settling time, to within 5%, is nearly equal to 0.26 s, which is consistent with the crossover frequency (5 Hz) of the voltage control loop. Therefore, $T_{SH} = 0.4$ s was selected. It is worth noting that the influence of the chosen value of T_{SH} on the performance of the proposed algorithm is such that, for a given K_I, the higher the T_{SH} the slower the tracking process under time-varying irradiance conditions and the higher the waste of available energy under steady-state irradiance conditions. Then, given the value of T_{SH}, the proper value of K_I can be identified on the basis of the maximum allowed steady-state variation, $\Delta V_{PV} = K_I \cdot T_{SH}$, of the PV operating point, and, in particular, the higher the value of ΔV_{PV}, the higher the steady-state maximum waste of available PV power. Since ΔV_{PV} was set equal to 10 V, it is $K_I = 25$ V/s. Moreover, the starting value, $V_{ref\_}$, of, V_{PV_ref}, was set equal to 450 V in order to ensure that, at the startup, the system works sufficiently far from the minimum allowed value of the DC input voltage of the full-bridge inverter that is equal to $\sqrt{2} \cdot 230$ V $= 325$ V. It is worth noting that because of the unit power stage efficiency of the power electronics interface of the ideal system implemented and simulated in PSIM, the maximum value of P_{grid} is only obtained when the PV source operates in its MPP.

4.1. Constant Irradiance

The first set of simulations were carried out in order to show the correct working of the proposed T4S MPPT technique and its ability to identify the maximum power point (MPP) under stationary irradiance conditions. In particular, a constant irradiance level, $S = 1000$ W/m^2, was considered. In such a condition the PV source is characterized by $P_{MPP} = 3058$ W and $V_{MPP} = 401.5$ V.

The following figures illustrate the behavior of the technique at the startup. In particular, in Figure 5 the PV extracted power is reported and compared with the available MPP power.

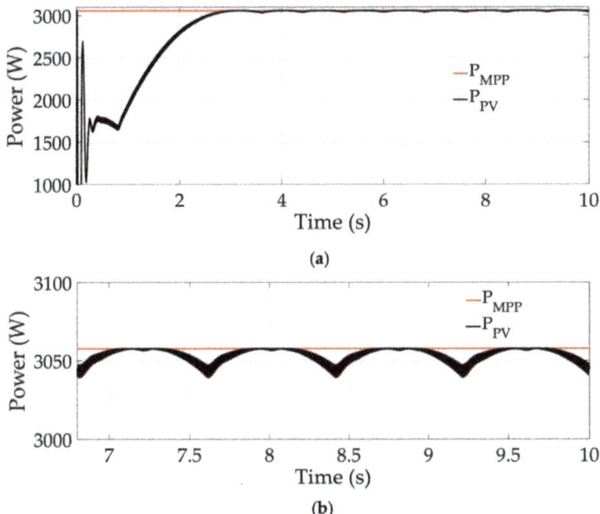

Figure 5. (**a**) Power extracted by adopting the T4S MPPT technique under a constant irradiance S = 1000 W/m^2 and (**b**) zoom of the steady-state behavior.

From the analysis of Figure 5a, the ability of the proposed technique to identify, after about 4 s, the MPP without errors is evident. In Figure 5b, a zoom of the steady-state behavior is reported. It is evident, as stated in the previous section, that the T4S technique leads to a periodic oscillation of the extracted power due to the periodic piecewise ramp profile of V_{PV_ref}, as shown in Figure 6a. Moreover, as predicted, the maximum waste of power due to such an oscillation around the MPP is equal to about 0.5%, i.e., (3058 W − 3042 W)/3042 W, as shown in Figure 5b, where 3058 W is the P_{MPP} and 3042 W represents the lowest level of the instantaneous extracted power. Moreover, from observations of Figure 6b, it is clear, as explained in the previous Section, that the voltage, V_{PV_ref}, is characterized by a period equal to $2 \cdot T_{SH} = 0.8$ s and a peak-peak amplitude equal to $K_I \cdot T_{SH} = 10$ V.

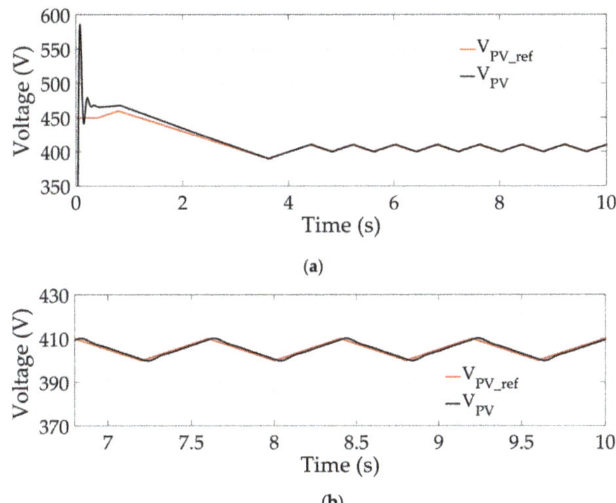

Figure 6. (**a**) Voltages, V_{PV_ref} and V_{PV}, obtained under a constant irradiance S = 1000 W/m^2 and (**b**) zoom of the steady-state behavior.

4.2. Time-Varying Irradiance

The second set of simulations were carried out in order to test the ability of the T4S MPPT technique to track the MPP under time-varying irradiance conditions. In particular, the operation of the T4S technique was tested by adopting the irradiance profile, as shown in Figure 7. The considered irradiance exhibits a step change from $S = 1000$ W/m^2 to $S = 500$ W/m^2 at $t = 5$ s and back from $S = 500$ W/m^2 to $S = 1000$ W/m^2 at $t = 10$ s. At $S = 1000$ W/m^2, it is $P_{MPP} = 3058$ W and $V_{MPP} = 401.5$ V and at $S = 500$ W/m^2, it is $P_{MPP} = 1598$ W and $V_{MPP} = 428$ V.

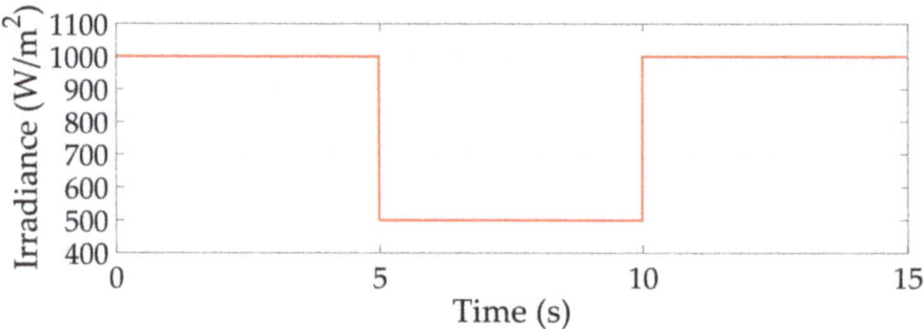

Figure 7. Time-varying irradiance profile.

The results of the numerical tests under the time-varying irradiance profile shown in Figure 7 are reported in the following figures. In particular, in Figure 8, the PV extracted power is reported and compared with the available MPP power. In Figure 9, the voltage, V_{PV_ref}, is reported together with the PV voltage, V_{PV}.

The previous figures show that the proposed MPPT technique is able to quickly track the MPP after any irradiance step variation.

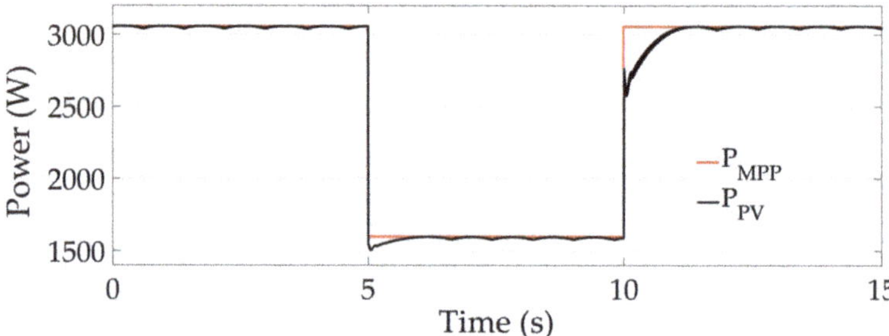

Figure 8. Power extracted by adopting the T4S technique under the irradiance profile of Figure 7.

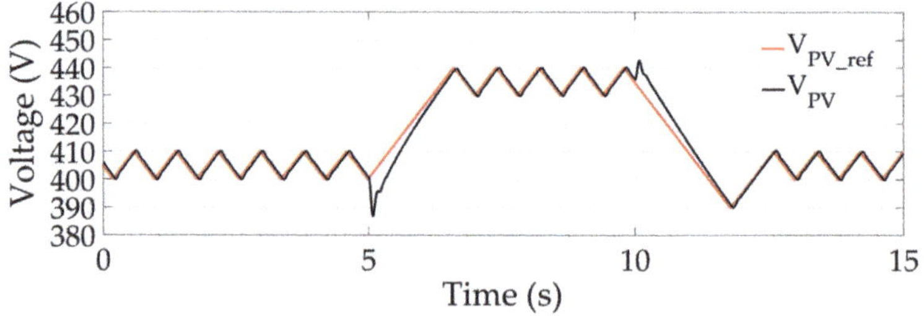

Figure 9. V_{PV_ref} and V_{PV}, T4S technique under the irradiance profile shown in Figure 7.

5. Comparison between the T4S MPPT Technique and the P&O MPPT Technique

The last set of numerical tests were carried out in order to compare the performance of the proposed T4S MPPT technique with that of the traditional P&O algorithm. to this aim, the system shown in Figure 1 was implemented and tested in a PSIM environment with the same values for the parameters as adopted in the previous section. The parameters of the P&O algorithm ($T_{P\&O}$, ΔV_{PV_ref}, and V_{ref_0}) were chosen on the basis of the guidelines reported in the literature [11] and in order to obtain the same steady-state maximum waste of power (0.5%) considered in the previous section, in particular, $T_{P\&O} = 0.4$ s, $\Delta V_{PV\_ref} = 10$ V, and $V_{ref\_0} = 450$ V. It is worth noting that the values of the parameters that have been chosen for the P&O technique fully ensure a fair comparison with the T4S MPPT technique.

In the following, the symbol $\kappa_{d\%}$ indicates the quantity $(P_{T4S\text{-}d}/P_{P\&O\text{-}d} - 1)\cdot100\%$ where $P_{T4S\text{-}d}$ is the average extracted power in dynamic conditions by adopting the T4S technique, and $P_{P\&O\text{-}d}$ is the average extracted power under the same dynamic conditions by adopting the P&O technique. "Dynamic conditions" mean working conditions taking place in the interval of time occurring between an event associated with a change of the irradiance conditions and the subsequent settling instant of steady-state periodic conditions. In particular, steady-state periodic conditions associated with the T4S technique are characterized by a piecewise ramp voltage behavior with period equal to $2\cdot T_{SH}$, as discussed in Section 3 and reported in Figure 4b. Instead, steady-state periodic conditions associated with the P&O technique are characterized by a three-level voltage behavior with period equal to $3\cdot T_{P\&O}$ [11]. It is worth noting that since the settling instant is, in general, different for the two MPPT techniques, the highest settling instant will be considered for the evaluation of $\kappa_{d\%}$. In addition, the symbol $\kappa_{ss\%}$ indicates the quantity $(P_{T4S\text{-}ss}/P_{P\&O\text{-}ss} - 1)\cdot100\%$, where $P_{T4S\text{-}ss}$ is the average extracted power in steady-state periodic conditions by adopting the T4S technique, for a given irradiance level, and $P_{P\&O\text{-}ss}$ is the average extracted power in steady-state periodic conditions by adopting the P&O technique for the same irradiance level.

5.1. Stationary Irradiance Conditions

The powers extracted by adopting the T4S and the P&O MPPT techniques under a constant irradiance ($S = 1000$ W/m^2) are reported together in Figure 10a. It is evident that both techniques need nearly the same time, at the start up, to reach the MPP. Such a startup can be seen as a step change of the irradiance level from 0 to 1000 W/m^2 and is characterized by a settling instant equal to 4 s, $P_{T4S\text{-}d} = 2519$ W, $P_{P\&O\text{-}d} = 2345$ W, and hence $\kappa_{d\%} = 7.4\%$. In addition, by looking at the zoom of Figure 10b, no remarkable difference in the steady-state efficiencies of the two MPPT techniques can be identified, although, indeed, the T4S performance is slightly better. In particular, $P_{T4S\text{-}ss} = 3053$ W, $P_{P\&O\text{-}ss} = 3048$ W, and hence $\kappa_{ss\%} = 0.16\%$.

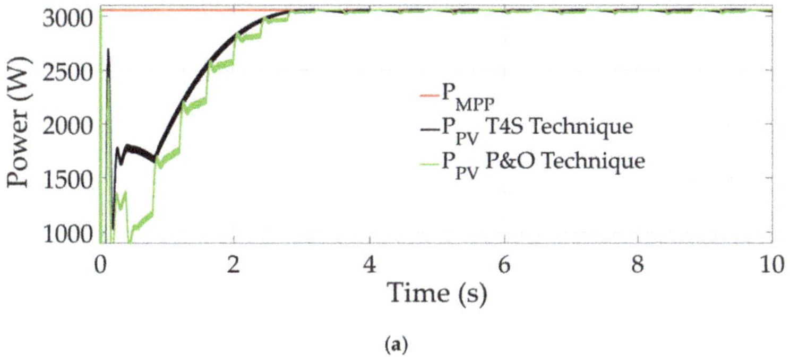

(a)

Figure 10. *Cont.*

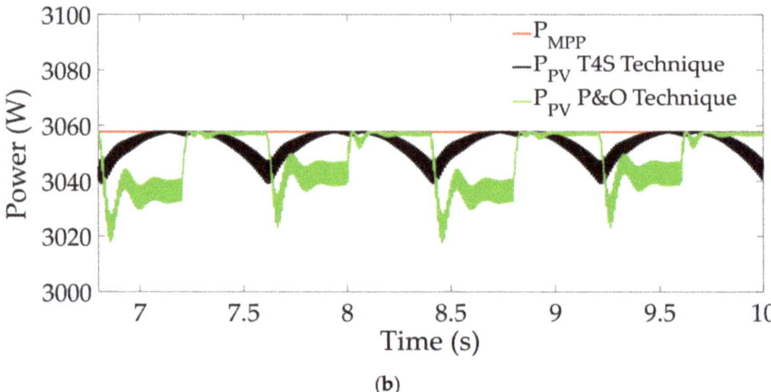

(b)

Figure 10. (**a**) Power extracted by adopting the T4s and the P&O MPPT techniques under a constant irradiance S = 1000 W/m^2 and (**b**) zoom of the steady-state behavior.

An interesting difference between the two MPPT techniques can be evidenced by observing Figure 11a and, in particular, its zoom in Figure 11b. In this figure, V_{PV_ref} and V_{PV} are reported for both of the MPPT techniques. It is clear for the P&O case, at steady-state, $V_{PV\_ref} \approx V_{PV}$ oscillates around the MPP by assuming, as expected, the following three values: V_{MPP} - ΔV_{PV}, V_{MPP}, and V_{MPP} + ΔV_{PV} [18]. Instead, in the T4S case, $V_{PV\_ref} \approx V_{PV}$ exhibits a piecewise linear behavior with values belonging to the range [V_{MPP} and V_{MPP} + ΔV_{PV}]. In other words, the T4S operating point is always located at the right of the MPP.

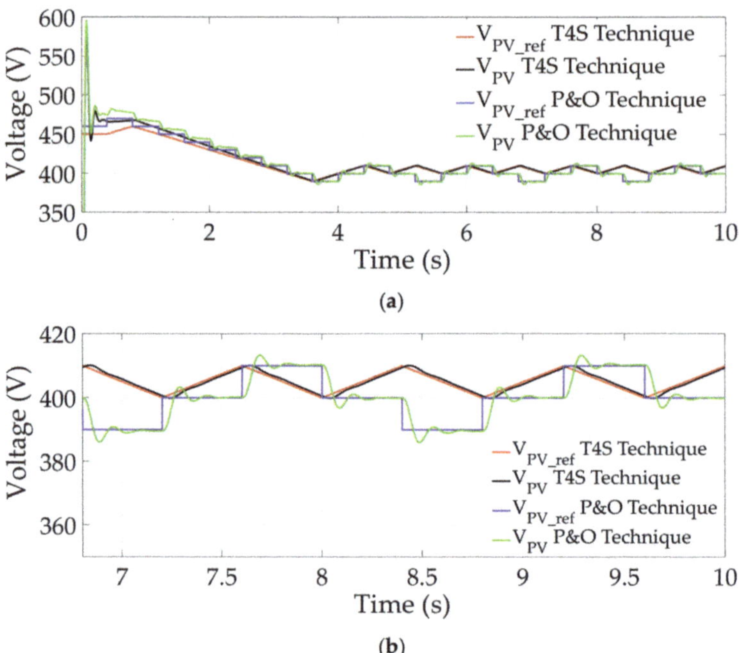

(a)

(b)

Figure 11. (**a**) V_{PV_ref} and V_{PV}, T4S and P&O MPPT techniques under a constant irradiance S = 1000 W/m^2 and (**b**) zoom of the steady-state behavior.

5.2. Time-Varying Irradiance

The performances of the two MPPT techniques were compared also under time-varying irradiance conditions. The adopted time-varying irradiance profile is shown in Figure 7. The powers extracted by adopting the T4S and the P&O MPPT techniques under such an irradiance profile are reported together in Figure 12. In Figure 13, V_{PV_ref} and V_{PV} are reported for both of the MPPT techniques. From the abovementioned figures, it is evident the P&O MPPT technique is outperformed by the T4S technique, at least in correspondence to the irradiance positive step.

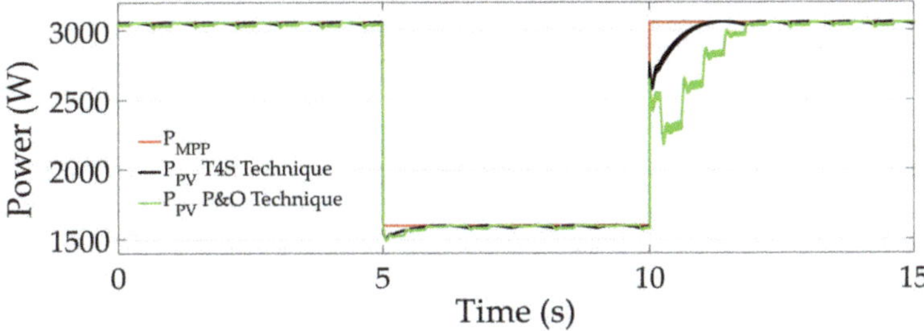

Figure 12. Power extracted by adopting the T4S and the P&O MPPT techniques under the irradiance profile of Figure 7.

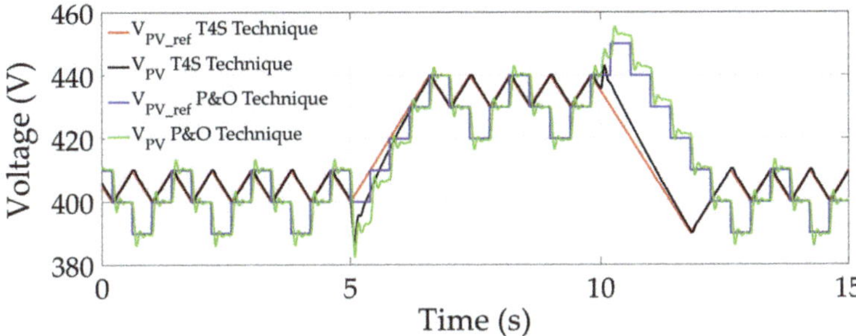

Figure 13. V_{PV_ref} and V_{PV}, T4S and P&O MPPT techniques under the irradiance profile shown in Figure 7.

From the above figures, it is evident the P&O MPPT technique is outperformed by the T4S technique, at least in correspondence to the irradiance positive step. In particular, in correspondence of the irradiance negative step occurring at $t = 5$ s the following values have been obtained: settling instant equal to 6.2 s, $P_{T4S\text{-}d} = 1571$ W, $P_{P\&O\text{-}d} = 1561$ W, and hence $\kappa_{d\%} = 0.64\%$. In correspondence to the irradiance positive step occurring at $t = 10$ s the following values have been obtained: settling instant equal to 12.2 s, $P_{T4S\text{-}d} = 2971$ W, $P_{P\&O\text{-}d} = 2721$ W, and hence $\kappa_{d\%} = 9.2\%$.

6. Conclusions

A novel MPPT technique, T4S (a technique based on the proper setting of the sign of the slope of the photovoltaic voltage reference signal), has been presented and discussed in this paper. Such a technique is specifically designed with reference to a single-stage grid-connected PV system. Its performance has been numerically compared to a benchmark MPPT technique, the P&O technique. The tracking efficiencies of the two techniques are more or less comparable, however, the T4S technique has a slightly better performance under both stationary and time-varying irradiance conditions. Furthermore, the T4S technique is characterized by the following two advantages with respect to the P&O technique: There is no need for PV current detection and the power stage efficiency profile is automatically taken into account. In other words, the T4S technique allows the tracking of the maximum average power injected into the grid rather than the maximum instantaneous power extracted by the PV source.

Author Contributions: Conceptualization, L.C. and M.V.; methodology, L.C. and M.V.; software, L.C. and M.V.; validation, L.C. and M.V.; formal analysis, L.C. and M.V.; investigation, L.C. and M.V.; resources, L.C. and M.V.; data curation, L.C. and M.V.; writing—original draft preparation, L.C. and M.V.; writing—review and editing, L.C. and M.V.; visualization, L.C. and M.V.

Funding: This research was partially funded by the "VALERE" research program of the Università della Campania "Luigi Vanvitelli".

Conflicts of Interest: The authors declare no conflict of interest. The funders had no role in the design of the study; in the collection, analyses, or interpretation of data; in the writing of the manuscript, or in the decision to publish the results.

References

1. Pathy, S.; Subramani, C.; Sridhar, R.; Thentral, T.M.T.; Padmanaban, S. Nature-Inspired MPPT Algorithms for Partially Shaded PV Systems: A Comparative Study. *Energies* **2019**, *12*, 1451. [CrossRef]
2. Balato, M.; Costanzo, L.; Vitelli, M. Chapter 5—DMPPT PV System: Modeling and Control Techniques. In *Advances in Renewable Energies and Power Technologies*; Yahyaoui, I., Ed.; Elsevier: Edinburgh, UK, 2018; pp. 163–205, ISBN 9780128129593. [CrossRef]
3. Saravanan, S.; Babu, N.R. Maximum power point tracking algorithms for photovoltaic system—A review. *Renew. Sustain. Energy Rev.* **2016**, *57*, 192–204. [CrossRef]
4. Balato, M.; Costanzo, L.; Marino, P.; Rubino, G.; Rubino, L.; Vitelli, M. Modified TEODI MPPT Technique: Theoretical Analysis and Experimental Validation in Uniform and Mismatching Conditions. *IEEE J. Photovolt.* **2017**, *7*, 604–613. [CrossRef]
5. Du, Y.; Yan, K.; Ren, Z.; Xiao, W. Designing Localized MPPT for PV Systems Using Fuzzy-Weighted Extreme Learning Machine. *Energies* **2018**, *11*, 2615. [CrossRef]
6. Yazdani, A.; Di Fazio, A.R.; Ghoddami, H.; Russo, M.; Kazerani, M.; Jatskevich, J.; Strunz, K.; Leva, S.; Martinez, J.A. Modeling Guidelines and a Benchmark for Power System Simulation Studies of Three-Phase Single-Stage Photovoltaic Systems. *IEEE Trans. Power Deliv.* **2011**, *26*, 1247–1264. [CrossRef]
7. Tan, C.W.; Green, T.C.; Hernandez-Aramburo, C.A. Analysis of perturb and observe maximum power point tracking algorithm for photovoltaic applications. In Proceedings of the 2008 IEEE 2nd International Power and Energy Conference, Johor Bahru, Malaysia, 1–3 December 2008; pp. 237–242. [CrossRef]
8. Carrasco, M.; Mancilla-David, F. Maximum power point tracking algorithms for single-stage photovoltaic power plants under time-varying reactive power injection. *Sol. Energy* **2016**, *132*, 321–331. [CrossRef]
9. Kouro, S.; Leon, J.I.; Vinnikov, D.; Franquelo, L.G. Grid-Connected Photovoltaic Systems: An Overview of Recent Research and Emerging PV Converter Technology. *IEEE Ind. Electron. Mag.* **2015**, *9*, 47–61. [CrossRef]
10. Li, L.; Wang, H.; Chen, X.; Bukhari, A.A.S.; Cao, W.; Chai, L.; Li, B. High Efficiency Solar Power Generation with Improved Discontinuous Pulse Width Modulation (DPWM) Overmodulation Algorithms. *Energies* **2019**, *12*, 1765. [CrossRef]
11. Femia, N.; Petrone, G.; Spagnuolo, G.; Vitelli, M. Optimization of perturb and observe maximum power point tracking method. *IEEE Trans. Power Electron.* **2005**, *20*, 963–973. [CrossRef]
12. Femia, N.; Fortunato, M.; Lisi, G.; Petrone, G.; Spagnuolo, G.; Vitelli, M. Guidelines for the Optimization of the P&O Technique in Grid-connected Double-stage Photovoltaic Systems. In Proceedings of the 2007 IEEE International Symposium on Industrial Electronics, Vigo, Spain, 4–7 June 2007; pp. 2420–2425. [CrossRef]
13. Rey-Boué, A.B.; Guerrero-Rodríguez, N.F.; Stöckl, J.; Strasser, T.I. Modeling and Design of the Vector Control for a Three-Phase Single-Stage Grid-Connected PV System with LVRT Capability according to the Spanish Grid Code. *Energies* **2019**, *12*, 2899. [CrossRef]
14. Jain, S.; Agarwal, V. Comparison of the performance of maximum power point tracking schemes applied to single-stage grid-connected photovoltaic systems. *IET Electr. Power Appl.* **2007**, *1*, 753–762. [CrossRef]
15. Wu, B.; Zhao, Z.; Liu, J. A Single-Stage Three-Phase Grid-Connected Photovoltaic System with Modified MPPT Method and Reactive Power Compensation. *IEEE Trans. Energy Convers.* **2007**, *22*, 881–886. [CrossRef]
16. Erickson, R.W.; Maksimovic, D. *Fundamentals of Power Electronics*, 2nd ed.; Kluwer: Norwell, MA, USA, 2001.

17. Su, M.; Luo, C.; Hou, X.; Yuan, W.; Liu, Z.; Han, H.; Guerrero, J.M. A Communication-Free Decentralized Control for Grid-Connected Cascaded PV Inverters. *Energies* **2018**, *11*, 1375. [CrossRef]
18. Balato, M.; Costanzo, L.; Vitelli, M. Maximum Power Point Tracking Techniques. In *Wiley Encyclopedia of Electrical and Electronics Engineering*; Wiley Online Library: Hoboken, NJ, USA, 2016; pp. 1–26, ISBN 9780471346081. [CrossRef]
19. Balato, M.; Costanzo, L.; Marino, P.; Rubino, L.; Vitelli, M. Dual implementation of the MPPT technique TEODI: Uniform and mismatching operating conditions. In Proceedings of the 2015 International Conference on Clean Electrical Power (ICCEP 2015), Taormina, Italy, 16–18 June 2015; pp. 422–429. [CrossRef]
20. Pearsall, N.M. (Ed.) 1—Introduction to photovoltaic system performance. In *The Performance of Photovoltaic (PV) Systems*; Woodhead Publishing: Cambridge, UK, 2017; pp. 1–19, ISBN 9781782423362. [CrossRef]
21. Demoulias, C. A new simple analytical method for calculating the optimum inverter size in grid-connected PV plants. *Electr. Power Syst. Res.* **2010**, *80*, 1197–1204. [CrossRef]
22. Zeb, K.; Khan, I.; Uddin, W.; Khan, M.A.; Sathishkumar, P.; Busarello, T.D.C.; Ahmad, I.; Kim, H.J. A Review on Recent Advances and Future Trends of Transformerless Inverter Structures for Single-Phase Grid-Connected Photovoltaic Systems. *Energies* **2018**, *11*, 1968. [CrossRef]
23. Li, H.; Wen, C.; Chao, K.-H.; Li, L.-L. Research on Inverter Integrated Reactive Power Control Strategy in the Grid-Connected PV Systems. *Energies* **2017**, *10*, 912. [CrossRef]
24. Sechilariu, M.; Locment, F. (Eds.) Chapter 1—Connecting and Integrating Variable Renewable Electricity in Utility Grid. In *Urban DC Microgrid*; Butterworth-Heinemann: Waltham, UK, 2016; pp. 1–33, ISBN 9780128037362. [CrossRef]
25. Silvestre, S. Chapter IIA-4—Review of System Design and Sizing Tools. In *Practical Handbook of Photovoltaics*, 2nd ed.; McEvoy, A., Markvart, T., Castañer, L., Eds.; Academic Press: Cambridge, MA, USA, 2012; pp. 673–694, ISBN 9780123859341. [CrossRef]
26. PSIM. Available online: https://powersimtech.com/ (accessed on 26 November 2019).

 *energies*

Article

A Novel Module Independent Straight Line-Based Fast Maximum Power Point Tracking Algorithm for Photovoltaic Systems

Anjan Debnath, Temitayo O. Olowu, Imtiaz Parvez, Md Golam Dastgir and Arif Sarwat *

Electrical and Computer Engineering, Florida International University, 10555 W Flagler St,
Miami, FL 33174, USA; adebn001@fiu.edu (A.D.); tolow003@fiu.edu (T.O.O.); iparv001@fiu.edu (I.P.);
mdast001@fiu.edu (M.G.D.)
* Correspondence: asarwat@fiu.edu

Received: 17 May 2020; Accepted: 12 June 2020; Published: 22 June 2020

Abstract: The maximum power point tracking (MPPT) algorithm has become an integral part of many charge controllers that are used in photovoltaic (PV) systems. Most of the existing algorithms have a compromise among simplicity, tracking speed, ability to track accurately, and cost. In this work, a novel "straight-line approximation based Maximum Power Point (MPP) finding algorithm" is proposed where the intersections of two linear lines have been utilized to find the MPP, and investigated for its effectiveness in tracking maximum power points in case of rapidly changing weather conditions along with tracking speed using standard irradiance and temperature curves for validation. In comparison with a conventional Perturb and Observe (*P&O*) method, the *Proposed* method takes fewer iterations and also, it can precisely track the MPP s even in a rapidly varying weather condition with minimal deviation. The *Proposed* algorithm is also compared with *P&O* algorithm in terms of accuracy in duty cycle and efficiency. The results show that the errors in duty cycle and power extraction are much smaller than the conventional *P&O* algorithm.

Keywords: linear approximation; MPPT algorithm; duty cycle; global horizontal irradiance; mathematical modeling

---

1. Introduction

Electricity generation from renewable energy sources has consistently increased over the past decade, with the largest percentage of renewable energy sources integrated being photovoltaic (PV) systems [1–5]. There has been a consistent increase in installed PV capacity globally which has lead to a corresponding increase in power generation from PV systems as shown in Figure 1 [6].

The massive integration of PV systems has also been aided by the declining cost of PV modules and improvement in their efficiencies [7]. Other than grid-tied PV systems, the use of PV modules has been extended to various applications such as rooftop solar power supply for residential homes, mobile charging systems, wearable devices, standalone PV arrays as car parks and electric vehicle charging stations, remote weather stations, international space station, and robots for space exploration.

In contrast to PV systems, Solar Thermal Plants (STPs) allow power to be generated by concentrating solar radiation which causes a very high temperature. The heat produced is subsequently used to convert water to steam which then used to turn power turbines for electricity generation. STPs are oriented to allow maximum tracking of the sun's insolation. The plants usually consist of reflectors and receivers. The reflectors (which are typically mirrors) are used to capture and concentrate the sun's light rays onto the receivers. The heat absorbed by the receivers is used to convert water to steam in order to drive the conventional steam turbine generators. It is of uttermost importance for the reflectors used in solar thermal plants to constantly track the movement of the sun in order to ensure maximum extraction of solar rays in order to maximize the efficiency of the plant [8–10]. In comparison with PV systems, Thermal Energy Storage (TES) technologies can be used to store energy for use at night, during severe cloud coverage, or periods with little or no sunlight. This allows STPs to be more dispatchable and achieved a higher capacity factor compared to PV. On the other hand, the decking cost of PV systems makes them cheaper than the STPs. A report by [11] suggests that STP technology might in the future help to increase the amount of PV penetration in the grid rather than being a direct competitor or alternative.

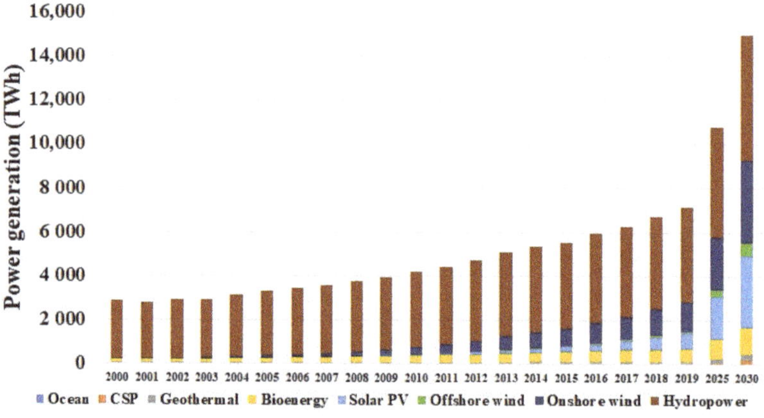

Figure 1. Power generation from renewable energy sources globally [6].

Due to the relatively low efficiency of PV systems compared to conventional power sources such as coal, diesel, or gas-fired plants, extraction of the maximum power per unit PV module becomes imperative. Maximum Power Point Tracking (MPPT) algorithms are developed to allow PV systems to operate at their maximum power under prevailing weather conditions [12–14]. The power generation from PV systems largely depends on the global horizontal irradiance and the atmospheric (consequently module) temperature. These parameters are stochastic in nature which means an accurate tracking system (simply called the Maximum Power Point Tracking (MPPT) system) to extract the maximum power out of the PVs is necessary. Several techniques and algorithms have been proposed for MPPT applications. These techniques and algorithms include the machine learning [15,16], hill climbing method [17–21], incremental conductance technique [22–25], Perturb and Observe [26–28], fractional open circuit voltage [29–31], fractional short circuit current [32–34], and fuzzy logic-based MPPT algorithms [35–38] amongst others. These algorithms aim to achieve high efficiency, fast-tracking speed, reduced steady-state oscillations, and reduced complexity in hardware implementation [38]. Out of these algorithms, the Perturb and Observe (P&O) algorithm, Incremental Conductance, and Fractional Open Circuit algorithm are most popular because of their simplicity. The *P&O* MPPT algorithm has dual shortcomings: enormous power loss due to large oscillations and deflection of tracking under rapid weather conditions. The large power loss can be minimized by choosing a small step size around the Maximum Power Point (MPP). However, the convergence speed has to be sacrificed [39]. The main drawbacks of Incremental and Conductance algorithms are the complex circuitry because of the derivative finding and choice of perturbation extent. The fractional open circuit based algorithm depends on the open circuit voltage of the panel, so at every change of irradiance, there is a temporary power loss due to the measurement of V_{OC}. Moreover, it is not module independent since the V_{OC} value has to be known prior to the application of the algorithm. Fractional short circuit current algorithm has the same issue as the fractional open circuit voltage. The algorithm is required to know the characteristics of the PV module and manufacturing specifications which makes it module dependent and less efficient. Machine learning (such as ANN) based MPPT algorithms are dependent on features (weather data) to generate the desired duty cycle for finding the MP point and one of the most common features is irradiance which requires costly sensors to measure. Though these algorithms are fast in finding MPPs, they are not cost-effective. Moreover, they require a lot of data to train which is the most important part of those models to demonstrate good accuracy. Therefore, it can not be generalized for any module. The PV module characteristics also change with time which necessitates the need for periodic training of the ANN-based controller to accurately track the MPP.

It is obvious that selecting an MPPT algorithm for many applications is usually a trade-off between efficiency, complexity, speed, cost, and ease of hardware implementation. This paper proposes a novel MPPT algorithm where the Maximum Power Point is acquired by the intersection of two linear lines which are tangents to the power vs voltage curve of the photovoltaic module. The major contributions of this paper are the following:

(i)   Number of iterations in reaching the MPP s are way less than conventional MPPT algorithm

(ii)   Reduction of oscillations around the MPP is much better than conventional, less than 0.03%

(iii)   Module independence, therefore the algorithm can be used in any PV modules without knowing the module data

(iv)   Highly efficient and easily implementable

MATLAB/SIMULINK environment is used for the simulation and verification of the *Proposed* algorithm. BP Solar BP SX 150S PV module is chosen in the MATLAB simulation model. The module is made of 72 multi-crystalline silicon solar cells in series and provides 150 W of nominal maximum power [40]. Table 1 shows the electrical parameters of the module:

Table 1. Electrical characteristics of PV Module Specifications.

Maximum Power (P_{MAX})	150 W
Voltage at $P_{MAX}(V_{MP})$	34.5 V
Current at $P_{MAX}(I_{MP})$	4.35 A
Open Circuit Voltage (V_{OC})	43.5 V
Short Circuit Current (I_{SC})	4.75 A

The rest of this paper is structured as follows: Section 2 presents the development of *Proposed* algorithm and description of the *P&O*; Section 3 represents simulation results using the *Proposed* algorithm under slow and fast varying weather conditions, a comparison between the *Proposed* algorithm with the *P&O* algorithm and validation of the *Proposed* algorithm using the CENELEC EN50530 standard test procedures; Section 4 concludes this paper.

2. Development of *Proposed* Algorithm and Description of the *P&O*

Several MPPT algorithms have been proposed in literature. The P&O, which is one of the widely used MPPT algorithms is selected as a baseline to compare the performance of the *Proposed* algorithm. The detailed formulation and modeling of the *Proposed* algorithm and a brief description of the *P&O* is as presented the following subsection.

2.1. Development of the Proposed Algorithm

Figure 2, the flowchart of the *Proposed* algorithm, depicts the construction of the algorithm. In developing the algorithm, a straight line approximation technique is utilized which is based on the P-V curve shapes. There is no evidence of significantly different P-V curve shape other than an inverted tilted '*V*'. So, the *Proposed* algorithm can be employed universally for a controller coupled with any kind of P-V panel without taking the panel's preset values. The details of the flowchart are depicted in the following:

1. At first, it senses voltage V_1 and I_1 by voltage and current sensors , respectively, and measured the power, P_1. Then the voltage is perturbed by a small amount, b and another voltage, V_2, is taken . The the power, P_2 is measured .

2. If $\Delta P = P_2 - P_1 > 0$, the operating point is at the left of MPP , else the operating point is at the right of the MPP.

3. Afterwards , the algorithm perturbs the voltage by a bigger step size c and senses the new voltage, V_3 followed by a small perturbation b. Then it measures voltage V_4 and calculates $\Delta P = (P_4 - P_3)$.

4. If $\Delta P < 0$, the desired four points (V_1, P_1), (V_2, P_2) and (V_3, P_3), (V_4, P_4) are found to draw the two straight lines. The algorithm will find then the intersecting point V_i by solving the equations. The intersecting point should be close to the MPP . If $\Delta P > 0$, step 3 needs to be repeated until $\Delta P < 0$ is fulfilled.

5. The point V_i is not the desired MPP, rather it is very close to MPP. To track the MPP , the algorithm will perturb the voltage by a small amount b and calculate ΔP again.

6. If $\Delta P < 0$, the operating point is at the right side of MPP, else the operating point is at the left side of the MPP.

7. If the operating point lies at the left side of MPP ($\Delta P > 0$), the algorithm will keep on perturbing the voltage by a small amount 'b' until $\Delta P < 0$ is satisfied. Once it reaches to that condition, the algorithm stops and P_{iii} will be the desired MPP. However, if it starts from the right side of MPP ($\Delta P < 0$), the algorithm will then keep on perturbing the voltage by a small amount 'b' to the left side. Once $\Delta P > 0$ is satisfied, the algorithm stops and P_{iii} is the desired MPP.

2.1.1. Mathematical Model of the *Proposed* Algorithm

In the developed method, MPP can be tracked very quickly and this operation is graphically illustrated in Figure 3. To understand the *Proposed* algorithm, let us take the operating point P_1 which is at the left side of MPP. By a small perturbation of voltage, another point P_2 can be found as:
$\Delta P = P_2 - P_1 > 0$

Now, by straight a line approximation, we can write an equation:

$$P = m_1 V + C_1 \tag{1}$$

where:

$$m_1 = (P_2 - P_1)/(V_2 - V_1) \quad \text{and}$$

$$C_1 = (V_1 P_2 - V_2 P_1)/(V_2 - V_1)$$

Now by taking another two points in the right side of MPP where $\Delta P = P_4 - P_3 < 0$ does match, another equation can be written also.

$$P = m_2 V + C_2 \tag{2}$$

where:

$$m_2 = (P_4 - P_3)/(V_4 - V_3) \quad \text{and}$$

$$C_2 = (V_3 P_4 - V_4 P_3)/(V_3 - V_4)$$

By solving Equations (1) and (2), the intersecting point V_A can be found as follows:

$$m_1 V + C_1 = m_2 V + C_2 \quad \text{and}$$

$$V_A = (C_2 - C_1)/(m_1 - m_2)$$

After getting V_A which is very near to MPP, the algorithm would reach MPP very fast as depicted in Figure 3.

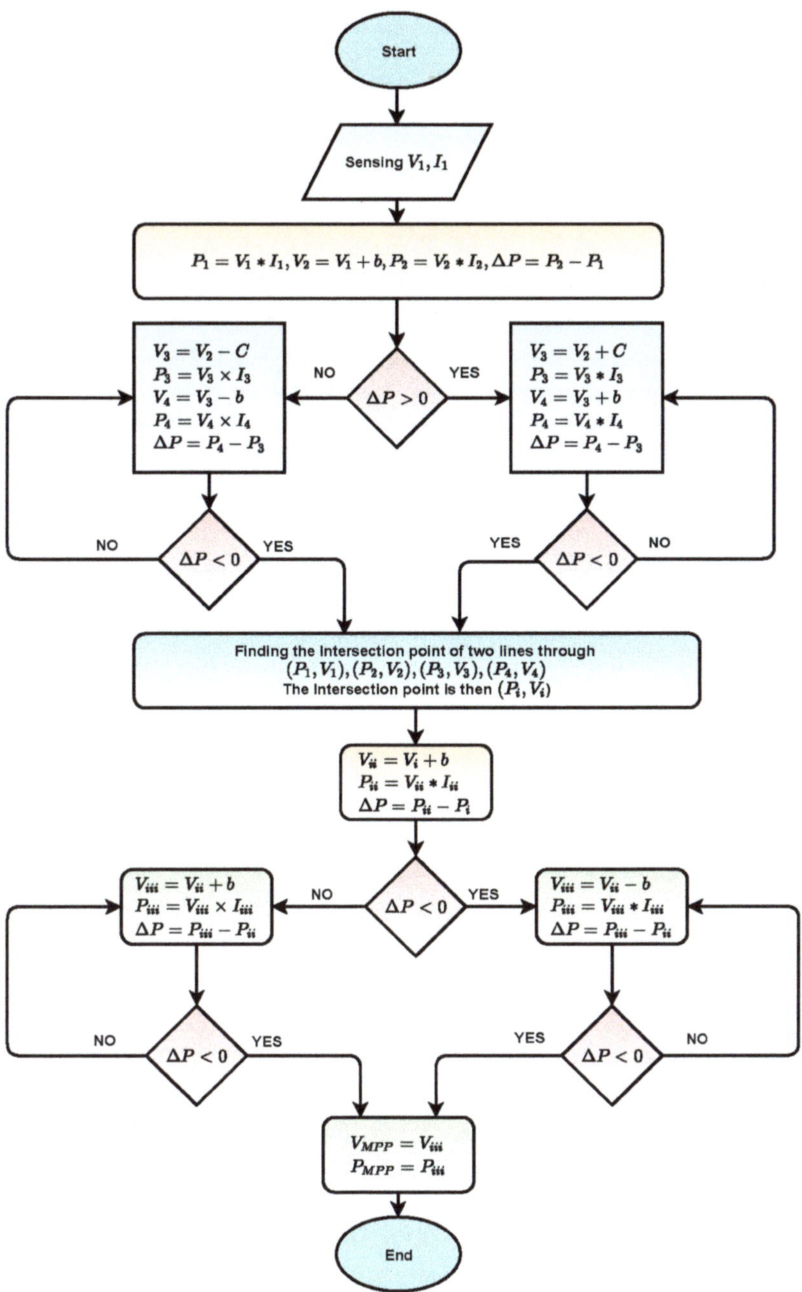

Figure 2. Flowchart of the *Proposed* algorithm.

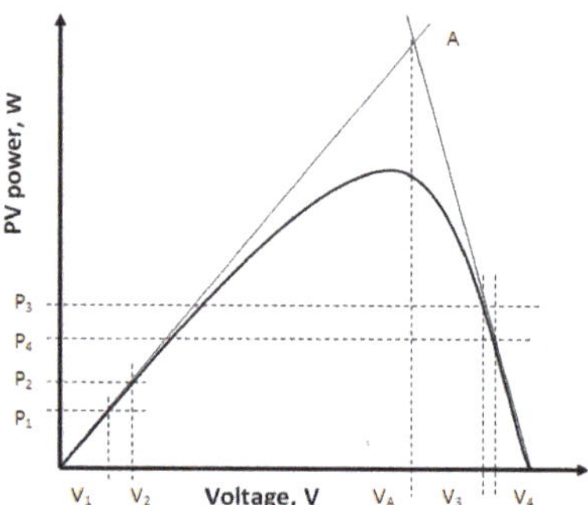

Figure 3. Straight line approach to find the MPP.

2.1.2. Response of the *Proposed* Algorithm When Irradiance Increases Rapidly

Here, the reaction towards rapidly changing weather conditions has been analyzed. In Figure 4, it is shown how the normal *P&O* method fails for a rapid variation in irradiance. Starting from an operating point located at P_1 (right side of MPP) we see, after a little perturbation, the new value is P_2. If the irradiance does not change during the perturbation, then P_2 will be lying on the same curve with G_1, I_1. So:

$$\Delta P = P_2 - P_1 < 0$$
$$\Delta V = V_k - V_{k-1} > 0$$

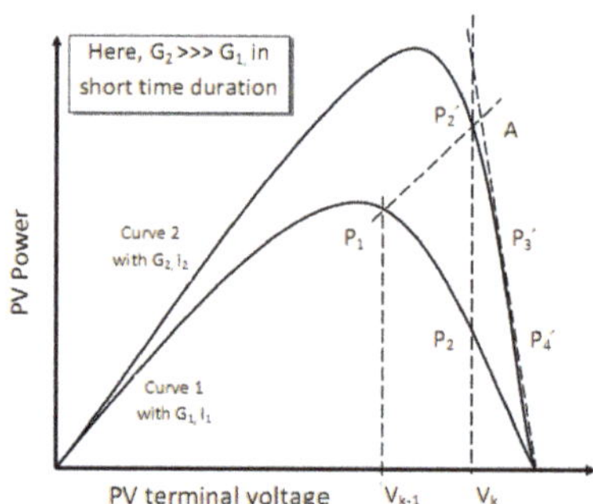

Figure 4. Response of *Proposed* method for rapidly increasing irradiance.

Consequently, the *P&O* algorithm will drive the operating point leftward which would be the correct direction. However, if the irradiance changes rapidly during the perturbation, then P_2 should lie on the G_2 curve at corresponding V_K. Let us call it P_2'.

$$\Delta P = P_2' - P_1 > 0$$
$$\Delta V = V_k - V_{k-1} > 0$$

Accordingly, the *P&O* algorithm would drive the operating point into the right side which is the wrong direction even though it should have driven it to the left, whereas, the developed method would draw the straight line passing through these two points P_1, P_2' and will take another two points P_3', P_4' on the I_2 curve and draw the straight line as depicted in Figure 4. After calculating the intersection point, V_A of these two straight lines, the algorithm would then drive the operating point to leftward to reach the MPP on G_2 curve.

2.2. The Conventional P&O Algorithm

In the conventional *P&O* algorithm, duty signal is perturbed till the operating point converges at the MPP. The algorithm compares the voltage and power at two consecutive samples. A small voltage perturbation is performed, if the change of power is positive, the perturbation is continued in the same direction and for a negative change in power, the perturbation is reversed in order to reach the MPP. By this approach, the whole P-V curve is checked using fixed perturbations to find the MPP. If the perturbation size is small, the algorithm takes a longer time in reaching the MPP with a high level of accuracy and vice versa. A large perturbation size could cause the algorithm to produce steady-state oscillations around the MPP [41]. Figure 5 illustrates the steps required in determining the MPP using the *P&O* approach.

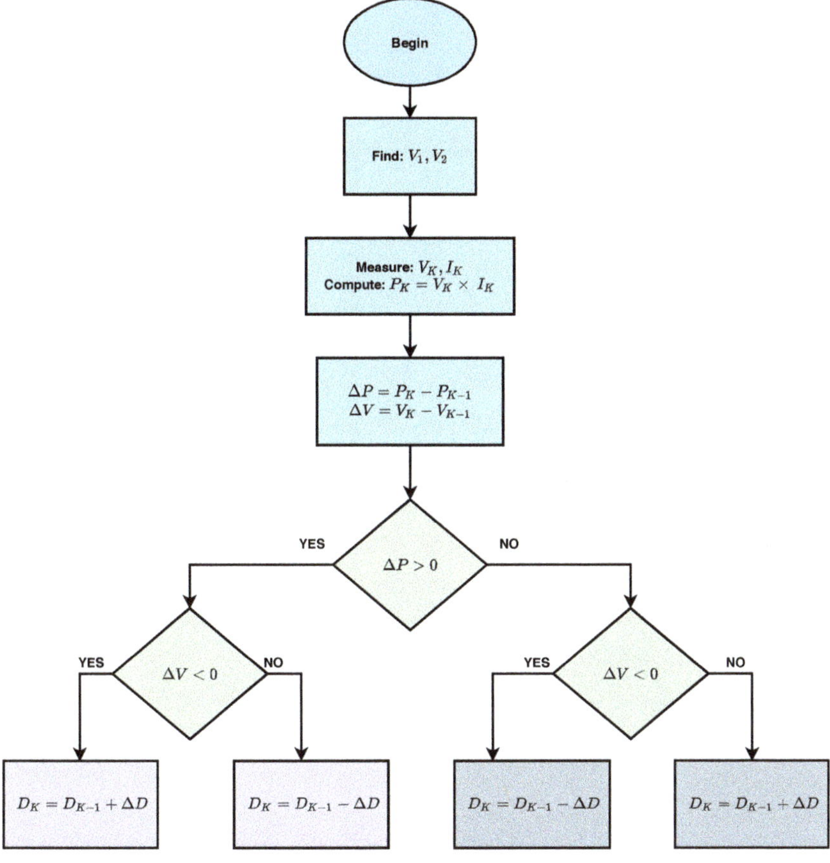

Figure 5. Flowchart of the conventional *P&O* Algorithm.

3. Simulation Results Using the *Proposed* Algorithm

In this section, simulation was conducted under normal and rapid weather conditions to test the response of the *Proposed* algorithm. Then comparison was made between *P&O* and *Proposed* algorithm in terms of the total number of iterations in reaching the MPP. After that, the percentage of error in duty signal and power corresponding to the MPPs was compared. Finally, the validation of the *Proposed* algorithm was performed using standard irradiance curves.

3.1. Simulation Results under Normal and Rapid Weather Conditions

In order to investigate the responses of the algorithm, practical irradiance data was taken a for a sunny day and a cloudy day for 12 h period from 6 a.m. to 6 p.m. This Global Horizontal Irradiance

(GHI) is plotted for a sunny day and cloudy day as shown in Figure 6a,b respectively in order to certify the efficacy of the *Proposed* MPPT algorithm.

Figure 7a–c, illustrate the accuracy of the *Proposed* algorithm for sunny and cloudy weather conditions. In Figure 7a,b, the module power vs module voltage graph is plotted for when irradiance was varying for a particularly sunny day and cloudy day respectively. The red dots shown in Figure 7a,b, are the exact MPPs and the green curve is the output of the developed algorithm. Itis very clear from those figures that the *Proposed* algorithm accurately identified the exact MPPs. The cloudy day irradiance shown in Figure 6a changed very rapidly; nonetheless, the *Proposed* algorithm was able to determine the MPPs (red dots) very successfully which is very clear from Figure 7b. Figure 7c shows how the module power could be varied just by varying the duty cycle which ensured that the operating point of the PV panel could be varied by changing the duty ratio.

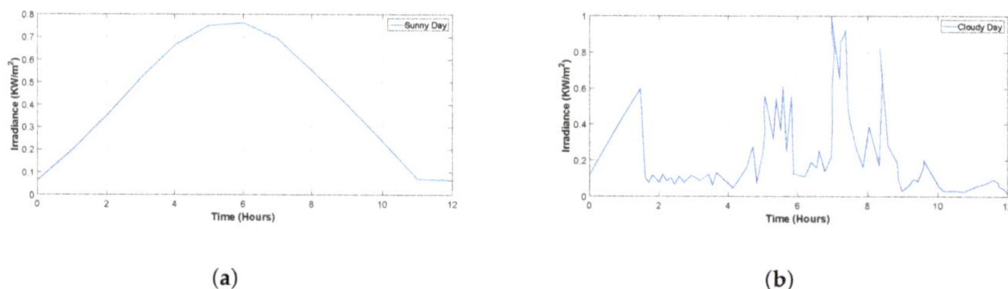

(a)

(b)

Figure 6. (**a**) Sunny Day Irradiance Data from 6 a.m. to 6 p.m. (**b**) Cloudy Day Irradiance Data from 6 a.m. to 6 p.m.

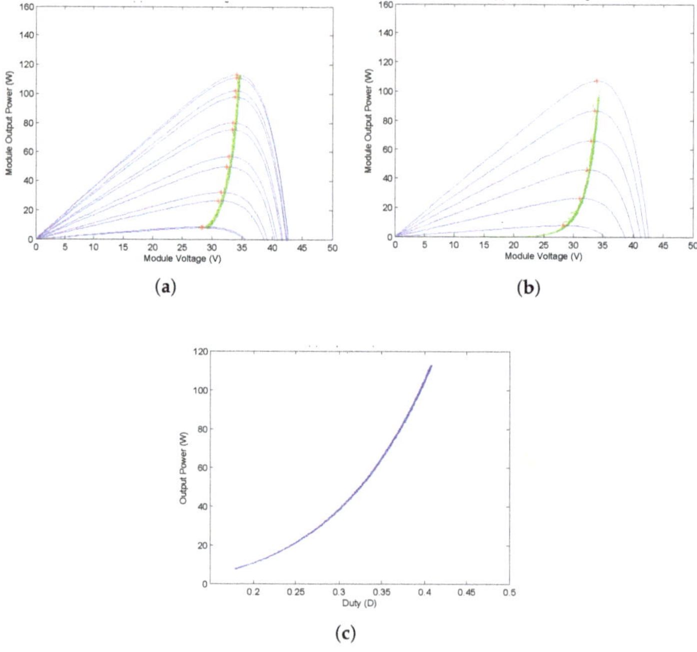

(a)

(b)

(c)

Figure 7. (**a**) Panel voltage vs. power for a sunny day (**b**) Panel voltage vs Power for a Cloudy Day (**c**) Duty vs. Output Power.

3.2. Comparison of the Proposed Algorithm with Conventional Perturb and Observe Algorithm

Here, the number of iterations to reach the MPP was compared between the developed method and Perturb and Observe method for the following conditions:

Irradiance, $G = 1\,\text{KW/m}^2$ Temperature, $T_a C = 25\,°C$

Operating point, Va = 5 V (left side of MPP)

When the operating point lay at the left of MPP, Table 2 shows a comparison between the number of iterations used by the *Proposed* and *P&O* algorithm, where N_{op-Ip} is the number of iterations from operating point to intersection point and N_{Ip-MPP} is the number of iterations from intersection point to MPP. The number of iterations used by the developed algorithm was much less than that of the *P&O* method. Initially, the developed method had big steps (2 V) in order to get the intersection point (32.2346 V) of two straight lines and took 15 iterations. The intersection point was close to the MPP as mentioned earlier. From MPP intersection point to the, the algorithm took 22 iterations since it went with small steps (0.1 V). The small steps after finding the intersection point not only made sure of small oscillations around MPP but also ensured high efficiency.

When the operating point lay at the right side of MPP, Table 3 data confirms again that the number of iterations of the *Proposed* algorithm was much smaller than the conventional *P&O* method.

Irradiance, $G = 1\text{KW/m}^2$ Temperature, $TaC = 25\,°C$ Operating point, Va = 40 V (right side of MPP).

Table 2. Number of iterations comparison between developed and *P&O* algorithm when the operating point lies at left of MPP.

	P&O (Step Size = 0.1)	Proposed (Step Size = 2, 0.1)	Exact Value
MPP Voltage	34.6	34.5346	34.5
MPP Power	149.9709	149.9825	149.9858
Intersection Voltage	NA	32.2346	NA
N_{op-Ip}	NA	15	NA
N_{Ip-MPP}	NA	22	NA
Total of iterations	296	37	NA

Table 3. No. of iterations comparison between developed and *P&O* when the operating point lies at right of MPP.

	P&O (Step Size = 0.1)	Proposed (Step Size = 2, 0.1)	Exact Value
MPP Voltage	34.6	34.3498	34.5
MPP Power	149.9709	149.9784	149.9858
Intersection Voltage	NA	32.2346	NA
N_{op-Ip}	NA	3	NA
N_{Ip-MPP}	NA	29	NA
Total of iterations	54	32	NA

3.3. Comparison between Proposed and Perturb and Observe Algorithm in Terms of Duty and Maximum Power Point

Here, the irradiance and temperature were varied to analyze the performance and compare the two algorithms. It is very apparent from Table 4 that the *Proposed* method obtained the duty cycles and maximum power points much closer to the exact duty cycle and maximum power points than the Perturb and Observe method under varying weather conditions which secured better efficiency.

The comparison in terms of percentage error in duty cycle and percentage error in finding MPPs are illustrated in Figure 8a,b. It is evident in Figure 8a that the *Proposed* algorithm had a very small error in finding the correct duty cycle under varying irradiance conditions. The error was less than $\pm0.5\%$ whereas the Perturb and Observe algorithm had more than $\pm1\%$ error in finding the correct duty cycle. The similar scenario was observed in Figure 8b as well where the percentage error in finding the Maximum Power Point under varying irradiance conditions was less than 0.03% for the *Proposed* algorithm whereas it was more than 0.23% in the case of the Perturb and Observe method.

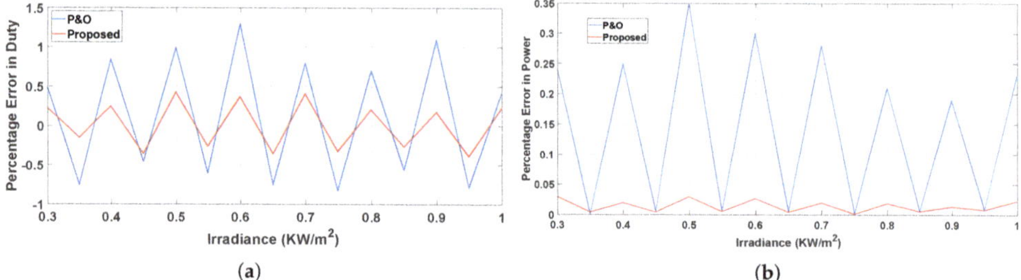

(a)   (b)

Figure 8. (**a**) Comparison of percentage of error in duty cycle under 25 °C between *Proposed* and conventional Perturb and Observe (P&O) algorithm (**b**) Comparison of percentage of error in Maximum Power Point under 25 °C between *Proposed* and conventional *P&O* algorithm.

Table 4. Comparison of duty cycle and maximum power between *Proposed* and *P&O* algorithm at different irradiance and temperature.

G and T(kW/m², °C)	D_{exact}	$D_{Proposed}$	$D_{P\&O}$	P_{exact}	$P_{Proposed}$	$P_{P\&O}$
(0.3, 0)	0.3177	0.3201	0.3100	46.8442	46.8059	46.4333
(0.3, 10)	0.3228	0.3200	0.3100	44.8187	44.7739	43.8932
(0.3, 20)	0.3275	0.3280	0.3310	42.7801	42.7802	42.7203
(0.5, 0)	0.3725	0.3720	0.3730	80.4209	80.4200	80.4165
(0.5, 10)	0.3774	0.3780	0.3790	77.1584	77.1546	77.1326
(0.5, 20)	0.3824	0.3820	0.3790	73.8728	73.8702	73.7657
(0.8, 10)	0.4314	0.4320	0.4270	126.4301	126.4219	126.1785
(0.8, 20)	0.4367	0.4350	0.4390	121.3375	121.3139	121.2492
(0.8, 30)	0.4415	0.4420	0.4390	116.2146	116.2139	116.1323
(1.0, 20)	0.4627	0.4620	0.4630	153.1360	153.1256	153.1357
(1.0, 30)	0.4683	0.4680	0.4690	146.8280	146.8288	146.8160
(1.0, 40)	0.4734	0.4720	0.4750	140.4862	140.4611	140.4538
(1.2, 20)	0.4848	0.4850	0.4870	184.8904	184.8890	184.7909
(1.2, 30)	0.4897	0.4890	0.4870	177.4167	177.4042	177.2795
(1.2,40)	0.4948	0.4950	0.4990	169.9001	169.9017	169.6730

3.4. Validation of the Proposed MPPT Algorithm

In order to validate the *Proposed* algorithm, the standard test irradiance profile (according to CENELEC EN50530) was used [42,43] which is shown in Figures 9a–12a. Figures 9a–12c show the performance of the developed algorithm under irradiation ramps with very steep gradients of 0.5, 5, 20, 50 and 100 W/m²/s. From these figures, it can be seen that the quality of the MPP tracking was very good and it was accurate with all gradients. The figures show that the current always tracked the irradiance changes accurately, whereby the voltage was either lagging or leading the MPP by a very small amount. This can be explained by the fact that the current was linearly proportional to irradiance and it was almost impossible to detect the exact MPP voltage because the algorithm had to use an epsilon(stopping condition when close to MPP) value which was less than a preset value (the difference between exact V_{MPP} and estimated V_{MPP} which was very close to exact MPP but not equal) to get very close to exact V_{MPP}. Therefore the detected V_{MPP} points stayed either before or after the exact V_{MPP} point, however very close.

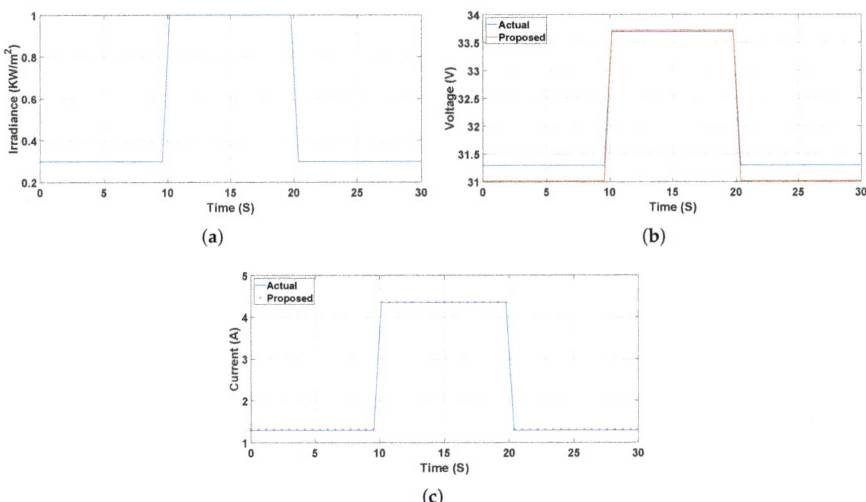

Figure 9. (**a**) Very steep gradient (**b**) accuracy of the *Proposed* algorithm in finding V_{mpp} at very steep gradient (**c**) accuracy of the *Proposed* algorithm in finding I_{mpp} at very steep gradient.

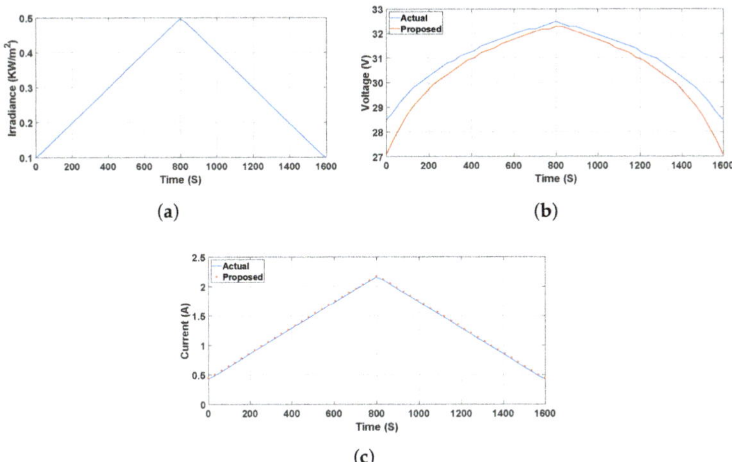

Figure 10. (**a**) Irradiance with gradient of 0.5 W/m^2/s (**b**) accuracy of *Proposed* algorithm in finding V_{MPP} at 0.5 W/m^2/s gradient (**c**) accuracy of the *Proposed* algorithm in finding I_{MPP} at 0.5 W/m^2/s gradient.

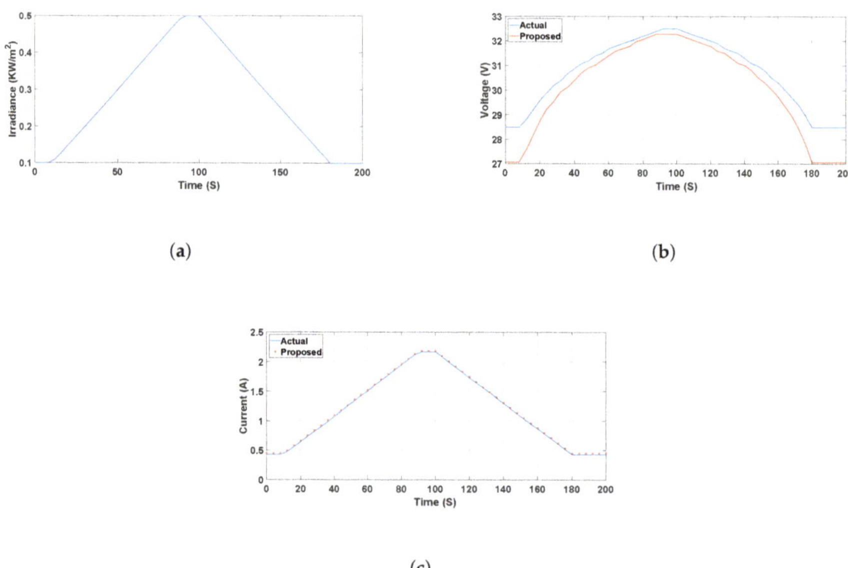

Figure 11. (**a**) Irradiance at a gradient of 5 W/m^2/s (**b**) accuracy of the *Proposed* algorithm in finding V_{mpp} at 5 W/m^2/s gradient (**c**) accuracy of the *Proposed* algorithm in finding I_{mpp} at 5 W/m^2/s gradient.

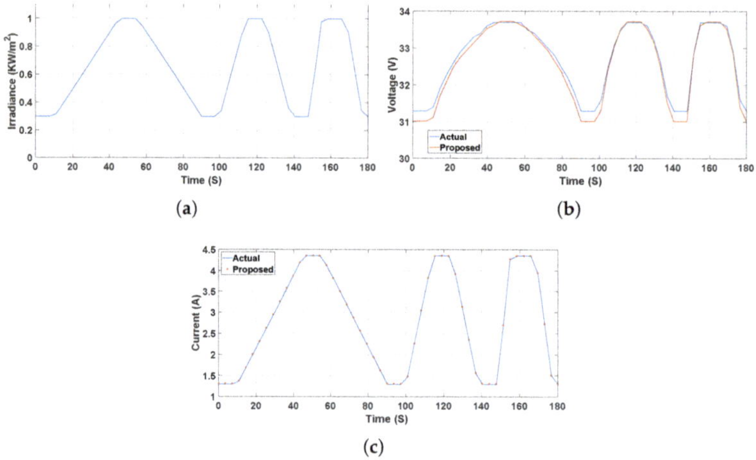

Figure 12. (**a**) Irradiance at gradients of 20, 50, 100 W/m^2/s (**b**) accuracy of the *Proposed* algorithm in finding V_{mpp} at gradients of 20, 50, 100 W/m^2/s (**c**) accuracy of the *Proposed* algorithm in finding I_{mpp} at Gradients of 20, 50, 100 W/m^2/s.

4. Conclusions

In this paper, a simple and accurate photovoltaic maximum power point tracking algorithm is proposed with its mathematical formulation. The *Proposed* MPPT algorithm is tested for its tracking speed, tracking accuracy, capability to rapid transition, and maximum power point efficiency and compared with the conventional hill climbing *P&O* method. The results show that the system reaches to MPP voltage in much fewer iterations and hence is very fast to converge to MPPs under fast-changing weather conditions. The duty cycle and maximum power point under various weather conditions have been analyzed also and compared with the *P&O* method. The results also show that the percentage of errors in finding duty cycles and maximum power points of the *Proposed* algorithm are much less than its counterpart (*P&O* method). Subsequently, the developed algorithm was validated according to the CENELEC EN50530 standard which stipulates how the efficiency of the MPPT algorithm should be measured. It is also worthy of note that the *Proposed* algorithm does not require the technical specifications of the PV module as an input to the algorithm.

As a future work, the algorithm will be implemented using a micro-controller and other necessary hardware for a PV-Battery microgrid system and voltage stability will be studied for varying load conditions along with other control-algorithms such as fuzzy logic, artificial intelligence, model predictive control, etc. PV power ramp rate control also will be studied with the *Proposed* algorithm for high penetration level of PV systems as future smart grid applications.

Author Contributions: Conceptualization, A.D.; Funding acquisition, A.S.; Investigation, A.D.; Methodology, A.D.; Project administration, I.P.; Supervision, I.P. and A.S.; Writing–original draft, T.O.O.; Writing–review & editing, T.O.O. and M.G.D. All authors have read and agreed to the published version of the manuscript.

Funding: This research was supported by National Science Foundation (NSF) under the grant number 1553494.

Conflicts of Interest: The authors declare no conflict of interest.

References

1. Olowu, T.O.; Sundararajan, A.; Moghaddami, M.; Sarwat, A.I. Future challenges and mitigation methods for high photovoltaic penetration: A survey. *Energies* **2018**, *11*, 1782. [CrossRef]
2. Jafari, M.; Olowu, T.O.; Sarwat, A.I.; Rahman, M.A. Study of Smart Grid Protection Challenges with High Photovoltaic Penetration. In Proceedings of the 2019 North American Power Symposium (NAPS), Wichita, KS, USA, 13–15 October 2019; pp. 1–6. [CrossRef]
3. Olowu, T.O.; Jafari, M.; Sarwat, A.I. A Multi-Objective Optimization Technique for Volt-Var Control with High PV Penetration using Genetic Algorithm. In Proceedings of the 2018 North American Power Symposium (NAPS), Fargo, ND, USA, 9–11 Septemper 2018; pp. 1–6. [CrossRef]
4. Olowu, T.O.; Sundararajan, A.; Moghaddami, M.; Sarwat, A.I. Fleet Aggregation of Photovoltaic Systems: A Survey and Case Study. In Proceedings of the 2019 IEEE Power Energy Society Innovative Smart Grid Technologies Conference (ISGT), Washington, DC, USA, 18–21 February 2019; pp. 1–5. [CrossRef]
5. Sundararajan, A.; Olowu, T.O.; Wei, L.; Rahman, S.; Sarwat, A.I. Case study on the effects of partial solar eclipse on distributed PV systems and management areas. *IET Smart Grid* **2019**, *2*, 477–490. [CrossRef]
6. Statistics, I.R.E. The International Renewable Energy Agency. July 2017. Available online: Http://www.irena.org/--/media/Files/IRENA/Agency/Publication/2017/ (accessed on 10 June 2020).
7. Fu, R.; Feldman, D.J.; Margolis, R.M. *US Solar Photovoltaic System Cost Benchmark: Q1 2018*; Technical Report; National Renewable Energy Lab. (NREL): Golden, CO, USA, 2018.
8. Rashid, K. Design, Economics, and Real-Time Optimization of a Solar/Natural Gas Hybrid Power Plant. Ph.D. Thesis, The University of Utah, Salt Lake City, UT, USA, 2019.
9. Alva, G.; Liu, L.; Huang, X.; Fang, G. Thermal energy storage materials and systems for solar energy applications. *Renew. Sustain. Energy Rev.* **2017**, *68*, 693–706. [CrossRef]
10. Rashid, K.; Safdarnejad, S.M.; Powell, K.M. Process intensification of solar thermal power using hybridization, flexible heat integration, and real-time optimization. *Chem. Eng. Process.-Process Intensif.* **2019**, *139*, 155–171. [CrossRef]
11. Sioshansi, R.; Denholm, P. The value of concentrating solar power and thermal energy storage. *IEEE Trans. Sustain. Energy* **2010**, *1*, 173–183. [CrossRef]
12. Debnath, A.; Huda, M.N.; Saha, C.; Roy, S. Determining Optimum Generator for South-East Coast of Bangladesh: Hybrid, Solar-Only or Wind-Only? In Proceedings of the 2018 10th International Conference on Electrical and Computer Engineering (ICECE), Dhaka, Bangladesh, 20–22 December 2018; pp. 165–168.

13. Saha, C.; Huda, M.N.; Mumtaz, A.; Debnath, A.; Thomas, S.; Jinks, R. Photovoltaic (PV) and thermo-electric energy harvesters for charging applications. *Microelectron. J.* **2020**, *96*, 104685. [CrossRef]

14. Debnath, A.; Parvez, I.; Dastgir, M.; Nabi, A.; Olowu, T.O.; Riggs, H.; Sarwart, A. Voltage Regulation of Photovoltaic System with varying Loads. In Proceedings of the 2020 SoutheastCon, Raleigh, NC, USA, 10 March 2020; pp. 1–7.

15. Parvez, I.; Sarwat, A.; Debnath, A.; Olowu, T.; Dastgir, M.G. Multi-layer Perceptron based Photovoltaic Forecasting for Rooftop PV Applications in Smart Grid. In Proceedings of the 2020 SoutheastCon, Raleigh, NC, USA, 12–15 March 2020;

16. Shahriar, M.H.; Haque, N.I. ; Rahman, M.A.; Alonso Jr, M. G-IDS: Generative Adversarial Networks Assisted Intrusion Detection System. In Proceedings of the 2020 International Computer Software and Applications Conference (COMPSAC), Madrid, Spain, 13–17 July 2020; pp. 1–10.

17. Xiao, X.; Huang, X.; Kang, Q. A Hill-Climbing-Method-Based Maximum-Power-Point-Tracking Strategy for Direct-Drive Wave Energy Converters. *IEEE Trans. Ind. Electron.* **2016**, *63*, 257–267. [CrossRef]

18. Motsoeneng, P.; Bamukunde, J.; Chowdhury, S. Comparison of Perturb Observe and Hill Climbing MPPT Schemes for PV Plant Under Cloud Cover and Varying Load. In Proceedings of the 2019 10th International Renewable Energy Congress (IREC), Sousse, Tunisia, 26–28 March 2019; pp. 1–6. [CrossRef]

19. Du, X.; Yin, H. MPPT control strategy of DFIG-based wind turbines using double steps hill climb searching algorithm. In Proceedings of the 2015 5th International Conference on Electric Utility Deregulation and Restructuring and Power Technologies (DRPT), Changsha, China, 26–29 November 2015; pp. 1910–1914. [CrossRef]

20. Tan, C.Y.; Rahim, N.A.; Selvaraj, J. Improvement of hill climbing method by introducing simple irradiance detection method. In Proceedings of the 3rd IET International Conference on Clean Energy and Technology (CEAT) 2014, Kuching, Malaysia, 24–26 November 2014; pp. 1–5. [CrossRef]

21. Fatemi, S.M.; Shadlu, M.S.; Talebkhah, A. Comparison of Three-Point P O and Hill Climbing Methods for Maximum Power Point Tracking in PV Systems. In Proceedings of the 2019 10th International Power Electronics, Drive Systems and Technologies Conference (PEDSTC), Shiraz, Iran, 12–14 February 2019; pp. 764–768. [CrossRef]

22. Deopare, H.; Deshpande, A. Fuzzy based incremental conductance algorithm for PV system. In Proceedings of the 2016 International Conference on Automatic Control and Dynamic Optimization Techniques (ICACDOT), Pune, India, 9–10 September 2016; pp. 683–687. [CrossRef]

23. Ananthi, C.; Kannapiran, B. Improved design of sliding-mode controller based on the incremental conductance MPPT algorithm for PV applications. In Proceedings of the 2017 IEEE International Conference on Electrical, Instrumentation and Communication Engineering (ICEICE), Karur, India, 27–28 April 2017; pp. 1–6. [CrossRef]

24. Anowar, M.H.; Roy, P. A Modified Incremental Conductance Based Photovoltaic MPPT Charge Controller. In Proceedings of the 2019 International Conference on Electrical, Computer and Communication Engineering (ECCE), Cox'sBazar, Bangladesh, 7–9 February 2019; pp. 1–5. [CrossRef]

25. Vineeth Kumar, P.K.; Manjunath, K. Analysis, design and implementation for control of non-inverted zeta converter using incremental conductance MPPT algorithm for SPV applications. In Proceedings of the 2017 International Conference on Inventive Systems and Control (ICISC), Coimbatore, India, 19–20 January 2017; pp. 1–5. [CrossRef]

26. Kumar, N.; Hussain, I.; Singh, B.; Panigrahi, B.K. Framework of Maximum Power Extraction From Solar PV Panel Using Self Predictive Perturb and Observe Algorithm. *IEEE Trans. Sustain. Energy* **2018**, *9*, 895–903. [CrossRef]

27. Abdel-Salam, M.; El-Mohandes, M.T.; Goda, M. An improved perturb-and-observe based MPPT method for PV systems under varying irradiation levels. *Sol. Energy* **2018**, *171*, 547–561. [CrossRef]

28. Nedumgatt, J.J.; Jayakrishnan, K.B.; Umashankar, S.; Vijayakumar, D.; Kothari, D.P. Perturb and observe MPPT algorithm for solar PV systems-modeling and simulation. In Proceedings of the 2011 Annual IEEE India Conference, Hyderabad, India, 16–18 December 2011; pp. 1–6. [CrossRef]

29. Ahmad, J. A fractional open circuit voltage based maximum power point tracker for photovoltaic arrays. In Proceedings of the 2010 2nd International Conference on Software Technology and Engineering, San Juan, Puerto Rico, 3–5 October 2010; Volume 1, pp. 247–250. [CrossRef]

30. Das, P. Maximum Power Tracking Based Open Circuit Voltage Method for PV System. *Energy Procedia* **2016**, *90*, 2–13. doi:10.1016/j.egypro.2016.11.165. [CrossRef]

31. Baimel, D.; Tapuchi, S.; Levron, Y.; Belikov, J. Improved Fractional Open Circuit Voltage MPPT Methods for PV Systems. *Electronics* **2019**, *8*, 321. [CrossRef]

32. Sher, H.A.; Murtaza, A.F.; Noman, A.; Addoweesh, K.E.; Chiaberge, M. An intelligent control strategy of fractional short circuit current maximum power point tracking technique for photovoltaic applications. *J. Renew. Sustain. Energy* **2015**, *7*, 013114. [CrossRef]

33. Husain, M.A.; Tariq, A.; Hameed, S.; Arif, M.S.B.; Jain, A. Comparative assessment of maximum power point tracking procedures for photovoltaic systems. *Green Energy Environ.* **2017**, *2*, 5–17. [CrossRef]

34. Sher, H.A.; Murtaza, A.F.; Noman, A.; Addoweesh, K.E.; Al-Haddad, K.; Chiaberge, M. A New Sensorless Hybrid MPPT Algorithm Based on Fractional Short-Circuit Current Measurement and P O MPPT. *IEEE Trans. Sustain. Energy* **2015**, *6*, 1426–1434. [CrossRef]

35. Youssef, A.; El Telbany, M.; Zekry, A. Reconfigurable generic FPGA implementation of fuzzy logic controller for MPPT of PV systems. *Renew. Sustain. Energy Rev.* **2018**, *82*, 1313–1319. [CrossRef]

36. Yang, B.; Zhong, L.; Zhang, X.; Shu, H.; Yu, T.; Li, H.; Jiang, L.; Sun, L. Novel bio-inspired memetic salp swarm algorithm and application to MPPT for PV systems considering partial shading condition. *J. Clean. Prod.* **2019**, *215*, 1203–1222. [CrossRef]

37. Rizzo, S.A.; Scelba, G. ANN based MPPT method for rapidly variable shading conditions. *Appl. Energy* **2015**, *145*, 124–132. [CrossRef]

38. Motahhir, S.; El Hammoumi, A.; El Ghzizal, A. The Most Used MPPT Algorithms: Review and the Suitable Low-cost Embedded Board for Each Algorithm. *J. Clean. Prod.* **2019**. [CrossRef]

39. Wilson, K.R.; Rao, Y.S. Comparative Analysis of MPPTT Algorithms for PV Grid Tied Systems: A Review. In Proceedings of the 2019 2nd International Conference on Intelligent Computing, Instrumentation and Control Technologies (ICICICT), Kannur, India, 5–6 July 2019; Volume 1, pp. 1105–1110.

40. BP SX 150. Available online: https://www.abcsolar.com/pdf/bpsx150.pdf/ (accessed on 12 June 2020).

41. Kamran, M.; Mudassar, M.; Fazal, M.R.; Asghar, M.U.; Bilal, M.; Asghar, R. Implementation of improved Perturb & Observe MPPT technique with confined search space for standalone photovoltaic system. *J. King Saud Univ.-Eng. Sci.* **2018**. doi.org/10.1016/j.jksues.2018.04.006. [CrossRef]

42. Andrejašič, T.; Jankovec, M.; Topič, M. Comparison of direct maximum power point tracking algorithms using EN 50530 dynamic test procedure. *IET Renew. Power Gener.* **2011**, *5*, 281–286. [CrossRef]

43. Bründlinger, R.; Henze, N.; Häberlin, H.; Burger, B.; Bergmann, A.; Baumgartner, F. prEN 50530–The new European standard for performance characterisation of PV inverters. In Proceedings of the 24th EU PV Conference, Hamburg, Germany, 21–25 September 2009.

 *energies*

Article

Application of Genetic Algorithm for More Efficient Multi-Layer Thickness Optimization in Solar Cells

Premkumar Vincent [1,†], Gwenaelle Cunha Sergio [1,†], Jaewon Jang [1], In Man Kang [1], Jaehoon Park [2], Hyeok Kim [3], Minho Lee [4] and Jin-Hyuk Bae [1,*]

[1] School of Electronics Engineering, Kyungpook National University, 80 Daehakro, Bukgu, Daegu 41566, Korea; 2014600014@knu.ac.kr (P.V.); gwena.cs@gmail.com (G.C.S.); j1jang@knu.ac.kr (J.J.); imkang@ee.knu.ac.kr (I.M.K.)

[2] College of Software, Hallym University, Chuncheon 24252, Korea; jaypark@hallym.ac.kr

[3] Department of Electrical and Computer Engineering, University of Seoul, 163 Seoulsiripdaero, Dongdaemun-gu, Seoul 02504, Korea; hyeok.kim@uos.ac.kr

[4] Department of Artificial Intelligence, Kyungpook National University, 80 Daehakro, Bukgu, Daegu 41566, Korea; mholee@knu.ac.kr

* Correspondence: jhbae@ee.knu.ac.kr

† Co-first authors with equal contribution.

Received: 9 March 2020; Accepted: 1 April 2020; Published: 4 April 2020

Abstract: Thin-film solar cells are predominately designed similar to a stacked structure. Optimizing the layer thicknesses in this stack structure is crucial to extract the best efficiency of the solar cell. The commonplace method used in optimization simulations, such as for optimizing the optical spacer layers' thicknesses, is the parameter sweep. Our simulation study shows that the implementation of a meta-heuristic method like the genetic algorithm results in a significantly faster and accurate search method when compared to the brute-force parameter sweep method in both single and multi-layer optimization. While other sweep methods can also outperform the brute-force method, they do not consistently exhibit 100% accuracy in the optimized results like our genetic algorithm. We have used a well-studied P3HT-based structure to test our algorithm. Our best-case scenario was observed to use 60.84% fewer simulations than the brute-force method.

Keywords: genetic algorithm; solar cell optimization; finite difference time domain; optical modelling

---

1. Introduction

Simulations of optoelectronic devices have helped to understand and design better optimized structures with efficiencies nearing the theoretical maximum. Lucio et al. analyzed the possibility of achieving the limits of c-Si solar cells through such simulations [1]. Simulations have reduced the time it takes for researchers to find optimized device structure. However, the most common way to obtain results over a large range of a parameter's values is through parameter sweep method. This brute-force method is ineffective in most cases where the user only requires the end optimized device structure. Genetic algorithm (GA) is an optimization algorithm in artificial intelligence based on Darwin's evolution and natural selection theory, in which the fittest outcome survives [2,3]. This algorithm sets an environment with a random population and a function, which is called the fitness function, that scores each individual of that population. The environment then selects individuals to become the parents of the next generation through a selection process. The next generation of individuals (children of the previous generation's parents) is obtained via a crossover method. Similar to natural mutation in genes of the offspring, the new generation's individuals can also suffer mutation in their genes. After several generations, the population converges to the individuals representing the optimal solution. Through the application of genetic algorithm, Jafar-Zanjani et al. designed a binary-pattern reflect-array for highly efficient beam steering [4] and Tsai et al. was able to beam-shape the laser to obtain up to 90% uniformity in intensity distribution [5]. Genetic algorithm has been used by Donald et al. to improve the focusing of light propagating through a scattering medium [6] and by Wen et al. for designing highly coherent optical fiber in the mid-infrared spectral range [7]. It has also been used for designing nanostructures to improve light absorption in solar cells. Chen et al. were able to surpass the Yablonovitch Limit using genetic algorithms to design light trapping nanostructures [8]. Rogério et al. used a genetic algorithm to design surface structures on a Si solar cell to increase the short-circuit current density obtained from it [9].

In this article, we have demonstrated the optimization of an organic solar cell through the optimization of the optical spacer layers. Traditionally, finite difference time domain (FDTD) method was used to simulate the ideal short-circuit current density (J_{sc}) of the solar cell through the Lumerical, FDTD solutions software similar to our previous reported study [10]. Parameter sweep or brute-force method was used to vary the thickness of the optical spacer layers of the solar cell. At the optimized layer thicknesses, the solar cell will be observed to have the highest J_{sc} output. Although not computationally intensive for a single-layer optimization, the number of simulations expands as in Equation (1) for multi-layer optimization problems.

$$N = n_1 \cdot n_2 \cdot n_3 ... NN \tag{1}$$

where N is the total number of simulations and n_1, n_2, and n_3 are the number of simulations performed for layers 1, 2, and 3, and NN is the total number of layers ($NN > 3$), respectively.

To alleviate the brute-force method's limitations, we propose the use of GA. This article then aims to heuristically assert the hypothesis that GA is a more efficient approach than brute-force algorithms in tasks such as optimizing optoelectronic device structures.

2. Methodology

Figure 1 shows the device structure that was constructed in Lumerical, FDTD solutions. It consists of a 150 nm indium tin oxide (ITO) similar to our previous study [11]. The Al thickness was set to 100 nm as light gets completely reflected from the Al electrode at this thickness. Any further increase in its thickness would not affect the outcome of our optical simulation. The active layer, poly (3-hexylthiophene) (P3HT): indene-C_{60} bisadduct (ICBA), was designed to be 200 nm as it exhibited good efficiency in our previous study [12]. The charge transport layers, zinc oxide (ZnO) and Molybdenum oxide (MoOx), also act as optical spacer layers and are variable quantities in our simulation. In order to qualify as an optical spacer, the layer material should have the refractive index properties to shift the light induced electric field inside the solar cell structure by varying its thickness. ZnO was already shown to be a good optical spacer in our previous research [11]. While a ITO/PEDOT:PSS/P3HT:PCBM/ZnO/Al is a more commonly used solar cell structure [13], PEDOT:PSS does not have the suitable refractive index to control the distribution of electric field for the wavelengths under consideration. MoOx was found to be a suitable hole transport layer substitute to the PEDOT:PSS by a previously reported study by Bohao et al. [14]. MoOx was also found to have good optical spacer properties in our simulations and so it was used as the other optical spacer layer.

The simulation was done in perpendicular illumination onto the solar cell as done for device simulations. This method does not simulate the real outdoor operation of the solar cell, and thus, the optimized thickness values calculated in this article would not be applicable in the view of energy yield optimization. However, our study was a comparison of the algorithms, and our conclusion should be consistent.The solar cell layers were stacked along the y-axis in the FDTD software. The incident light was set up as a plane wave with the spectral intensity of AM1.5G. The plane wave's propagation direction was the same axis along which the solar cell's layers were stacked. It was incident through the ITO electrode. Periodic boundaries were set along the x-axis and perfectly matched layers were used along the y-axis to set the FDTD boundary conditions. The device was meshed good enough for the results to converge. Further simulation setup details are published elsewhere [11,12]. To simulate the ideal J_{sc}, 100% internal quantum efficiency was assumed.

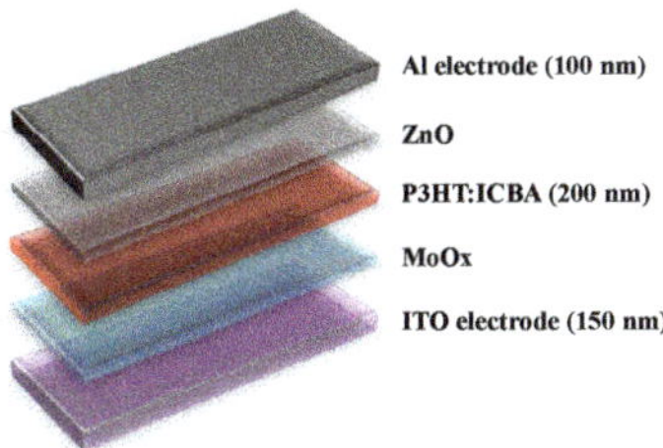

Figure 1. Solar cell device structure. The electron transport layer is ZnO, while the hole transport layer is MoOx. Both also act as optical spacer layers as they affect the distribution of light inside the device. We optimized these optical spacer layers for maximizing photon absorption inside the active layer of the solar cell.

2.1. Brute Force

Using the software's parameter sweep option, we simulated our device structure according to three sections. The first section optimized only the ZnO layer, while keeping the MoOx layer at 10 nm thickness. The second simulation section consists of optimizing only the MoOx layer, while the ZnO layer was fixed at 30 nm. Our final section optimized both optical spacer layers together. The results of the brute-force method are provided in Figure 2.

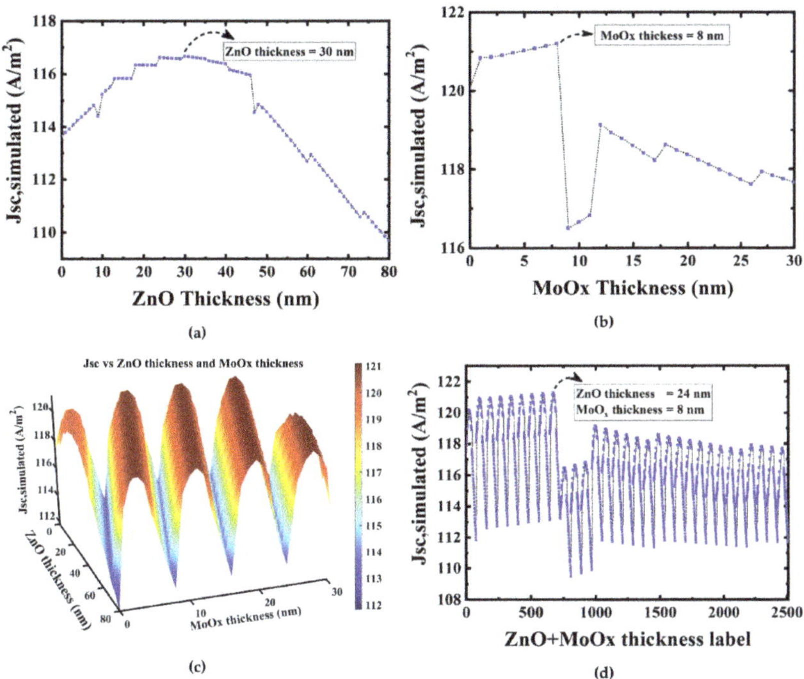

Figure 2. Brute-force method results: (**a**) single ZnO layer optimization, (**b**) single MoOx layer optimization, (**c**) multiple ZnO and MoOx layers optimization, (**d**) 2D data representation of (**c**) using labels pointing to the ZnO and MoOx layer thickness combinations for the ease of computation.

Using parameter sweep to simulate every possible combination is time-consuming. Figure 2 contains multiple local maxima and minima points. This makes it tedious to extrapolate the optimal thickness from fewer simulation points. Due to this, every possible simulation point is required to find the optical result. This way of obtaining the optimal result is termed as brute-force method henceforth. In order to make a 2-layer optimization problem similar to that of the single-layer optimization, we replaced the optical spacers' thickness combinations with a label in Figure 2d, effectively converting 3D data to a 2D data. The label number is given by *label number* = *max*(ZnO *thickness*) × *MoOx thickness* + *MoOx thickness* + 1.

2.2. Genetic Algorithm

To alleviate the computational and time inefficiencies of the brute-force method, we found inspiration in Darwin's natural selection theory and proposed the use of a genetic algorithm [3] for our optimization problem. Consider that there is a random population and their adaptability to the environment is given by a fitness function. In our optimization problem, the fitness function was to maximize J_{sc} output from the FDTD simulation. Each population contains several individual chromosomes, which in turn consists of an array of bits called genes. Different selection methods are used to choose certain chromosomes in each generation (see Section 2.2.1) in order to reproduce off-springs using a crossover method (see Section 2.2.2). To further mimic biology, there is also a probability that a chromosome might suffer mutation, which is provided by the mutation rate, which allows the algorithm to escape local minima in the data. In the end, the fittest members of the population prevail, meaning that the algorithm converges to the optimal solution.

A step-by-step of the works of the genetic algorithm is shown in Algorithm 1. The initial population *pop* of size *p* is randomly selected from the search space, which in our case is the maximum and minimum thickness of the optical spacer layers. The first iteration of the algorithm then starts. For each population value, the GA calls the FDTD software to simulate and extract the J_{sc} result. The fitness function (J_{sc}) is applied to all individuals in the population and ranked from the highest J_{sc} to the lowest. A selection method is then applied taking into consideration each chromosome and their respective fitness score. After selecting the parents responsible for the next generation, they reproduce to obtain the next generation, a step detailed in Algorithm 2. Please note that in our algorithm, the fittest individual in the current generation is cloned to be part of the next generation. The current generation is then updated with the next generation, with the algorithm continuing until the maximum number of generations has been reached.

Algorithm 1: Genetic Algorithm

    Data: *max_generation, population_size (p), mutation_prob, search_space* (range of population)
    Result: Best individual and fitness value (optimal solution)
    pop ← random initial population of size *p* from *search_space*;
    while *generation* <= *max_generation* **do**
        fitness_score ← Fitness(*pop*);
        fitness_score_sorted, pop_sorted ← Rank(*fitness_score, pop*);
        new_pop ← **Cloning** of the *fittest*;
        next_parents ← Selection(*fitness_score_sorted, pop_sorted*);
        new_pop ← [*new_pop*, Reproduction(*p* − 1, *next_parents, mutation_prob*)];
        pop ← *new_pop*;
        generation += 1;
    Save *pop_sorted* and *fitness_score_sorted* in *.mat*;

As for the reproduction algorithm, detailed in Algorithm 2, the first step is to allocate an empty array for the new population. The second step is to select two parents to produce *C* children. Due to the crossover method used and mutation ratio, the same parents can reproduce different children. The new population is then updated with the new children obtained through crossover and mutation.

Algorithm 2: Reproduction Algorithm

Data: *next_parents, p, mutation_prob*
Result: New population
new_pop ← [];
for *i=1 to p / 2* **do**
    parent1 ← *next_parents*(i);
    parent2 ← *next_parents*(len(*next_parents*) - i + 1);
    Create C children from *parent1* and *parent2* with Crossover and Mutations;
    new_pop ← *new_pop* + new children

2.2.1. Selection Methods

Four selection methods were examined in this work: random, tournament, roulette wheel, and breeder. For better clarity of the following explanations, the fittest individual means the individual with highest fitness score, since our problem is a maximization problem.

Random

This is the simplest selection method since it does not incorporate selection criteria. This method consists of randomly selecting individuals to be the next generation's parents, with no regards to the fitness function. Because of this Monte-Carlo-like approach, it can take very long for the algorithm to converge.

Tournament

Tournament Selection [15] samples k individuals with replacement from a population of p and applies the fitness function to those individuals in order to select the one with best fitness score, also known as the fittest individual. One can think of this method as a battle of the fittest, where k individuals face each other in tournament fashion to decide the fittest. The fittest individuals from each tournament round will then constitute the parents responsible for forming the next generation.

Roulette Wheel

Roulette-Wheel Selection (RWS) [16] is a popular way of parent selection in which individuals have a fitness-proportionate probability of being selected. In that way, if an individual is very fit, it has a higher chance of being chosen, otherwise their chance is lower.

Breeder

Breeder Selection [17] follows the same strategy used for breeding animals and plants, where the goal is to preserve certain desired properties from the parents in their children. This is achieved by conserving the genetic material from the fittest individuals while still giving some mutation leeway by adding a few random individuals (lucky few) to the mix of parents for the next generation.

2.2.2. Crossover

Crossover is the reproduction method in genetic algorithms, and it consists of choosing the parts of each parent that will be present in their child.

Uniform

A uniform crossover is shown in Figure 3. In this type of crossover, each bit is chosen from one of the parents with probability of 0.5. The advantage of this method is that the same parents can form many children with a more diverse set of genes.

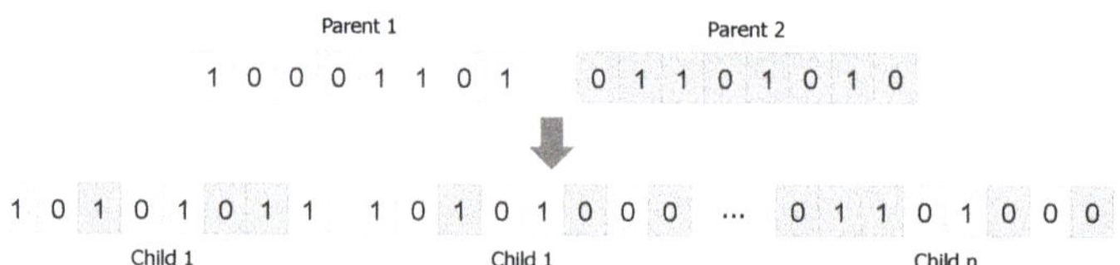

Figure 3. Uniform crossover: bitwise reproduction.

k-Point

In a *k*-point crossover, each parent is divided into *k* segments. These segments are then chosen with equal probability to compose the new chromosomes for their children. This crossover, shown in Figure 4, can be a one-point crossover ($k = 1$), or a multi-point crossover ($1 < k < N_C$, where N_C is the length of a chromosome). The disadvantage of the one-point crossover is that it is only able to generate two distinct children.

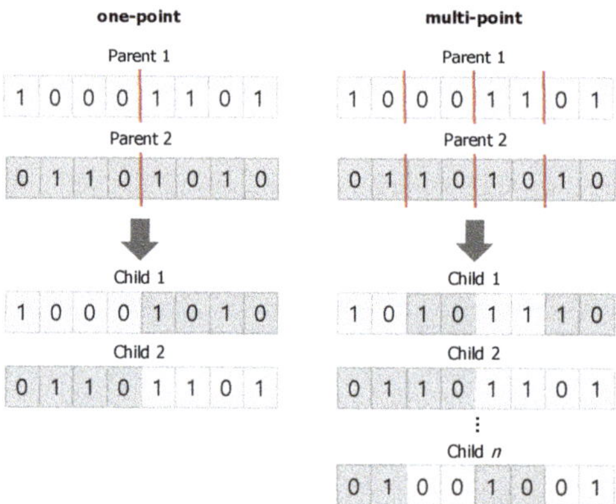

Figure 4. *k*-point crossover: blockwise reproduction.

2.2.3. Mutation

This genetic operator is used to ensure genetic diversity within a group of individuals and to ensure the algorithm does not converge to a local minimum. The mutation operator works by flipping bits in a chromosome according to a mutation probability, as shown in Figure 5.

Figure 5. Mutation of a bit in a chromosome.

3. Complexity Analysis

In this section, we aim to further explain our algorithm's suitability with respect to its complexity. In brute force, the total simulation count increases with respect to the solar cell's layers which are thickness-optimized, as shown previously in Equation (1). In Big-O notation, which represents the upper bound for time complexity in an algorithm, the brute-force method has algorithmic complexity as in Equation (2):

$$O_{bruteforce} = O(n^l) \tag{2}$$

where l is the number of layers and n is the number of fitness function evaluations for a layer. In other words, the complexity increases exponentially with the number of layers, and that can be very expensive as that number grows.

Our goal is to efficiently optimize layer thickness in devices composed of a single layer and multiple layers alike. Hence, we used genetic algorithm for our global optimization requirement. GA can converge to an optimal solution by evaluating less individuals than the brute-force method. However, in the classical approach, the same individual might be evaluated multiple times. To prevent this redundancy from happening, we implement GA with dynamic programming. In this approach, there is a dynamic dictionary, also called lookup table, which serves as memory to store the already evaluated individuals and their respective fitness scores. The dictionary is said to be dynamic because it may change size if the algorithm receives an individual whose fitness score has not been calculated yet. This approach is illustrated with an example in Figure 6, where in every generation, or iteration, the algorithm searches the memory for a desired individual. If this individual has already been computed, its fitness score can simply be used by the algorithm. Otherwise, the fitness function is evaluated, and the lookup table is updated.

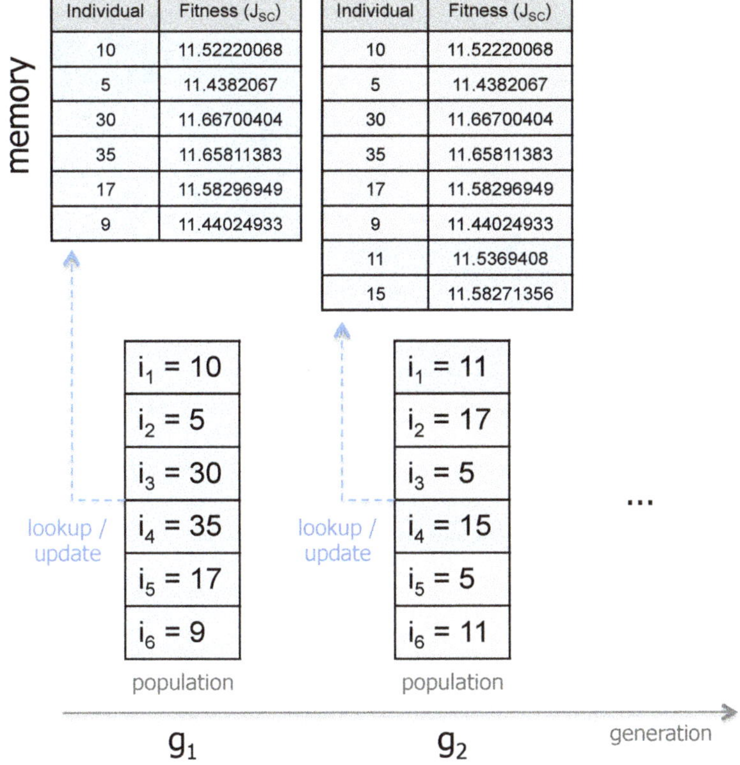

Figure 6. Genetic algorithm with dynamic programming, where blue represents values that need to be added to the lookup table in order to update it and grey represents values that have already been evaluated and need only to be copied when necessary.

4. Results and Discussion

Since GA is a stochastic algorithm, we calculated the average number of simulations required by the GA and its standard deviation from 5000 repeated runs per section. The accuracy of the data discussed below are all 100%, which means that all the 5000 runs converged to the optimal solution. We have discussed three sections below: single layer—ZnO thickness optimization; single layer—MoOx thickness optimization; and multiple layers—concurrent ZnO and MoOx thickness optimization.

4.1. Single Layer

4.1.1. ZnO Optical Spacer Layer

For the single ZnO optical spacer layer optimization, we fixed the MoOx layer thickness as 10 nm. We applied the brute force and the genetic algorithm to our task and solved it within the same thickness limits of 0 to 80 nm. The best J_{sc}, which is also the fitness value, was obtained when the ZnO thickness was optimized to 30 nm (as shown in Figure 2a). We have compared the total number of simulations and their respective accuracies for different selection methods. The initial population size, generations count, and mutation probability were the factors that were iterated in order to find the conditions that would use the least number of simulations to optimize the device structure. The population size was varied from 10 to 80 in increments of 10, the generation count from 10 to 100 in increments of 10, and the mutation probability from 5 to 100 in increments of 5. While the brute-force method required 81 simulations in total, the number of simulations required by the genetic algorithm was dependent on the selection method and initialization parameters used. The initialization parameters which provided the least average number of simulations from the 5000 simulations, while keeping the accuracy at 100%, is provided in Table 1.

Table 1. Comparison results for ZnO single-layer optimization.

	Brute-Force Method: Number of Simulations = 81 Optimized ZnO Thickness = 30 nm Selection Method			
Parameter	**Random**	**Roulette**	**Tournament**	**Breeder**
Population	20	80	70	60
Generation	40	10	30	10
Mutation prob (%)	80	15	60	75
Mean (simulations)	78.42 ± 1.82	80.47 ± 0.50	$\mathbf{78.16 \pm 1.65}$	79.17 ± 1.37

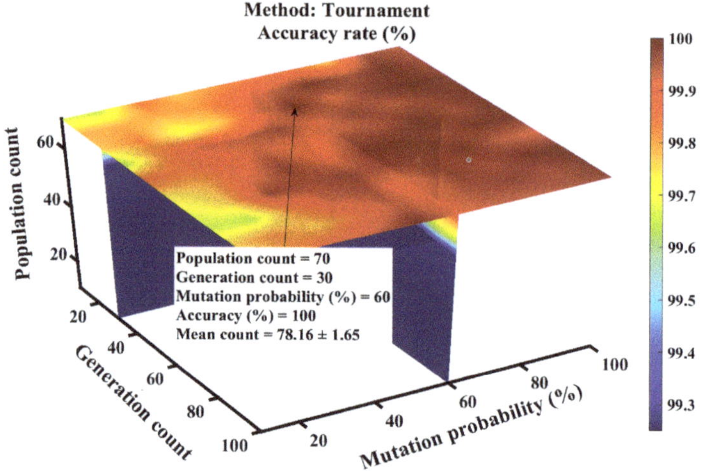

Figure 7. Single ZnO layer optimization using GA with tournament selection: accuracy (%) distribution data sliced at population size of 70, generation count of 30, and mutation probability of 60%.

Figure 7 presents the accuracy distribution over different initialization parameters. It was observed that the best result was obtained while using tournament selection model with the population size of 70, generation count of 30, and mutation probability of 60%. It required 78.16 ± 1.65 simulations to reach the optimal solution. Although the GA algorithm was observed to produce only a reduction of

2.26 $\pm$ 2.04% in the number of simulations required, this was mainly due to two optimal result points in the data. Since the device with a 24 nm ZnO layer thickness exhibited a J_{sc} of 116.62 A/m^2 and the one with the optimal 30 nm ZnO layer thickness exhibited a near same J_{sc} of 116.67 A/m^2, the algorithm took longer to converge at the optimal structure. In a practical scenario, however, if both the above-mentioned structures were regarded as optimal, the algorithm would converge with high accuracy with much lesser number of simulations.

4.1.2. MoOx Optical Spacer Layer

For the MoOx optical spacer single-layer optimization, the ZnO thickness was fixed at 30 nm. The brute-force method required 31 simulations to determine the optimized layer thickness of 8 nm (Figure 2b). For finding the best initialization parameter combination, we varied the population size from 5 to 20 in increments of 5, the generation count from 10 to 100 in increments of 10, and the mutation probability from 5 to 100 in increments of 5. The best case results for each selection model is provided in Table 2.

Table 2. Comparison results for MoOx single-layer optimization.

	Brute-Force Method: Number of Simulations = 31 Optimized MoOx Thickness = 8 nm Selection Method			
Parameter	**Random**	**Roulette**	**Tournament**	**Breeder**
Population	15	5	15	15
Generation	20	100	30	80
Mutation prob (%)	80	75	75	80
Mean (simulations)	30.91 $\pm$ 0.31	**13.05 $\pm$ 3.24**	30.97 $\pm$ 0.16	30.97 $\pm$ 0.18

It was observed that while the roulette method was able to use 57.9 $\pm$ 10.45% fewer simulations to determine the optimal MoOx thickness, the other methods required nearly the same number of simulations as the brute-force method. We hypothesize that the roulette method's preferential weighing of the fittest model aided in the convergence to the optimal solution faster. Figure 8 presents the accuracy distribution for the optimum initializing parameter values.

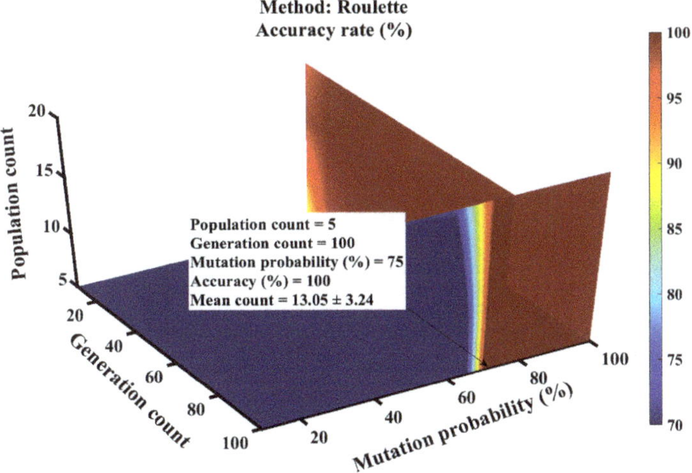

Figure 8. Single MoOx layer optimization using GA with roulette selection: accuracy (%) distribution data sliced at population size of 5, generation count of 100, and mutation probability of 75%.

4.2. Multi-Layer: ZnO + MoOx

As mentioned earlier, multi-layer optimization can take a large number of simulations. The computational and time cost required to run these simulations are expensive. GA can be used to refine the optimization process to take as less simulations as required. The ZnO layer thickness was incremented by 1 nm from 0 to 80 nm, while the same thickness increment was done from 0 to 30 nm for the MoOx layer. The brute-force method used 2511 simulations to find the optimized optical spacer layer thicknesses of 24 nm ZnO and 8 nm MoOx. We converted the 3D Figure 2c to a 2D Figure 2d by applying labels for each ZnO and MoOx thickness combination. Thus, we have 2511 labels and the GA algorithm was applied to determine the label pointing to the optimal result. The population size that was varied from 500 to 1500 in increments of 500, the generation count from 10 to 100 in increments of 10, and the mutation probability from 10 to 100 in increments of 10. Table 3 presents the result from multi-layer optimization.

Table 3. Comparison results for multi-layer optimization.

	Brute-Force Method: Number of Simulations = 2511 Optimized ZnO Thickness = 24 nm, Optimized MoOx Thickness = 8 nm Selection Method			
Parameter	**Random**	**Roulette**	**Tournament**	**Breeder**
Population	500	1000	500	500
Generation	90	90	80	90
Mutation prob (%)	10	90	20	50
Mean (simulations)	2391.34 ± 38.13	**1758.77** ± 39.75	2428.84 ± 34.57	2256.80 ± 70.15

Figure 9 presents a 29.96 ± 1.58% reduction in the number of simulations required by the roulette selection method to obtain the optimal solution. As the complexity became higher, having more randomness in the population through a high mutation probability rate aided constructively to reduce the number of simulations required. Due to this, the roulette selection method was able to show a best average number of simulation count of 1758.77 ± 39.75 from 5000 repetitive runs.

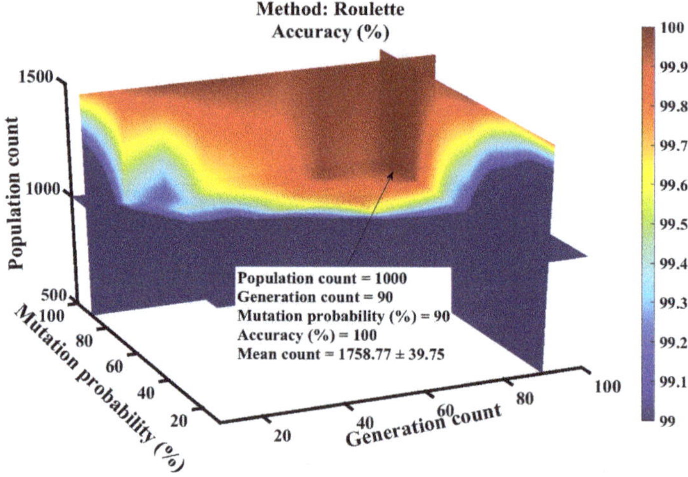

Figure 9. Multi-layer optimization using GA with roulette selection: accuracy (%) distribution data sliced at population size of 1000, generation count of 90, and mutation probability of 90%.

4.3. Performance Comparison: Uniform vs K-Point Crossover Methods

The roulette method demonstrates satisfactory performance in the ZnO layer thickness optimization problem and the best performances for both MoOx and multi-layer thickness optimization problems. These results are obtained using the uniform crossover genetic operator, whereas performances with the k-point crossover method remain uncharted. This section aims to fill that gap and provide a performance comparison between uniform and k-point crossover methods. We use k values of 1, 2, and 4 for our k-point crossover method.

Table 4 shows the average number of simulations required to solve the optimization problem when using uniform versus when using k-point crossover. It also compares the result obtained through the various crossover methods to the chromosome's binary bit size. In the MoOx case with $k = 1$ and $bits = 12$, the chromosome is divided into segments of 6 bits. Since the maximum number of bits required to define the MoOx layer thicknesses is 5 bits, one of the parents does not pass on its chromosome to the child. Due to this, there is no genetic variation between the parents and children, essentially meaning that the children are copies of one parent, hence the solution did not converge. We also observed that in most cases using k-point crossover, the best results were obtained when the chromosome was segmented into segments of 4 bits in length. The exception to this were the MoOx case with 5-bit chromosomes and the ZnO case with 12-bit chromosomes. Uniform crossover and 4-point crossover proved better in these cases, respectively. We hypothesize that the poorer performance at other k-point values was linked to the amount of variation in the chromosome. Lower k-point values meant fewer segmentations in the parent chromosome, which in turn meant fewer genetic variations occurred, leading to a slower convergence of the result. However, at the same time, higher k-point value meant high genetic variation which also lead to slower convergence.

Table 4. Comparison results for uniform and k-point crossover methods.

Layers	Uniform	1-point	2-point	4-point	Bits
	Average Number of Simulations				
	Crossover Methods				
MoOx	**13.05 ± 3.24**	17.83 ± 2.54	20.89 ± 2.84	14.73 ± 3.06	5
MoOx	**13.05 ± 3.24**	13.83 ± 1.69	19.88 ± 3.74	20.99 ± 3.72	8
MoOx	13.05 ± 3.24	- *	**12.14 ± 1.26**	19.05 ± 2.13	12
ZnO	80.47 ± 0.50	**70.76 ± 0.84**	75.72 ± 1.44	74.00 ± 1.96	8
ZnO	80.47 ± 0.50	74.04 ± 1.58	76.37 ± 1.56	**73.90 ± 1.51**	12
ZnO-MoOx	1758.77 ± 39.75	1701.78 ± 15.34	**1140.06 ± 37.11**	1355.23 ± 68.35	12

* Solution did not converge.

The best initialization parameters for the optimization problems discussed in this article are tabulated in Table 5.

Table 5. Initialization parameters for best performances.

Layers	Average Number of Simulations	Crossover Method	Bits	Population	Generation Count	Mutation Rate
					Initialization Parameters	
MoOx	12.14 ± 1.26	2-point	12	10	30	10
ZnO	70.76 ± 0.84	1-point	8	70	10	15
MoOx + ZnO	1140.06 ± 37.11	2-point	12	500	70	70

5. Conclusions

We have demonstrated that the genetic algorithm can perform better than the conventional parameter sweep used in simulations. In our best-case scenario, it exhibited no loss in accuracy, while outperforming the brute-force method by up to 57.9% with the correct initialization parameters. In the worst-case scenario, the GA utilized the same number of simulations as the brute-force method, demonstrating that it cannot be outperformed by brute force. We found that the best selection method was the roulette-wheel selection. For uniform crossover method, it exhibited a satisfactory performance for ZnO layer optimization and an outstanding performance for the MoOx and 2-layer optimization problems. When using k-point crossover, the roulette method was able to further decrease the number of simulations required to converge to the optimal result. In conclusion, we were able to reduce the average number of simulations required for MoOx layer optimization to 12.14 (brute force = 31), ZnO layer optimization to 70.76 (brute force = 81), and ZnO-MoOx layers optimization to 1140.06 (brute force = 2511). The GA is dependent on its initialization parameters and the selection method chosen. This article does not discuss an automated way to assign these parameters as it is not in its research scope. However, the results suggest that there is possibility for greatly refining the parameter sweep method using genetic algorithms as shown with both single and multi-layer optimization of the solar cell structure. Code and additional results are at https://github.com/gcunhase/GeneticAlgorithm-SolarCells.

Author Contributions: Conceptualization, P.V.; Methodology, P.V. and G.C.S.; Software, P.V., and G.C.S.; Validation, J.-H.B., J.J., I.M.K., J.P., H.K., and M.L.; Formal analysis, P.V., G.C.S., J.-H.B., J.J., H.K., and J.P.; Investigation, P.V., and G.C.S.; Resources, H.K., and J.-H.B.; Data curation, P.V., and G.C.S.; Writing—original draft preparation, P.V., and G.C.S.; Writing—review and editing, P.V., G.C.S., J.-H.B.; Visualization, P.V., and G.C.S.; Supervision, J.-H.B., and M.L.; Project administration, J.-H.B.; Funding acquisition, J.-H.B. All authors have read and agreed to the published version of the manuscript.

Funding: This research was supported by Basic Science Research Program through the National Research Foundation of Korea (NRF) funded by the Ministry of Science and ICT (2018R1A2B6008815), and also by the BK21 Plus project funded by the Ministry of Education, Korea (21A20131600011).

Conflicts of Interest: The authors declare no conflict of interest.

Abbreviations

The following abbreviations are used in this manuscript:

GA	Genetic Algorithm
FDTD	Finite Difference Time Domain
ZnO	Zinc Oxide
MoOx	Molybdenum Oxide
RWS	Roulette-Wheel Selection

References

1. Andreani, L.C.; Bozzola, A.; Kowalczewski, P.; Liscidini, M.; Redorici, L. Silicon solar cells: Toward the efficiency limits. *Adv. Phys. X* **2019**, *4*, 1548305. [CrossRef]
2. Darwin, C. *On the Origin of Species, 1859*; Routledge: Abingdon, UK, 2004.
3. Man, K.F.; Tang, K.S.; Kwong, S. Genetic algorithms: Concepts and applications [in engineering design]. *IEEE Trans. Ind. Electron.* **1996**, *43*, 519–534. [CrossRef]
4. Jafar-Zanjani, S.; Inampudi, S.; Mosallaei, H. Adaptive Genetic Algorithm for Optical Metasurfaces Design. *Sci. Rep.* **2018**, *8*, 11040. [CrossRef] [PubMed]
5. Tsai, C.M.; Fang, Y.C.; Lin, C.T. Application of genetic algorithm on optimization of laser beam shaping. *Opt. Express* **2015**, *23*, 15877–15887. [CrossRef] [PubMed]
6. Conkey, D.B.; Brown, A.N.; Caravaca-Aguirre, A.M.; Piestun, R. Genetic algorithm optimization for focusing through turbid media in noisy environments. *Opt. Express* **2012**, *20*, 4840–4849. [CrossRef] [PubMed]
7. Zhang, W.Q.; Monro, T.M. A genetic algorithm based approach to fiber design for high coherence and large bandwidth supercontinuum generation. *Opt. Express* **2009**, *17*, 19311–19327. [CrossRef] [PubMed]
8. Wang, C.; Yu, S.; Chen, W.; Sun, C. Highly Efficient Light-Trapping Structure Design Inspired By Natural Evolution. *Sci. Rep.* **2013**, *3*, 1025. [CrossRef] [PubMed]
9. Gouvea, R.A.; Moreira, M.L.; Souza, J.A. Evolutionary design algorithm for optimal light trapping in solar cells. *J. Appl. Phys.* **2019**, *125*, 043105. [CrossRef]

10. Kim, J.Y.; Vincent, P.; Jang, J.; Jang, M.S.; Choi, M.; Bae, J.H.; Lee, C.; Kim, H. Versatile use of ZnO interlayer in hybrid solar cells for self-powered near infra-red photo-detecting application. *J. Alloy. Compd.* **2020**, *813*, 152202. [CrossRef]
11. Vincent, P.; Song, D.S.; Jung, J.H.; Kwon, J.H.; Kwon, H.B.; Kim, D.K.; Choe, E.; Kim, Y.R.; Kim, H.; Bae, J.H. Dependence of the hybrid solar cell efficiency on the thickness of ZnO nanoparticle optical spacer interlayer. *Mol. Cryst. Liq. Cryst.* **2017**, *653*, 254–259. [CrossRef]
12. Vincent, P.; Shin, S.C.; Goo, J.S.; You, Y.J.; Cho, B.; Lee, S.; Lee, D.W.; Kwon, S.R.; Chung, K.B.; Lee, J.J.; et al. Indoor-type photovoltaics with organic solar cells through optimal design. *Dyes Pigment.* **2018**, *159*, 306–313. [CrossRef]
13. Jouane, Y.; Colis, S.; Schmerber, G.; Kern, P.; Dinia, A.; Heiser, T.; Chapuis, Y.A. Room temperature ZnO growth by rf magnetron sputtering on top of photoactive P3HT: PCBM for organic solar cells. *J. Mater. Chem.* **2011**, *21*, 1953–1958. [CrossRef]
14. Li, B.; Ren, H.; Yuan, H.; Karim, A.; Gong, X. Room-Temperature, Solution-Processed MoOx Thin Film as a Hole Extraction Layer to Substitute PEDOT/PSS in Polymer Solar Cells. *ACS Photonics* **2014**, *1*, 87–90. [CrossRef]
15. Fang, Y.; Li, J. A review of tournament selection in genetic programming. In Proceedings of the International Symposium on Intelligence Computation and Applications, Wuhan, China, 22 October 2010; pp. 181–192.
16. Jebari, K.; Madiafi, M. Selection methods for genetic algorithms. *Int. J. Emerg. Sci.* **2013**, *3*, 333–344.
17. Mühlenbein, H.; Schlierkamp-Voosen, D. Predictive models for the breeder genetic algorithm i. continuous parameter optimization. *Evol. Comput.* **1993**, *1*, 25–49. [CrossRef]

Article

A Side-Absorption Concentrated Module with a Diffractive Optical Element as a Spectral-Beam-Splitter for a Hybrid-Collecting Solar System

An-Chi Wei [1,2,*], Wei-Jie Chang [2] and Jyh-Rou Sze [3]

[1] Graduate Institute of Energy Engineering, National Central University, Taoyuan 320, Taiwan
[2] Department of Mechanical Engineering, National Central University, Taoyuan 320, Taiwan; gmformax@gmail.com
[3] Instrument Technology Research Center, National Applied Research Laboratories, Hsinchu 300, Taiwan; sze@narlabs.org.tw
* Correspondence: acwei@ncu.edu.tw; Tel.: +886-3-4267-378

Received: 2 November 2019; Accepted: 27 December 2019; Published: 1 January 2020

Abstract: In this paper, we propose a side-absorption concentrated module with diffractive grating as a spectral-beam-splitter to divide sunlight into visible and infrared parts. The separate solar energy can be applied to different energy conversion devices or diverse applications, such as hybrid PV/T solar systems and other hybrid-collecting solar systems. Via the optimization of the geometric parameters of the diffractive grating, such as the grating period and height, the visible and the infrared bands can dominate the first and the zeroth diffraction orders, respectively. The designed grating integrated with the lens and the light-guide forms the proposed module, which is able to export visible and infrared light individually. This module is demonstrated in the form of an array consisting of seven units, successfully out-coupling the spectral-split beams by separate planar ports. Considering the whole solar spectrum, the simulated and measured module efficiencies of this module were 45.2% and 34.8%, respectively. Analyses of the efficiency loss indicated that the improvement of the module efficiency lies in the high fill-factor lens array, the high-reflectance coating, and less scattering.

Keywords: solar concentrator; spectral beam splitting; diffractive optical element; diffractive grating

---

1. Introduction

Solar technologies have drawn significant attention due to the Earth's extreme climate and energy crises, and these technologies have recently made great progress. Mostly, those solar technologies transform sunlight into electricity or thermal power. Although one of the dominating technologies for generating electricity from sunlight is photovoltaics, this technology entails significant energy loss, including thermal loss and spectral loss. To enhance or extend the energy usage of photovoltaic systems (PV), researchers have collected waste thermal energy, such as cascading photovoltaic and thermal modules [1], integrated a PV-powered air-conditioning unit with a boiler [2], reduced the PV temperature for higher efficiency by using air-based hybrid photovoltaic/thermal systems (PV/T), water-based PV/T, and refrigerant-based PV/T [3–5], and even extended the operating bands by means of multi-junction photovoltaics, spectral-beam-splitters (SBS), and so on [6–9]. Among the systems belonging to the PV/T regime, SBS has the following advantages. Photovoltaic cells are no longer used as thermal receivers, so over-heating can be avoided. Moreover, their relatively low operation temperatures have led to high efficiency. The temperature of the heat transfer fluid (HTF) of thermal modules can be unrestricted by the operating temperature of the cells, resulting in a broad range of thermal applications [8]. Because of the characteristic of spectral splitting, SBS can be applied not only to the PV/T but also other hybrid-collecting solar systems, such as a dual-photovoltaic system with its cells operating in different energy-conversion bands [10,11]. In terms of configurations, the technique of SBS includes the following categories: dichroic filtering, liquid absorption, diffraction, and others [8,12]. Diffractive type of SBS can be sorted mainly by the alignment method of the sunlight receivers. The first type of alignment adopts a common optical axis for different receivers, such as photovoltaic cells and thermal tubes [13–15]. Thus, the shadow effect becomes unavoidable. The second kind of alignment arranges multiple receivers with different optical axes [16–19]. In this arrangement, the sunlight receivers can be photovoltaic cells with different absorption bands and lateral arrangements [12,13]. The third kind is proposed as a planar concentrator with a zig-zag optical axis, with the receivers located at different sides of the concentrator [20,21]. Herein, we propose a novel SBS configuration (of the third kind of alignment), consisting of a zig-zag optical axis, with integration of the diffractive optical element (DOE) and other planar optics to form a side-absorption concentrated module. In addition to the design principle of the module constructed by a single unit, we present and discuss a practical demonstration of seven units. Notably, because of its side-absorption structure, the system's thickness, complexities in component alignment, and wire connections can be reduced [22–24]. In brief, the proposed side-absorption concentrated module inherits the benefits of SBS, such as the improved operating efficiency of photovoltaics, more probable thermal-applications, etc., while its novel configuration facilitates a compact PV/T system with a simplified packaging process because of its side-absorption structure.

2. Principles

2.1. Side-Absorption Concentrated Module for Spectral-Beam Splitting

The proposed side-absorption concentrated module utilizes lenses, DOEs, and a light-guide as the condenser, the spectral beam-splitter and the out-coupling adapter, respectively. The structure of the proposed module is illustrated in Figure 1. Assume that the whole module is allocated on a solar tracker and that sunlight is regarded as normally incident to the module. When sunlight illuminates the positive lens, the light will be condensed and will pass through a DOE (considered a diffractive grating, herein). Then, the grating will diffract the light according to its wavelength. In this study, a structured light-guide with its top surface as the entrance and two side surfaces as two output ports will receive the diffracted waves, while the two ports export the sunlight in the visible and infrared bands individually. By means of this light-guide, the diffracted waves will be directed toward different output ports according to their spectra, resulting in the module splitting sunlight spectrally. Meanwhile, the proposed module inherits the advantages of the planar concentrator such that the output ports lie in the same horizontal plane, thereby allowing for planar outputs and saving the system volume. Afterwards, this module can be applied to a hybrid solar system with different energy-conversion mechanisms. As for the issues and challenges of the proposed module, they are discussed in Section 5.3 along with the comparisons to other techniques.

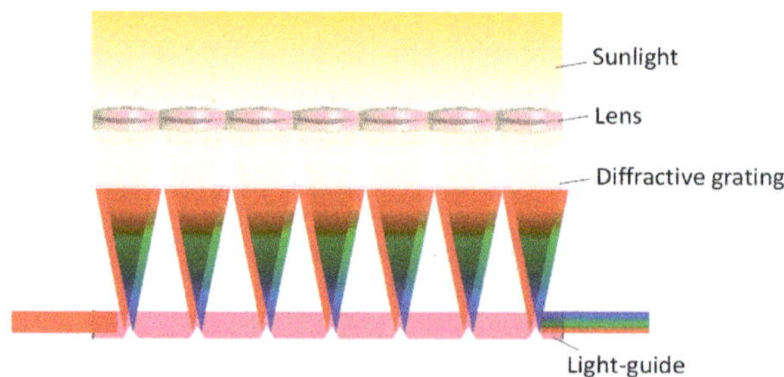

Figure 1. Schematic of the side-absorption concentrated module for spectral-beam splitting.

2.2. Performance Evaluation

The evaluation of the proposed side-absorption concentrated module consists of two stages. The primary stage evaluates the diffraction performance when the light passes through the grating, and the secondary stage evaluates the out-coupling performance when the light propagates through all components and couples out from the different output ports of the light-guide. Compared to the lens, the grating is more dominant during the first stage because its dispersion affects the performance of beam-splitting, and its high-order diffraction results in a loss. Accordingly, the lens is regarded as ideal to simplify the grating analyses in the primary stage. In this study, we evaluate grating by its diffractive efficiency in the visible band, the infrared band, and the whole solar spectrum. Meanwhile, the grating is designated to diffract the visible portion of sunlight toward the first-order diffraction and the infrared portion toward the zeroth-order diffraction. The efficiency of the grating in the visible band, $\eta_{grating}^{vis}$, can be then considered as the ratio of the optical power of the first-order visible output to that of the visible sunlight:

$$\eta_{grating}^{vis} = \frac{\int_{\lambda_s^{vis}}^{\lambda_e^{vis}} \eta_{diff,grating}^1(\lambda) \cdot S(\lambda)d\lambda}{\int_{\lambda_s^{vis}}^{\lambda_e^{vis}} S(\lambda)d\lambda}, \tag{1}$$

where $\eta_{diff,grating}^1(\lambda)$ denotes the first-order diffraction efficiency of the grating under a source with wavelength λ, and S represents the solar spectrum. λ_s^{vis} and λ_e^{vis} are the starting and ending wavelengths of the considered visible spectrum, respectively. Similarly, the grating efficiency in the infrared band, $\eta_{grating}^{IR}$, relates to the optical power of the zeroth-order infrared output:

$$\eta_{grating}^{IR} = \frac{\int_{\lambda_s^{IR}}^{\lambda_e^{IR}} \eta_{diff,grating}^0(\lambda) \cdot S(\lambda)d\lambda}{\int_{\lambda_s^{IR}}^{\lambda_e^{IR}} S(\lambda)d\lambda}, \tag{2}$$

where $\eta_{diff,grating}^0(\lambda)$ denotes the zeroth-order diffraction efficiency of the grating, and λ_s^{IR} and λ_e^{IR} are the start and the end wavelengths of the considered infrared spectrum, respectively. Likewise, all out-coupling waves from the designated diffraction orders are counted to derive the grating efficiency within the whole solar spectrum:

$$\eta_{grating}^{total} = \frac{\int_{\lambda_s^{vis}}^{\lambda_e^{vis}} \eta_{diff,grating}^1(\lambda) \cdot S(\lambda)d\lambda + \int_{\lambda_s^{IR}}^{\lambda_e^{IR}} \eta_{diff,grating}^0(\lambda) \cdot S(\lambda)d\lambda}{\int_{\lambda_s^{total}}^{\lambda_e^{total}} S(\lambda)d\lambda}, \tag{3}$$

where λ_s^{total} and λ_e^{total} are the start and the end wavelengths of the considered solar spectrum, respectively.

At the second stage, the synergetic effects of all components, including the lens, the grating, and the light-guide, are taken into account. The module efficiency for the visible band relates to the optical power of visible light out-coupled from the visible port of the light-guide:

$$\eta_{module}^{vis} = \frac{\int_{\lambda_s^{vis}}^{\lambda_e^{vis}} \eta_{lens}(\lambda) \cdot \eta_{diff,grating}^1(\lambda) \cdot \eta_{lightguide}(\lambda) \cdot S(\lambda)d\lambda}{\int_{\lambda_s^{vis}}^{\lambda_e^{vis}} S(\lambda)d\lambda}, \tag{4}$$

where $\eta_{lens}(\lambda)$ and $\eta_{lightguide}(\lambda)$ are the wavelength-dependent transmittance of the lens and the guiding efficiency of the light-guide, respectively. Similarly, the module efficiency for the infrared band and the whole solar spectrum can be respectively expressed as follows:

$$\eta_{module}^{IR} = \frac{\int_{\lambda_s^{IR}}^{\lambda_e^{IR}} \eta_{lens}(\lambda) \cdot \eta_{diff,grating}^{0}(\lambda) \cdot \eta_{lightguide}(\lambda) \cdot S(\lambda)d\lambda}{\int_{\lambda_s^{IR}}^{\lambda_e^{IR}} S(\lambda)d\lambda} \tag{5}$$

and

$$\eta_{module}^{total} = \frac{\int_{\lambda_s^{vis}}^{\lambda_e^{vis}} \eta_{lens}(\lambda) \cdot \eta_{diff,grating}^{1}(\lambda) \cdot \eta_{lightguide}(\lambda) \cdot S(\lambda)d\lambda + \int_{\lambda_s^{IR}}^{\lambda_e^{IR}} \eta_{lens}(\lambda) \cdot \eta_{diff,grating}^{0}(\lambda) \cdot \eta_{lightguide}(\lambda) \cdot S(\lambda)d\lambda}{\int_{\lambda_s^{total}}^{\lambda_e^{total}} S(\lambda)d\lambda}. \tag{6}$$

According to Equations (1)–(6), the spectral-splitting performance of the grating and that of the entire side-absorption concentrated module can be calculated and evaluated. In the following sections, these equations will be utilized for the design and optimization of the optical components in the proposed module, as well as for evaluation of the whole module.

3. Modeling

In order to facilitate the demonstration of a single unit and an assembled array of the proposed module, we considered the commercial specifications of lenses and then designated a lens as the condenser with a diameter of 25.4 mm and a distance of 30 mm between the lens and the grating. Since a lens with a long focal length produces a thick module, and that with a short focal length brings about large spherical aberrations, a lens with an appropriate focal length was plotted, and a commercial lens (model: AC254-080-A) with a focal length of 80 mm was selected. Notably, the selected lens uses achromatic coating to reduce chromatic aberration, which can simplify the optical properties of the waves incident to the grating and facilitate the grating design.

3.1. Optimization of Diffractive Grating

There are several considerations for designing the diffractive grating in the proposed module, including the grating period, the geometric shape, the material, and the diffraction orders in use. Because of the available fabrication processes and the requirement to achieve duplication with sufficient efficiency, the shape and material of the grating were designed as blazed and polyethylene terephthalate (PET), respectively. Meanwhile, the zeroth and the first diffraction orders were utilized. According to the grating equation, the diffractive beams can be characterized by their diffractive angles. For a normally incident light source, the diffractive angle of the mth-order diffraction can be expressed as in [25]:

$$\theta_m = \sin^{-1}\left(\frac{m\lambda}{d}\right), \tag{7}$$

where d is the grating period. By differentiation, the angular dispersion of the grating is derived as in [26]:

$$\frac{\partial \theta_m}{\partial \lambda} = \frac{m}{d \times \cos \theta_m}. \tag{8}$$

Equation (8) indicates that diffraction order m and grating period d are the determinative factors of angular dispersion. In the proposed module, the diffractive angle of every spectral wave influenced the overlapping of the diffractive spots on the following light-guide. Therefore, the effects of the grating period were analyzed based on Equation (7) for every wavelength. By means of the optical tool, LightTools, the diffractive spots from a broad-band source were simulated. The results show that when the grating period was less than 16 μm, the overlapping of the diffractive spots was eliminated. With consideration of the fabrication accuracy, the grating period was optimized as 15 μm.

Next, since the designated shape of the grating was blazed, the blaze angle was determined in terms of its diffraction efficiency, which is the merit-function to determine the blaze angle when the grating period is given [27]. Further, because of the wide spectral range of the light source, the diffraction efficiency was unable to be calculated accurately via scalar diffraction theory. Thus, the simulation was executed by means of the rigorous-coupled-wave-analysis (RCWA) software, Gsolver, to derive the diffraction efficiency of the blazed grate with different blaze angles. According to Equations (1)–(3), the grating efficiencies for the considered visible band (380–780 nm), infrared band (780~2520 nm), and whole solar band (280~2520 nm) were calculated, as shown in Figure 2. To maximize the grating efficiency within the whole solar spectrum, the optimized blaze angle was 3.51°, which is equivalent to a grating height of 0.92 μm. Then, the maximum grating efficiency for the whole solar spectrum reached 63.3%, while the corresponding efficiencies for the visible and infrared spectra were 73.8% and 54.4%, respectively. Additionally, the spectral-splitting performance illustrated in Figure 3 reveals that the unused diffraction orders were greatly suppressed.

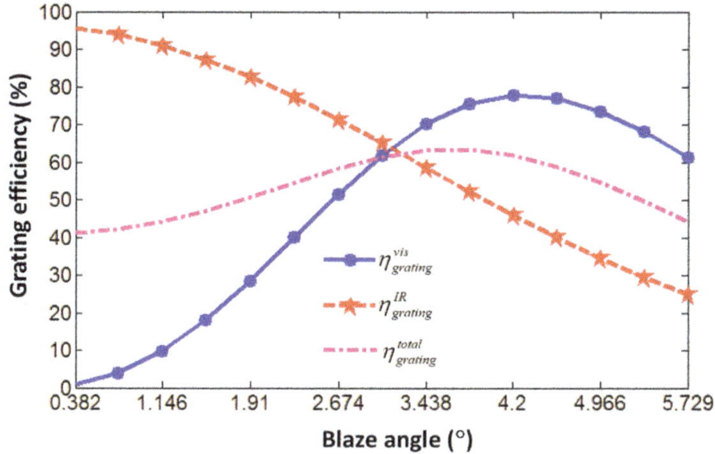

Figure 2. Relations between the grating efficiency and the blazed angle for different spectral bands.

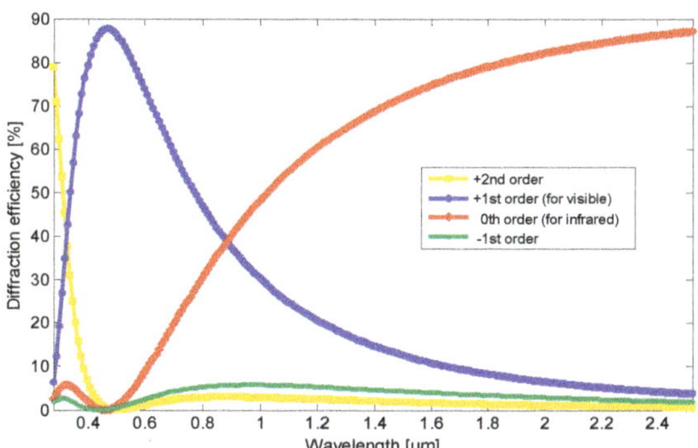

Figure 3. Diffraction efficiency for different diffraction orders in the spectral range from 280 to 2520 nm.

3.2. Design of the Module

After the lens and the grating had been designed, the following step was to determine the parameters of the light-guide. Herein, a single-unit module was considered for the design. An isosceles v-groove was assumed to exist on the bottom side of the light-guide with reflective coating to direct the spectral-split beams toward different output ports. The ray-tracing diagram for a collimated light source impinging on the proposed module is illustrated in Figure 4. This diagram shows that, for the first-order diffraction, the spot on the light-guide shifts along the +y axis when the wavelength increases, but this y-shift phenomenon does not occur for the zeroth-order light. Accordingly, adjusting the length, vertex angle, and center of the reflective v-groove, one can direct the first-order diffractive light, mainly in the visible spectrum, toward the visible output port and allow the on-axis zeroth-order diffractive light (mainly infrared) to be reflected toward the infrared output port. In order to maintain efficient propagation, the v-groove must make the light propagating within the light-guide to fulfill the condition of total internal reflection (TIR). Then, the v-groove was assigned as an isosceles triangle with a vertex angle of 120° to facilitate fabrication. On the other hand, when the proposed module was constructed as an array to enlarge the solar insolation, the greater number of units resulted in longer optical paths for the rays reflected by the v-groove and a higher possibility for them to encounter other v-grooves. Subsequently, the propagation inclination of these rays changed, and the probability of rays exiting the light-guide through the entrance surface increased. In this way, more loss was induced, and the module efficiency was reduced. Thus, the geometries of the v-groove and light-guide were optimized by maximizing the module efficiency of the whole solar spectrum, as defined in Equation (6). The material of the light-guide was assigned as polymethylmethacrylate (PMMA). The simulation was performed using the software LightTools with consideration of the antireflective coating on the lens, the Fresnel loss at every interface, and the half-angle subtended by the sun, while other factors, such as the reflectance of the v-groove, were regarded as ideal. Moreover, the sizes of the grating and the light-guide in the simulation were designed according to the dimensions of the lenses, as well as the required borders and output area compatible with the receiver.

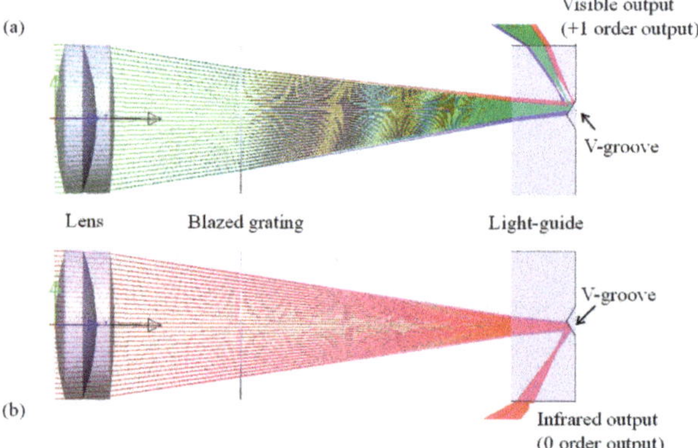

Figure 4. Single unit of the proposed module with the ray-tracing results for (**a**) the first-order and (**b**) the zeroth-order diffractive beams exiting through the visible and the infrared output ports, respectively.

Based on Equation (6), the module efficiency was counted from the simulated output power. Based on the results shown in Figure 5a, although the reflectance of the v-groove was assigned as unity, a longer v-groove length (as illustrated in Figure 5b) did not always contribute to higher module efficiency. The reason for this result is that an increased v-groove length enlarges the probability of rays encountering the second or successive grooves, resulting in more possible rays being reflected out of the light-guide through the entrance surface and leading to greater possible loss. Based on these simulated results, the geometric parameters of the v-groove were determined after the number of units for a module was given. The designed parameters of the v-groove and the light-guide will be presented in the next section along with those of the other components.

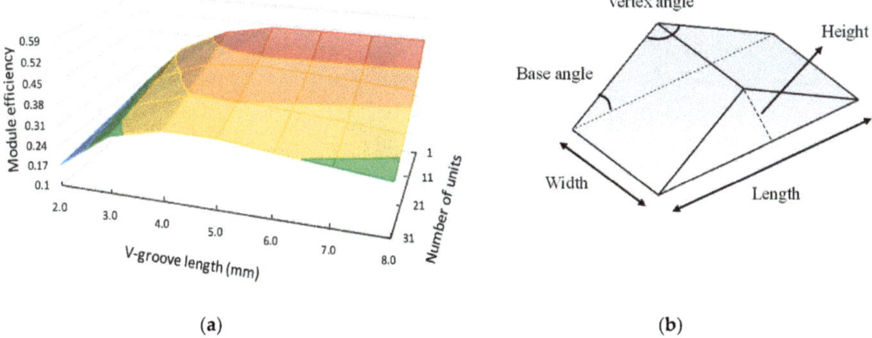

(**a**) (**b**)

Figure 5. (**a**) Relation between the module efficiency for the whole solar spectrum, the length of the v-groove, and the number of units based on the simulation results and (**b**) structure of the v-groove.

4. Demonstration and Experiment

The proposed module was demonstrated in the form of an array comprising seven units. Each unit consisted of three components, including a commercial achromatic lens (AC254-080-A), a blaze grating made of PET, and a PMMA light-guide with a carved v-groove. Based on the aforementioned design considerations, the parameters of the components were determined under the condition of a seven-unit module, as listed in Table 1. It is worth noting that the gratings of all units can be integrated into one device to simplify fabrication, and the light-guide has similar properties. Thus, the models of these two components were constructed as monolithic. Using Equations (4)–(6), the simulated module efficiency of the seven-unit module for the whole solar spectrum was 45.2%, and the efficiency for the visible and the infrared output ports was 53.8% and 37.5%, respectively. The grating sheet was fabricated via a roll-to-roll process. Meanwhile, the structured light-guide was fabricated by ultra-precision diamond machining, and then the carved v-grooves were coated with a reflective aluminum film. Moreover, in order to assemble all the components of the module, a fixture was designed and fabricated. The assembled module is shown in Figure 6.

After the assembly of all components with the fixture, the whole module was mounted on a dual-axis solar tracker for measurement. A customized integrating sphere, a power meter (NOVA II, Ophir Optronics), a visible spectrometer (SE1020C-025-VNIR, OtO Photonics Inc.), and an infrared spectrometer (SW2830S-050-NIRA, OtO Photonics Inc.) were utilized to measure the solar irradiance along with the out-coupling power and spectra from both the output ports. According the specifications of these instruments, the guaranteed power accuracy and the spectral accuracies of the visible and infrared bands were ±3%, less than 0.4 nm, and less than 1 nm, respectively. A photo of the experimental setup is shown in Figure 7.

Table 1. Designed parameters of the components for each unit in the 7-unit module.

Lens	Diameter		Focal Length		Material *
	25.4 mm		80 mm		N-BK7, N-SF5
Grating	**Period**		**Blaze Angle**		**Material**
	15 μm		3.51°		PET
Light-guide	**Dimensions (Width × Length × Thickness)**				**Material**
	25.4 mm × 25.4 mm × 10.8 mm				PMMA
V-groove	**Vertex Angle**	**Vertex Location**	**Length**	**Width**	**Coating**
	120°	Off-axis 0.82 mm	5 mm	5 mm	Al

* A commercial sample treated with achromatic coating was utilized.

Figure 6. Photo of the demonstrated module.

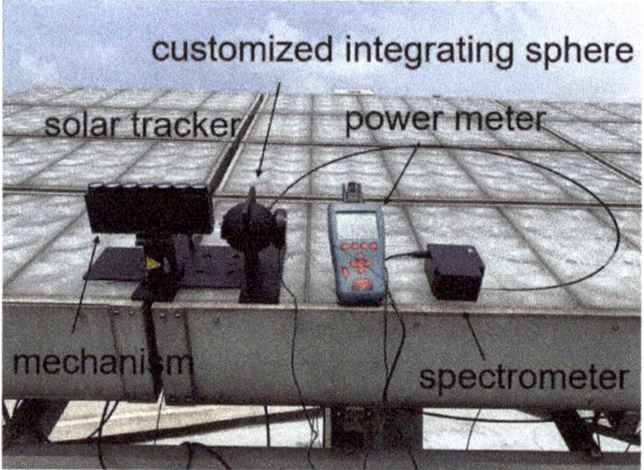

Figure 7. Assembled module with the measuring instruments allocated on the solar tracker.

The experiment was performed on a clear sunny day with the whole module fixed on the solar tracker. The irradiance of sunlight was measured as 606 W/m^2, and the incident solar power was counted as 2149.5 mW. The measured power in the visible spectrum from the visible output port was 523.7 mW, and that in the infrared spectrum from the infrared output port was 225.28 mW. Based on Equations (4)–(5), the module efficiencies were 44.8% and 24.6% for the visible and the infrared spectra, respectively. The experimental module efficiency for the whole solar spectrum was then counted as 34.8%, according to Equation (6).

When the proposed side-absorption concentrated module is connected to different receivers that are energy conversion devices, the system efficiency will vary. Ideally, the visible and the infrared output ports will be linked to the devices with complete energy conversion in the visible and infrared bands, respectively. Because of its proper absorption band and low cost, a silicon solar cell is an economic and efficient choice to link the visible output port. For the infrared port, the successive receiver for energy conversion can be either infrared solar cells or high-efficiency thermal modules, such as a solar water heater coupled with a phase-change material [28]. The system efficiencies of several feasible configurations were analyzed, and their details are reported in the following section.

5. Analyses and Discussions

5.1. System Efficiency

As mentioned above, different energy conversion devices for the proposed module will result in different system efficiencies. For example, the visible port of the proposed module can be connected to a commercial mono-crystalline silicon solar cell, with the external quantum efficiency (EQE) coarsely calculated as 75% [29], and the infrared port can be connected to a common water heater with a thermal efficiency of 70% [30,31]. Along with the measured data of the proposed module, the system efficiency is then estimated to be 25.9%.

Preferably, high-performance devices will be utilized for the proposed module. An example of such a device is an inter-digitated back contact (IBC) silicon solar cell with a high EQE [32]. This type of potential device will preferably be coupled to the visible port of the proposed module. On the other hand, a commercial germanium cell with a high EQE is one possible choice for the infrared port [33]. According to the reported EQE data for these cells and the simulated spectral responses of the proposed module, as illustrated in Figure 8, the system efficiency is calculated as 29.5%. The proposed module using this combination of silicon and germanium cells is comparable in efficiency to other double-cell techniques (for example, advancing tandem cells made of silicon and perovskite [34,35]).

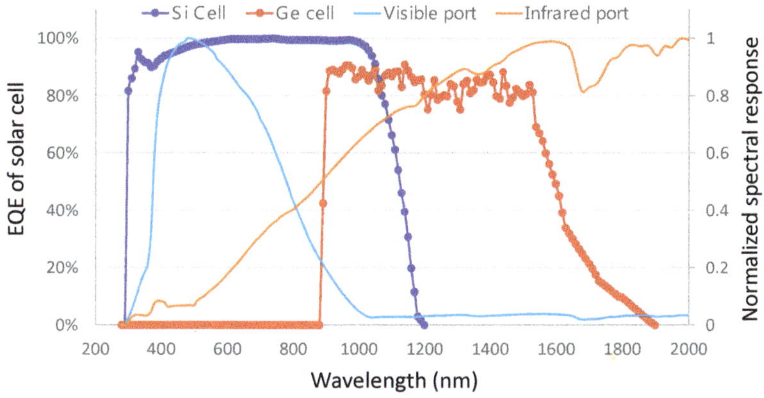

Figure 8. External quantum efficiency (EQE) data for the aforementioned silicon and germanium cells [32,33] and the normalized spectral responses of the visible and infrared ports in the proposed module.

5.2. Loss Analyses and Efficiency Enhancement

In order to investigate the discrepancy between the simulated and measured module efficiency, we performed loss analyses via simulation. The pertinent factors are described as follows. Unlike the original design with the lenses integrated compactly, the fixture of the fabricated module produced certain spaces for mounting the lenses, as shown in Figure 9. The modeling results show that such incompactness reduced the module efficiency. Thus, a lens array designed with a high-fill-factor is preferred. Meanwhile, the reflectance of the v-groove was not as ideal as unity. Thus, the reflectance spectrum of the aluminum-coated film, as plotted in Figure 10, is not negligible. In addition, a laser beam was utilized as a testing source to examine the surface quality of the v-groove, and scattering due to the surface roughness was inspected. Based on these factors, the module efficiency was analyzed, and the results are listed in Table 2. This analysis shows that the effects of these fabrication factors on the module efficiency for the whole solar spectrum are close (each around 3% to 4%), and their synergy dominates a 10% loss of such efficiency.

According to the above root cause analysis of the efficiency loss, improvement of module efficiency lies in the high fill-factor for the lens array, the high-reflectance coating (such as a silver coating), and the decreased scattering through the advanced polishing process. Furthermore, through modeling, we found that the Fresnel loss occurring at each refractive interface is also a dominant factor. When all Fresnel losses are eliminated, the module efficiency over the whole solar spectrum can be increased theoretically from 45.2% to 51.2%. Although producing a module without any Fresnel loss is difficult, a feasible approach for such a module would be to process two significant surfaces—the entrance surfaces of the grating and the light-guide—with an antireflective coating. Considering this feasible antireflective-coating, the module efficiency over the whole solar spectrum is 49.3%, based on the simulation results.

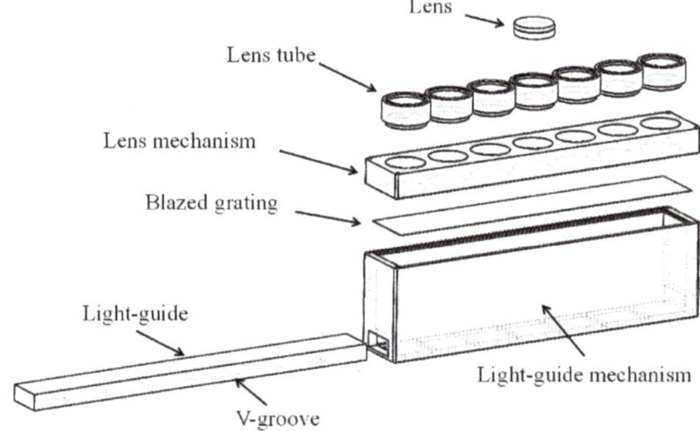

Figure 9. Exploded drawing of the practical module.

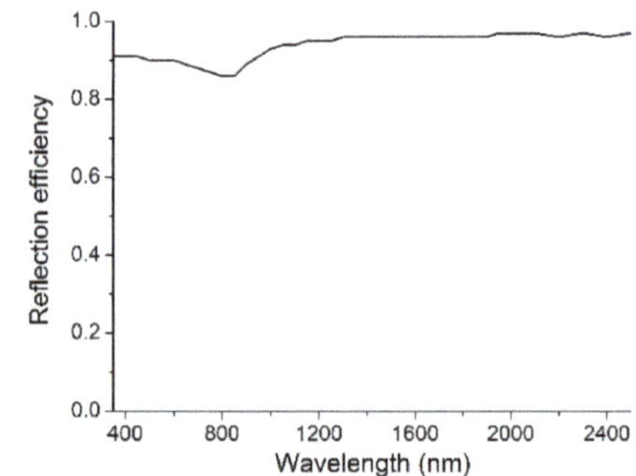

Figure 10. Spectral reflectance of the aluminum-coated film.

Table 2. Analyses of the module efficiency for a module with specific fabrication factors.

No.	Factors	η_{module}^{vis}	η_{module}^{IR}	η_{module}^{total}
(1)	Primary model [a]	53.8%	37.5%	45.2%
(2)	(1) + Bolder between lenses	52.9%	31.0%	42.0%
(3)	(2) + Reflectance of Al-coated film	47.4%	28.8%	38.1%
(4)	(3) + Scattering effect	43.6%	26.3%	34.9%
	Measurement	44.8%	24.6%	34.8%
	Expected system efficiency [b]	42.8%	14.6%	29.5%

[a] The primary model considered the geometry and material of each component, the antireflective coating of lens, the Fresnel loss at every refractive interface, and the half angle subtended by the sun. [b] The expected system efficiency was calculated on the basis of model No. (4) for an embodiment with the proposed module linking assumptively to an inter-digitated back contact (IBC) silicon solar cell and a commercial germanium cell [32,33].

5.3. Techniques Comparisons

The proposed side-absorption concentrated module is a zig-zag type of diffractive SBS. Although SBS possesses the aforementioned electrical and thermal advantages, there are different issues and challenges for different categories of SBS. The dichroic filtering category requires a dichroic filter to perform spectral beam splitting [12]; however, filters made of metal suffer from a considerable absorption loss, while dielectric filters made from multiple thin layers generally decrease performance after receiving solar insolation for a period of time. The liquid-absorption category uses liquid to absorb the thermal portion of sunlight and cool the photovoltaics [36]. However, the flow of liquid may affect the stability of the output power of the photovoltaics. The diffractive category can realize a relatively compact structure. However, the fabrication accuracy of the diffractive element is a critical factor and relates highly to the cost.

Among the available diffractive SBS techniques, the type using a common-axis possesses a simple structure but suffers from shadowing effect and its resultant loss. The type with multiple optical axes results in a compressed system by reducing the effective length of the optical axis. Nevertheless, the energy loss occurring at the spacing between its spectrally adjacent receivers is unavoidable. The type with a zig-zag axis contributes to another compact system by laterally (rather than longitudinally) allocating the electric or thermal modules. However, the longer zig-zag path may induce more loss due to the more probability of rays encountering the successive grooves. The performance of each representative diffractive-SBS technique is listed in Table 3 for comparison. This comparison shows that the performance of the proposed module is acceptable.

Table 3. Comparisons between the different diffractive-spectral-beam-splitters (SBS) techniques.

Category	Characteristics of Technique	Module Efficiency [a] or System Efficiency	Reference
Common axis	• A reflective hologram pasted on the quadric surface to form a spectral-beam-splitter	NA	[13]
	• A hologram mounted above a broadband receiver, whose center has an opening for another spectrally selective receiver	System efficiency = 21.4% for a PV/T system [a]	[14]
Multiple optical axis	• A nonuniform diffractive-grating as a spectral-beam-splitter	System efficiency = 34.7% for a dual-cell (InGaP/GaAs) system [b]	[17]
	• Micro-prism arrays as a spectral-beam-splitter	System efficiency = 46.05% for a triple-cell (InGaP/GaAs/InGaAs) system [c]	[18]
Zig-zag axis	• An integrated diffractive/refractive optical element as a spectral-beam-splitter, and a waveguide with engraved microstructures beneath the spectral beam splitter	Module efficiency ≤ 55% [b]	[20]
	• Lenses, diffractive grating, and a light-guide as the condenser, the spectral-beam-splitter, and the out-coupling adapter, respectively	Module efficiency = 34.8% [d], system efficiency = 29.5% for a dual-cell (Si/Ge) system [b]	Proposed

[a] Equivalent to "Optical efficiency" in some other literatures", [b] simulation data, [c] estimated results from the measured optical efficiency, [d] measured data.

6. Conclusions

In this study, we proposed a side-absorption concentrated module as a spectral-beam-splitter to separate the visible and infrared bands of sunlight for different energy-conversion applications. The proposed module integrates diffractive grating, lenses, and a light-guide to split sunlight according to spectral bands and to export the spectrum-split rays to individual planar ports. A design example with seven units has been provided for demonstration. In the simulation, the grating efficiency and module efficiency of this module were counted as 63% and 45.2%, respectively, for the whole solar spectrum. Experimentally, the components of the proposed module were fabricated and assembled; the module efficiency for the whole solar spectrum was then measured as 34.8%. Using this type of module with its visible and infrared ports linked to high-EQE silicon and germanium solar cells, respectively, a system efficiency of 29.5% is expected (based on the simulation results and the released data of solar cells). Accordingly, this system is comparable in efficiency to other double-cell systems. Moreover, the details of efficiency loss were analyzed for the proposed module, while the approaches to enhance module efficiency were discussed and presented. This analysis shows that when all Fresnel losses and the pertinent factors, including fixture limitation, reflectance of Al-coated film, and scatting effect are well-controlled, a module efficiency of 51.2% is expected for the whole solar spectrum. As for the energy gain in a PV/T system, the worthiness of the hardware development, and the payback time, they relate highly to PV materials, thermal system designs, the number of units of the proposed module, the climate and weather conditions of the demonstration site, and so on. Since some of them, such as the climate conditions, generally need sufficient time to acquire, collecting the necessary data, optimizing the performance, and lowering the cost of the proposed module will be our future researches for system realization.

Author Contributions: Conceptualization, A.-C.W.; formal analysis, A.-C.W.; funding acquisition, J.-R.S.; investigation, W.-J.C.; methodology, J.-R.S.; project administration, A.-C.W.; resources, A.-C.W.; software, W.-J.C.; supervision, A.-C.W.; validation, W.-J.C.; writing—original draft, W.-J.C.; writing—review and editing, A.-C.W. All authors have read and agreed to the published version of the manuscript.

Funding: This research was funded by the Ministry of Science and Technology, Taiwan, R.O.C., grant number MOST 107-2221-E-008-093 and 107-2221-E-492-026-MY3 by Academia Sinica, Taiwan, R.O.C., grant number AS-SS-108-04-02.

Acknowledgments: The authors thank Pi-Cheng Tung, Department of Mechanical Engineering, National Central University, Taiwan, R.O.C., and his group for the technical support on solar tracker.

Conflicts of Interest: The authors declare no conflict of interest.

References

1. Tian, Y.; Zhao, C.-Y. A review of solar collectors and thermal energy storage in solar thermal applications. *Appl. Energy* **2013**, *104*, 538–553. [CrossRef]
2. Nizetic, S.; Coko, D.; Marasovic, I. Experimental study on a hybrid energy system with small-and medium-scale applications for mild climates. *Energy* **2014**, *75*, 379–389. [CrossRef]
3. Zhang, X.; Zhao, X.; Smith, S.; Xu, J.; Yu, X. Review of R & D progress and practical application of the solar photovoltaic/thermal (PV/T) technologies. *Renew. Sustain. Energy Rev.* **2012**, *16*, 599–617.
4. Zhao, X.; Zhang, X.; Riffat, S.; Su, X. Theoretical investigation of a novel PV/e roof module for heat pump operation. *Energy Convers Manag.* **2011**, *52*, 603–614. [CrossRef]
5. Zondag, H.; De Vries, D.; Van Helden, W.; Van Zolingen, R.; Van Steenhoven, A. The yield of different combined PV-thermal collector designs. *Sol. Energy* **2003**, *74*, 253–269. [CrossRef]
6. Yamaguchi, M.; Takamoto, T.; Araki, K.; Ekins-Daukes, N. Multi-junction III–V solar cells: Current status and future potential. *Sol. Energy* **2005**, *79*, 78–85. [CrossRef]
7. Philipps, S.P.; Dimroth, F.; Bett, A.W. High-efficiency III–V multijunction solar cells. In *McEvoy's Handbook of Photovoltaics*; Elsevier: Amsterdam, The Netherlands, 2018; pp. 439–472.
8. Ju, X.; Xu, C.; Han, X.; Du, X.; Wei, G.; Yang, Y. A review of the concentrated photovoltaic/thermal (CPVT) hybrid solar systems based on the spectral beam splitting technology. *Appl. Energy* **2017**, *187*, 534–563. [CrossRef]
9. Jiang, S.; Hu, P.; Mo, S.; Chen, Z. Optical modeling for a two-stage parabolic trough concentrating photovoltaic/thermal system using spectral beam splitting technology. *Sol. Energy Mater. Sol. Cells* **2010**, *94*, 1686–1696. [CrossRef]

10. Froehlich, K.; Wagemann, E.U.; Frohn, B.; Schulat, J.; Stojanoff, C.G. Development and fabrication of a hybrid holographic solar concentrator for concurrent generation of electricity and thermal utilization. In Proceedings of the Optical Materials Technology for Energy Efficiency and Solar Energy Conversion XII, San Diego, CA, USA, 22 October 1993; pp. 311–319.

11. Fraas, L.M. Infrared Photovoltaics (IR PV) for Combined Solar Lighting and Electricity for Buildings. In *Low-Cost Solar Electric Power*; Springer: Berlin/Heidelberg, Germany, 2014; pp. 127–134.

12. Crisostomo, F.; Taylor, R.A.; Surjadi, D.; Mojiri, A.; Rosengarten, G.; Hawkes, E.R. Spectral splitting strategy and optical model for the development of a concentrating hybrid PV/T collector. *Appl. Energy* **2015**, *141*, 238–246. [CrossRef]

13. Meckler, M. Fixed Solar Concentrator-Collector-Satellite Receiver and Co-Generator. U.S. Patent 4,490,981, 1 January 1985.

14. Stojanoff, C.G.; Schulat, J.; Eich, M. Bandwidth-and angle-selective holographic films for solar energy applications. In Proceedings of the Solar Optical Materials XVI, Denver, CO, USA, 11 October 1999; pp. 38–49.

15. Vorndran, S.; Russo, J.M.; Wu, Y.; Gordon, M.; Kostuk, R. Holographic diffraction-through-aperture spectrum splitting for increased hybrid solar energy conversion efficiency. *Int. J. Energy Res.* **2015**, *39*, 326–335. [CrossRef]

16. Fröhlich, K.; Wagemann, E.; Schulat, H.; Schütte, H.; Stojanoff, C. Fabrication and test of a holographic concentrator for two color PVoperation. In Proceedings of the Optical Materials Technology for Energy Efficiency and Solar Energy Conversion XIII, Freiburg, Germany, 9 September 1994; pp. 812–821.

17. Stefancich, M.; Zayan, A.; Chiesa, M.; Rampino, S.; Roncati, D.; Kimerling, L.; Michel, J. Single element spectral splitting solar concentrator for multiple cells CPV system. *Opt. Express* **2012**, *20*, 9004–9018. [CrossRef] [PubMed]

18. Erim, M.N.; Erim, N.; Kurt, H. Spectral splitting for an InGaP/GaAs parallel junction solar cell. *Appl. Opt.* **2019**, *58*, 4265–4270. [CrossRef] [PubMed]

19. Li, D.; Michel, J.; Hu, J.; Gu, T. Compact spectrum splitter for laterally arrayed multi-junction concentrator photovoltaic modules. *Opt. Lett.* **2019**, *44*, 3274–3277. [CrossRef] [PubMed]

20. Kostuk, R.K.; Rosenberg, G. Analysis and design of holographic solar concentrators. In Proceedings of the High and Low Concentration for Solar Electric Applications III, San Diego, CA, USA, 9 September 2008; p. 70430I.

21. Michel, C.; Blain, P.; Clermont, L.; Languy, F.; Lenaerts, C.; Fleury-Frenette, K.; Décultot, M.; Habraken, S.; Vandormael, D.; Cloots, R. Waveguide solar concentrator design with spectrally separated light. *Sol. Energy* **2017**, *157*, 1005–1016. [CrossRef]

22. Karp, J.H.; Tremblay, E.J.; Ford, J.E. Planar micro-optic solar concentrator. *Opt. Express* **2010**, *18*, 1122–1133. [CrossRef] [PubMed]

23. Karp, J.H.; Tremblay, E.J.; Hallas, J.M.; Ford, J.E. Orthogonal and secondary concentration in planar micro-optic solar collectors. *Opt. Express* **2011**, *19*, A673–A685. [CrossRef]

24. Wei, A.-C.; Chen, Z.-R.; Sze, J.-R. Planar solar concentrator with a v-groove array for a side-absorption concentrated photovoltaic system. *Optik* **2016**, *127*, 10858–10867. [CrossRef]

25. Hecht, E. *Optics*; Pearson Education: London, UK, 2017.

26. Sze, J.-R.; Wei, A.-C. Crossed Czerny–Turner Spectrometer with Extended Spectrum Using Movable Planar Mirrors. *Appl. Spectrosc.* **2018**, *72*, 776–786. [CrossRef]

27. Sinzinger, S.; Jahns, J. *Microoptics*; Wiley-VCH Verlag GmbH: Darmstadt, Germany, 1999.

28. Sun, L.; Xiang, N.; Yuan, Y.; Cao, X. Experimental Investigation on Performance Comparison of Solar Water Heating-Phase Change Material System and Solar Water Heating System. *Energies* **2019**, *12*, 2347. [CrossRef]

29. Solar Cell Products of United Renewable Energy. Available online: http://www.nsp.com/upload/product/20190606145526_battery_pdf_1.pdf (accessed on 29 October 2019).

30. Jaisankar, S.; Ananth, J.; Thulasi, S.; Jayasuthakar, S.; Sheeba, K. A comprehensive review on solar water heaters. *Renew. Sustain. Energy Rev.* **2011**, *15*, 3045–3050. [CrossRef]

31. Jamar, A.; Majid, Z.; Azmi, W.; Norhafana, M.; Razak, A. A review of water heating system for solar energy applications. *Int. Commun. Heat Mass Transf.* **2016**, *76*, 178–187. [CrossRef]

32. Haase, F.; Hollemann, C.; Schaefer, S.; Merkle, A.; Rienaecker, M.; Krügener, J.; Brendel, R.; Peibst, R. Laser contact openings for local poly-Si-metal contacts enabling 26.1%-efficient POLO-IBC solar cells. *Sol. Energy Mater. Sol. Cells* **2018**, *186*, 184–193. [CrossRef]

33. Solar Cell Products of Arima. Available online: http://www.arima.com.tw/group1-detail.php?index_m1_id=1&index_m2_id=6&index_id=3#parentHorizontalTab3 (accessed on 29 October 2019).

34. Bisquert, J.; Juarez-Perez, E.J. The Causes of Degradation of Perovskite Solar Cells. *J. Phys. Chem. Lett.* **2019**, *10*, 5889–5891. [CrossRef] [PubMed]

35. Zheng, J.; Mehrvarz, H.; Ma, F.-J.; Lau, C.F.J.; Green, M.A.; Huang, S.; Ho-Baillie, A.W. 21.8% efficient monolithic perovskite/homo-junction-silicon tandem solar cell on 16 cm^2. *ACS Energy Lett.* **2018**, *3*, 2299–2300. [CrossRef]

36. Matsuoka, H.; Tamura, T. Design and evaluation of thermal-photovoltaic hybrid power generation module for more efficient use of solar energy. *NTT Docomo Tech. J.* **2007**, *12*, 61–67.

Article

Parameters Extraction of Photovoltaic Models Using an Improved Moth-Flame Optimization

Huawen Sheng, Chunquan Li *, Hanming Wang, Zeyuan Yan, Yin Xiong, Zhenting Cao and Qianying Kuang

School of Information Engineering, Nanchang University, Nanchang 330029, China;
huawensheng@outlook.com (H.S.); wanghanmingcsw@outlook.com (H.W.); yzyilia@outlook.com (Z.Y.);
xiongyin@ncu.edu.cn (Y.X.); zhentingcao125@gmail.com (Z.C.); QianYingKuangK@outlook.com (Q.K.)
* Correspondence: lichunquan@ncu.edu.cn

Received: 1 August 2019; Accepted: 6 September 2019; Published: 13 September 2019

Abstract: Photovoltaic (PV) models' parameter extraction with the tested current-voltage values is vital for the optimization, control, and evaluation of the PV systems. To reliably and accurately extract their parameters, this paper presents one improved moths-flames optimization (IMFO) method. In the IMFO, a double flames generation (DFG) strategy is proposed to generate two different types of target flames for guiding the flying of moths. Furthermore, two different update strategies are developed for updating the positions of moths. To greatly balance the exploitation and exploration, we adopt a probability to rationally select one of the two update strategies for each moth at each iteration. The proposed IMFO is used to distinguish the parameter of three test PV models including single diode model (SDM), double diode model (DDM), and PV module model (PMM). The results indicate that, compared with other well-established methods, the proposed IMFO can obtain an extremely promising performance.

Keywords: moth-flame optimization; parameter extraction; photovoltaic model; double flames generation (DFG) strategy

---

1. Introduction

Owing to fuel depletion, global warming, and environmental pollution, the renewable energy sources such as wave, tidal, wind, geothermal, and biomass have received more attention in recent years [1–4]. Solar energy is considered as an extremely promising renewable energy source owing to its usability and cleanliness [5–7]. Furthermore, solar energy is also widely adopted to generate power via photovoltaic (PV) systems [8]. However, PV systems are susceptible to external environmental factors such as temperature and global irradiance, affecting the use efficiency of solar energy [9]. Therefore, it is crucial to accurately establish current-voltage models for optimizing and controlling PV systems [10]. The current-voltage models mainly include single diode model (SDM) and double diode model (DDM) [11], which are widely utilized to indicate the connection between current and voltage. The accuracy and reliability of the PV models are primarily determined by their parameters. However, because of unstable operating cases such as faults and aging, the PV models' parameters are generally not available and vary. Thus, developing an effective method to accurately extract these parameters becomes especially critical.

Recently, a large number of algorithms have been proposed for identifying the parameters of PV models. They are mainly divided into the following three categories: analytical [12], deterministic [13], and heuristic methods [14]. The first category methods identify the PV models' parameters via analyzing a set of mathematical formulas [15,16], which accelerates and simplifies the calculation. However, the reliability and accuracy of the solution are poor due to some postulations that need to be made before the analysis [17,18]. The second category methods are greatly susceptible to initial assumptions and tend to fall into local optimum. In addition, they depend on the model's differentiability or convexity. However, the PV models are usually non-linear and multimodal, resulting in a set of poor solutions when using the second category methods. The third category methods are heuristic methods and can properly overcome the defects of analytical and deterministic methods. Therefore, they have become potential alternatives to PV models' parameters extraction.

Until now, various heuristic algorithms have been proposed for the parameter extraction of the PV models. Reference [14] presents a genetic algorithm (GA) for identifying the SDM parameters of a photovoltaic panel. Reference [19] proposes a new method adopting particle swarm optimization (PSO) with inverse barrier constraint to determine the uncertain parameters of a PV model. Reference [20] applies the artificial bee colony (ABC) to identify the optimal parameters of a silicon solar cell. Reference [21] presents an improved adaptive differential evolution (IADE) algorithm, which utilizes a simple structure based on the feedback of fitness value in the evolutionary process. Reference [22] uses the bird mating optimizer (BMO) to estimate the optimal parameters of silicon solar cell. Reference [23] proposes the chaotic whale optimization algorithm (CWOA), which adopts the chaotic maps to calculate and automatically adapt the internal parameters of the optimization algorithm. Reference [24] presents an improved ant lion optimizer (IALO), which arranges the initial positions of individuals by chaotic sequence to improve the uniformity and ergodicity of population. Reference [25] adopts the bacterial foraging algorithm (BFA) to predict the solar PV characteristics accurately. Reference [26] proposes an improved JAYA (IJAYA) algorithm, in which a self-adaptive weight is proposed to regulate the trend of moving towards the best result, getting out of the worst result at distinct search stages. Reference [27] uses a teaching learning–based optimization (TLBO) algorithm to extract all five parameters of a solar cell. Reference [28] proposes flower pollination algorithm (FPA) to extract the optimal parameters of different PV models.

Furthermore, as a population-based heuristic algorithm, MFO [29] is inspired by moths' behaviors and acquires great performance when solving non-convex and multimodal optimization problems. In MFO, each of the moths represents a solution for problems. Specifically, the algorithm's most striking feature is that each moth has its own flame and flies toward the flame in a spiral curve. Consequently, the flight curves of the moths are always oriented at a fixed angle to the remote target flames, known as the transverse orientation. In MFO, the flames' number is gradually decreased with the iterations' number raising, which allows MFO to obtain fast convergence speed. Note that reference [29] demonstrates MFO's excellent optimization performance in relation to other well-known methods such as GA [18], FPA [28], and PSO [19]. Besides, MFO has been commonly applied to various PV-related practical problems. In reference [30], MFO is utilized to estimate parameters of the three diode models for multi-crystalline solar cells/modules. In reference [31], MFO is used for PV MPPT and partial shading problem. In reference [32], MFO is applied for AI-based MPPT that is used for partial-shaded grid-connected PV plant.

Nevertheless, in MFO, the fast information exchanges among moths are produced owing to all the moths flying toward their best target flames. Once the global optimal value is close to the local optimum, the moths can easily fall into the local optimum of the search area. To overcome the shortcoming of MFO, various MFO variants have been proposed. Reference [33] presents a novel opposition-based moths-flames optimization (OMFO) that uses an opposition-based learning strategy in MFO to improve convergence speed. Reference [34] proposes an efficient hybrid algorithm based on water cycle and MFO algorithms (WCMFO), which combines the Levy-flight operator and spiral movement of MFO into the Water Cycle (WC) method to enhance the WC's ability of exploration and exploitation, respectively. Reference [35] presents a moth-flame optimization algorithm based on chaotic crisscross operator (CCMFO) that uses the chaotic operator and crisscross strategy in MFO to enhance the convergence of premature. However, the above MFO variants have no attempts for solving the PV models' parameters extraction problems.

This paper proposes an improved MFO (IMFO) for accurately estimating the parameters of the PV models. In IMFO, a double-flames generation (DFG) strategy is presented to generate two different kinds of target flames for effectively guiding the flying of moths. By adopting corresponding local best moths, a type of flames called local flame (LF) is produced. Another type of flames is called global flame (GF) and is generated by combining all the moths' current personal best positions (pbests). Furthermore, two different update strategies are created for updating the positions of moths, and each moth only adopts one of the two different evolution update strategies at each iteration. To ensure the effective compromise between global exploration and local exploitation, a probability is utilized to rationally select the two different update strategies for each moth. To demonstrate the superior performance of IMFO, we compare the proposed method with eight well-established algorithms on extracting the parameters of three different PV models, i.e., SDM, DDM, and PMM. The experimental results show that IMFO can achieve a highly competitive performance as well as can reliably and accurately extract the parameters of the three different PV models.

The main contributions of our work are as follows:

1. A novel IMFO algorithm is presented to extract the parameters of different PV models. Different from most existing MFO methods, IMFO develops a double-flames generation (DFG) strategy to generate two kinds of high-performance flames. Furthermore, in IMFO, two different update strategies of moths are developed to effectively tackle exploration and exploitation. Besides, a probability is applied to rationally select one of two different update strategies for each moth's update, contributing the efficient compromise between the global exploration and local exploitation.

2. To the best of our knowledge, there are no MFO methods for solving the parameters extraction on the PMM. Specifically, most existing MFO variants such as OMFO and WCMFO have never been used for the parameter extraction on the three PV models including SDM, DDM, and PMM. Although MFO [30] has been used for the DDM, it has never been adopted for the parameter on the SDM or PMM. However, the proposed IMFO fully considers the parameter extraction on the SDM, DDM, and PMM.

3. Comparisons are given between IMFO and eight well-established algorithms on extracting the parameters of three PV models. The results verify that IMFO can achieve better performance than the eight algorithms on the extraction of the three PV models' parameters.

4. The rest of the paper can be arranged as follows. Section 2 introduces the problem formulation of PV models. The original MFO algorithm is illustrated in Section 3. Section 4 gives the detail of the IMFO algorithm. The experimental results and analysis of different PV models are given in Section 5. Furthermore, the discussion is given in Section 6. Finally, the conclusion is displayed in Section 7.

2. Formulation of PV Models

The single diode model (SDM), double diode model (DDM), and PV module model (PMM) are generally adopted to indicate the relationship between current and voltage of solar cells and PV systems. In this section, the SDM, the DDM, and the PMM are given as follows. Furthermore, the objective function for extracting the parameters of the three models is described as follows.

2.1. Single Diode Model

The SDM has been commonly used to illustrate the static characteristics of solar cells owing to its simple structure and accuracy. As shown in Figure 1a, the SDM is composed of a diode, a current source, a shunt resistor, and a series resistor. Note that the shunt resistor and series resistor are used for indicating leakage current and denoting load current loss, respectively. The output current I_{o} of this diode model can be computed as follows:

$$I_{\mathrm{o}} = I_{\mathrm{ph}} - I_{\mathrm{d}} - I_{\mathrm{sh}} \tag{1}$$

where I_{sh}, I_d, and I_{ph} represent the shunt resistor current, the diode current, and the photo produced current, respectively; I_o denotes the calculated output current. The calculation formulas of I_d and I_{sh} are shown as follows:

$$I_d = I_{sd}\left[\exp\left(\frac{q(V_t + I_t R_s)}{mkT}\right) - 1\right] \tag{2}$$

$$I_{sh} = \frac{I_t R_s + V_t}{R_{sh}} \tag{3}$$

where I_{sd} indicates the saturation current of the diode; I_t and V_t respectively represent the tested output current and voltage; R_s means the series resistance; q and k are the electron charge ($1.60217646 \times 10^{-19}$C) and Boltzmann constant ($1.3806503 \times 10^{-23}$ J/K), respectively; m is the ideal factor of diode; T is the cell temperature (K); R_{sh} denotes the shunt resistance. Consequently, we can rewrite Formula (1) as follows:

$$I_o = I_{ph} - I_{sd}\left[\exp\left(\frac{q(V_t + I_t R_s)}{mkT}\right) - 1\right] - \frac{I_t R_s + V_t}{R_{sh}} \tag{4}$$

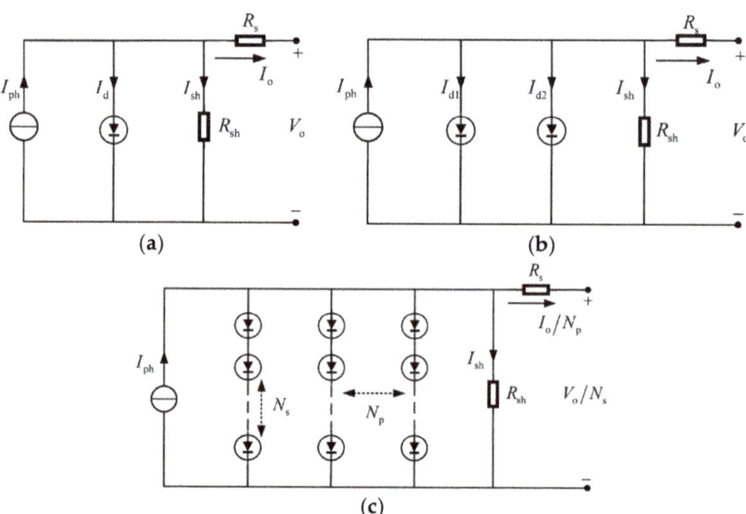

Figure 1. Equivalent circuit diagram of the (**a**) SDM, (**b**) DDM, and (**c**) PMM.

Note that Formula (4) has five undetermined parameters (I_{ph}, I_{sd}, R_s, R_{sh}, and m). Accurate and reliable extraction of the five parameters is important for solar cells, which can be realized by the proposed IMFO algorithms.

2.2. Double Diode Model

The DDM has been developed [36,37] due to this fact that the SDM has not considered the influence of composite current loss of the depletion area. As shown in Figure 1b, the DDM contains a current source, two diodes in parallel with a shunt resistance and a resistance in series with the current source. One of the diodes is applied to simulate the charge-recombination current with the non-ideal factors, and the other one is set as the rectifier. The output current I_o of this model can be acquired as follows:

$$I_o = I_{ph} - I_{d1} - I_{d2} - I_{sh}$$

$$= I_{ph} - I_{sd1}\left[\exp\left(\frac{q(V_t + I_t R_s)}{m_1 kT}\right) - 1\right] - I_{sd2}\left[\exp\left(\frac{q(V_t + I_t R_s)}{m_2 kT}\right) - 1\right] - \frac{I_t R_s + V_t}{R_{sh}} \tag{5}$$

where I_{d1} and m_1 denote the current of diffusion and the ideality factor, respectively; m_2 and I_{d2} indicate the composite diode ideality factor and saturation current, respectively. Different from the SDM, the DDM includes seven undetermined parameters (I_{ph}, I_{sd1}, I_{sd2}, R_s, R_{sh}, m_1, and m_2) that need to be accurately extracted.

2.3. PV Module Model

The PMM typically embodies multiple solar cells connected in parallel and/or in series. The equivalent circuit diagram of the single diode PMM is presented in Figure 1c. The output current of this model can be calculated by the following Formula (6):

$$I_o = N_p\left\{I_{ph} - I_{sd}\left[\exp\left(\frac{q\left(V_t/N_s + I_tR_s/N_p\right)}{mkT}\right) - 1\right] - \frac{I_tR_s/N_p + V_t/N_s}{R_{sh}}\right\} \tag{6}$$

where N_s and N_p represent the number of solar cells in series and parallel, respectively. Same as the SDM, PMM also has five undetermined parameters (I_{ph}, I_{sd}, R_s, R_{sh}, and m) to be accurately extracted.

2.4. Objective Function

For the extraction problem of the undetermined parameters in PV models, the main task is to explore the undetermined parameter optimal value for narrowing the difference between the calculated and tested current values. The errors on each set of calculated and tested current values of the SDM and DDM can be obtained by the following Formulas (7) and (8), respectively.

$$\begin{cases} f_x(V_t, I_t, X) = I_{ph} - I_{sd}\left[\exp\left(\frac{q(V_t + I_tR_s)}{mkT}\right) - 1\right] - \frac{I_tR_s + V_t}{R_{sh}} - I_t \\ X = \{I_{ph}, I_{sd}, R_s, R_{sh}, m\} \end{cases} \tag{7}$$

$$\begin{cases} f_x(V_t, I_t, X) = I_{ph} - I_{sd1}\left[\exp\left(\frac{q(V_t + I_tR_s)}{m_1kT}\right) - 1\right] - I_{sd2}\left[\exp\left(\frac{q(V_t + I_tR_s)}{m_2kT}\right) - 1\right] - \frac{I_tR_s + V_t}{R_{sh}} - I_t \\ X = \{I_{ph}, I_{sd1}, I_{sd2}, R_s, R_{sh}, m_1, m_2\} \end{cases} \tag{8}$$

In this work, to evaluate the entire difference between the calculated and tested current values, the root mean square error (RMSE) is utilized as objective functions as shown in Formula (9), which has been commonly adopted in literature [24,26,27]. Particularly, the smaller the RMSE(X) value, the more accurate the extraction of PV models' parameters.

$$\text{RMSE}(X) = \sqrt{\frac{\sum_{x=1}^{N} f_x(V_t, I_t, X)^2}{N}} \tag{9}$$

where N. denotes the number of tested current values, X is the decision vector that contains the undetermined parameters to be estimated.

3. Moth-Flame Optimization Algorithm

As a novel population-based heuristic algorithm, MFO is proposed by Mirjalili in 2015 [29]. The algorithm mainly contains two mechanisms: flame generation mechanism and the spiral flight search mechanism.

3.1. Flame Generation Mechanism

In MFO, the whole moth swarm' population size and the individual dimensional size are set as N and D, respectively. The whole moth swarm can be expressed as the following matrix $\boldsymbol{AM}$:

$$\boldsymbol{AM} = \begin{bmatrix} AM_1 \\ AM_2 \\ \vdots \\ AM_N \end{bmatrix} = \begin{bmatrix} am_{11} & am_{12} & \cdots & \cdots & am_{1D} \\ am_{21} & am_{22} & \cdots & \cdots & am_{2D} \\ \vdots & \vdots & \vdots & \vdots & \vdots \\ am_{N1} & am_{N2} & \cdots & \cdots & am_{ND} \end{bmatrix} \tag{10}$$

where $AM_i = [am_{i1}, am_{i2}, \cdots, am_{iD}]$ represents the state of the i-th moth, $i \in \{1, 2, \cdots, N\}$, and am_{id} indicates the d-th dimensional position of i-th moth, $d \in \{1, 2, \cdots, D\}$. When MFO is applied for the estimating a PV model's parameters, each dimension of $AM_i = [am_{i1}, am_{i2}, \cdots, am_{iD}]$ corresponds to a set of extracted parameters of the PV model; D stands for the number of the estimated parameters. For example, for the SDM with its five estimated parameters, each AM_i, $i \in \{1, 2, \cdots, N\}$ has five dimensions, written as $AM_i = [am_{i1}, am_{i2}, \cdots, am_{i5}]$; am_{i1}, am_{i2}, am_{i3}, am_{i4}, and am_{i5} represent five estimated parameters I_{ph}, I_{sd}, R_s, R_{sh}, and m, respectively.

$F[*]$ denotes the fitness value function. Therefore, for the i-th moth AM_i, its fitness value is indicated as $F[AM_i]$. Similarly, the fitness value vector of $\boldsymbol{AM}$ can be represented as follows:

$$F[\boldsymbol{AM}] = \begin{bmatrix} F[AM_1] \\ F[AM_2] \\ \vdots \\ F[AM_N] \end{bmatrix} \tag{11}$$

Similarly, for the estimating a PV model's parameters, $F[*]$ is defined as objective functions of the PV model.

Due to each moth has its own target flame, the number of the initial flames is equivalent to moths. The matrix $\boldsymbol{AF}$ is assumed to be the size of initial flames as follows:

$$\boldsymbol{AF} = \begin{bmatrix} AF_1 \\ AF_2 \\ \vdots \\ AF_N \end{bmatrix} = \begin{bmatrix} af_{11} & af_{12} & \cdots & \cdots & af_{1D} \\ af_{21} & af_{22} & \cdots & \cdots & af_{2D} \\ \vdots & \vdots & \vdots & \vdots & \vdots \\ af_{N1} & af_{N2} & \cdots & \cdots & af_{ND} \end{bmatrix} \tag{12}$$

where AF_i is corresponding to the AM_i, $i \in \{1, 2, \cdots, N\}$. D and N are the flames' dimensional number and population size, respectively.

Similarly, for the i-th flame AF_i, its fitness value is described as $F[AF_i]$. The fitness value vector of $\boldsymbol{AF}$ is expressed as below:

$$F[\boldsymbol{AF}] = \begin{bmatrix} F[AF_1] \\ F[AF_2] \\ \vdots \\ F[AF_N] \end{bmatrix} \tag{13}$$

The maximum number of iterations and the current number of iterations are set as K and k, respectively, $k \in \{1, 2, \cdots, K\}$. The flame generation mechanism is introduced as follows:

- Step one: when the current iterations' number k is one, the moths $\{AM_1, AM_2, \cdots, AM_N\}$ are sorted according to their corresponding fitness value $\{F[AM_1], F[AM_2], \cdots, F[AM_N]\}$ from small to large, and the i-th moth after sorting is set as AM_{si}, $AM_{si} \in \{AM_{s1}, AM_{s2}, \cdots, AM_{sN}\}$. Then, the AM_{si}, $i \in \{1, 2, \cdots, N\}$ means the i-th flame AF_i. Thus, the $\boldsymbol{AF}$ and its corresponding fitness value $F[\boldsymbol{AF}]$ can be initialed based on the sorted $\boldsymbol{AM}$ and $F[\boldsymbol{AM}]$, respectively.

- Step two: when the current number of iterations k is greater than one but less than or equal to K. We sort $AM(k-1)$ and $AF(k-1)$ from small to large according to their corresponding fitness values $F[AM(k-1)]$ and $F[AF(k-1)]$. Next, the top N individuals are selected as new flames $AF(k)$ from $AM(k-1)$ and $AF(k-1)$. The $F[AM(k-1)]$ and $AM(k-1)$ represent the fitness value vector and the position vector of moths at the $(k-1)$-th iterations, respectively. Similarly, the fitness value vector and position vector of flames at the $(k-1)$-th iterations are $F[AF(k-1)]$ and $AF(k-1)$, respectively.

From the above demonstration, AF denotes the essence of $AM(k-1)$ and $AF(k-1)$. Therefore, for the estimating a PV model's parameters, each dimension of $AF_i = [af_{i1}, af_{i2}, \cdots, af_{iD}]$ corresponds to an estimated parameters of the PV model; D stands for the number of the estimated parameters.

3.2. Spiral Flight Search Mechanism

Furthermore, since the moths fly toward their respective target flame in spiral, the spiral flight search mechanism has been developed to simulate the moth's logarithmic spirally flying curve, as below:

$$AM_i(k+1) = \begin{cases} D_i \cdot e^{bt} \cdot \cos(2\pi t) + AF_i(k), & i \le F_{nu} \\ D_i \cdot e^{bt} \cdot \cos(2\pi t) + AF_{F_{nu}}(k), & i > F_{nu} \end{cases} \tag{14}$$

where $AM_i(k+1)$ indicates the position of the i-th moth AM_i after renewing, $i \in \{1, 2, \cdots, N\}$; $D_i = |AF_i - AM_i|$ denotes the distance between the i-th flame AF_i and the AM_i; b is a constant for the logarithmic spiral shape; t is a random number of $[-1, 1]$, which denotes the distance between the next position of the moth and the flame ($t = -1$ is the smallest distance between AM_i and AF_i, and $t = 1$ is the farthest between AM_i and AF_i). For diverse t values, the spirally flying trajectory in one dimension is displayed in Figure 2.

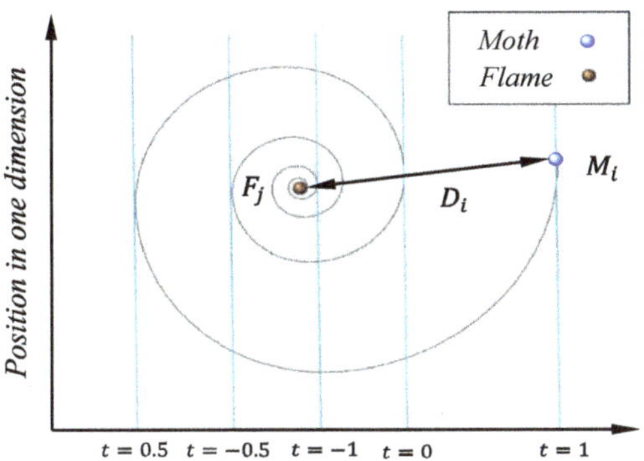

Figure 2. Geometric trajectory diagram of flying moth catching flame around logarithmic spiral curve.

F_{nu} is the number of the flames during the iteration process. Note that as the number of iterations increases, F_{nu} gradually decreases and eventually equals to 1. The process of adaptively reducing the number of flames can be expressed in Formula (15) as below:

$$F_{nu} = round\left[N - k \times \frac{N-1}{K}\right] \tag{15}$$

where N and K are the maximum number of flames and iterations, respectively; k denotes the current iterative number. Besides, $round[x^*]$ is the function that can make x^* be rounded to its nearest integer. Note that with $k = 0$, $F_{nu} = N$; when $k = K$, $F_{nu} = 1$ denotes that moths utilize the best flame to update their own positions at the last iteration in Formula (14).

In general, the flames surrounded by their corresponding moths can be randomly distributed in the search area. Consequently, the MFO can obtain the good balance between the local exploitation and the global exploration [29].

4. Improve Moth-Flame Optimization Algorithm

In the original MFO [29], the moths always flying towards their corresponding best target flames, resulting in the fast communication between moths. Therefore, the moths are easily trapped into local optimum when facing complex multi-modal problems. To address this problem, this paper develops an improved MFO (IMFO) algorithm. In the IMFO, we propose a double-flames generation (DFG) strategy to generate two different types of target flames for guiding moths flying. One type of flames is named local flame (LF), which can be created by dividing the entire swarm into several sub-swarms and then select the local best moths in a corresponding sub-swarm; another one is called global flame (GF) that is generated by combining all the moths' current personal best positions (pbests). Moreover, two different update strategies are developed for updating the positions of moths; each moth only adopts one of the two update strategies at each iteration. In addition, to ensure a great balance between the exploitation and exploration, we develop a probability to rationally select one of the two update strategies for each moth at each iteration. The above process is detailed below.

4.1. Double Flames Generation Strategy

In this strategy, two types of target flames are produced by adopting different moths' information. The first type of flames is the LFs, which can be generated by the following description:

1. Assume the population size of moths is N, AM_i and $F[AM_i]$ are the position and the fitness value of i-th moth, respectively, $i \in \{1, 2, \cdots, N\}$.
2. Divide all the moths in the entire swarm into m sub-swarms, and each sub-swarm has n moths. In this case, the α-th sub-swarm is composed of from number $(\alpha - 1) \times n + 1$ to number $\alpha \times n$, $\alpha \in \{1, 2, \cdots, m\}$.
3. According to the moths' fitness values, the best moth of each sub-swarm is select as LF_α, which denotes the LF of the α-th sub-swarm.

The second type is GF, which is generated by combining all the moths' pbests. The calculation formula of GF is displayed in Formula (16), we can see from the formula that the GF carries the information about all moths.

$$GF = \frac{1}{N} \sum_{i=1}^{N} pbest_i \tag{16}$$

4.2. Update Formulas

Furthermore, two different update strategies are created for generating new moths. Particularly, the two update strategies correspond to two types of flames, respectively as follows:

$$AM_i(k+1) = \left| LF_\alpha(k) - AM_i(k) \right| \times e^{bt} \times \cos(2\pi t) + LF_\alpha(k) \; (\phi > P) \tag{17}$$

$$AM_i(k+1) = \left| GF(k) - AM_i(k) \right| \times e^{bt} \times \cos(2\pi t) + GF(k) \; (\phi \leq P) \tag{18}$$

where $AM_i(k+1)$ and $AM_i(k)$ respectively represent the updated position of the i-th moth at the k-th and $(k+1)$-th iteration, $i \in \{1, 2, \cdots, N\}$; b indicates a constant for the logarithmic spiral shape; $LF_\alpha(k)$ and $GF(k)$ denote two type of flames; t is a number that randomly distributed in the interval of $[-1,1]$.

From the Formula (17), we can observe that as the local best flame in the α-th sub-swarm, $\alpha \in \{1, 2, \cdots, m\}$, $LF_\alpha(k)$ can effectively guide its corresponding moth $AM_i(k)$ to perform the local search and explore a new potential place in the solution area. On the other hand, we can see from the Formula (18) that the moths in the entire swarm can fly towards $GF(k)$ for improving the accuracy of solutions. Particularly, instead of using the best flame, Formula (18) adopts $GF(k)$, namely the average value of all the moths' pbests to avoid the fast communication between moths and improve the premature of the moths. In general, the two different update strategies make moths obtain the promising compromise between global exploration and local exploitation.

In addition, we develop a probability to rationally select one of the above two update strategies at each iteration. Specifically, a random number ϕ distributed in [0,1] is created during each iterative process. If the value of ϕ is larger than P, the Formula (17) can be selected to update the i-th moth AM_i; otherwise, the Formula (18) be done for updating the AM_i. Therefore, it is significant to acquire a suitable P value for balancing the exploitation and exploration abilities. The setting of P value is displayed in Section 6.

4.3. Process of IMFO

The pseudo code of IMFO is expressed in the Algorithm 1, where k denotes the number of the current iteration; LF_α denotes the local best flame of the α-th sub-swarm, $\alpha \in \{1, 2, \cdots, m\}$, where m indicates the number of sub-swarms. It can be seen from the Algorithm 1 that the complexity of IMFO is comparable to the original MFO. Moreover, the flow chart of IMFO is described in Figure 3, where the DFG strategy with two different update strategies are adopted to seek the best solution in the search space.

Algorithm 1: IMFO

1: /*Initialization*/
2: Initialize the position AM_i of moths, $i \in \{1, 2, \cdots, N\}$;
3: Calculate the fitness value $F[AM_i]$ of moths, initialize the pbest$_i$, gbest;
4: /* DFG strategy*/
5: $k = 1$;
6: **for** $\alpha = 1$ **to** m **do**
7: subswarm$_\alpha = \left[AM_{(\alpha-1)*n+1}, AM_{(\alpha-1)*n+2}, \cdots, AM_{\alpha*n} \right]$;
8: Select the best moth of sub $-$ swarm$_\alpha$ as $LF_\alpha(k)$;
9: **end for**
10: Calculate the $GF(k)$ with formula (16);
11: **while** (condition of termination not met) **do**
12: $k = k + 1$;
13: **for** $i = 1$ **to** N **do**
14: Create a random number ϕ;
15: **if** $\phi > P$
16: Update the $AM_i(k)$ with formula (17);
17: **else**
18: Update the $AM_i(k)$ with formula (18);
19: **end if**
20: Update the pbest$_i$, LF_α, GF, gbest, $i \in \{1, 2, \cdots, N\}$;
21: **end for**
22: **end while**

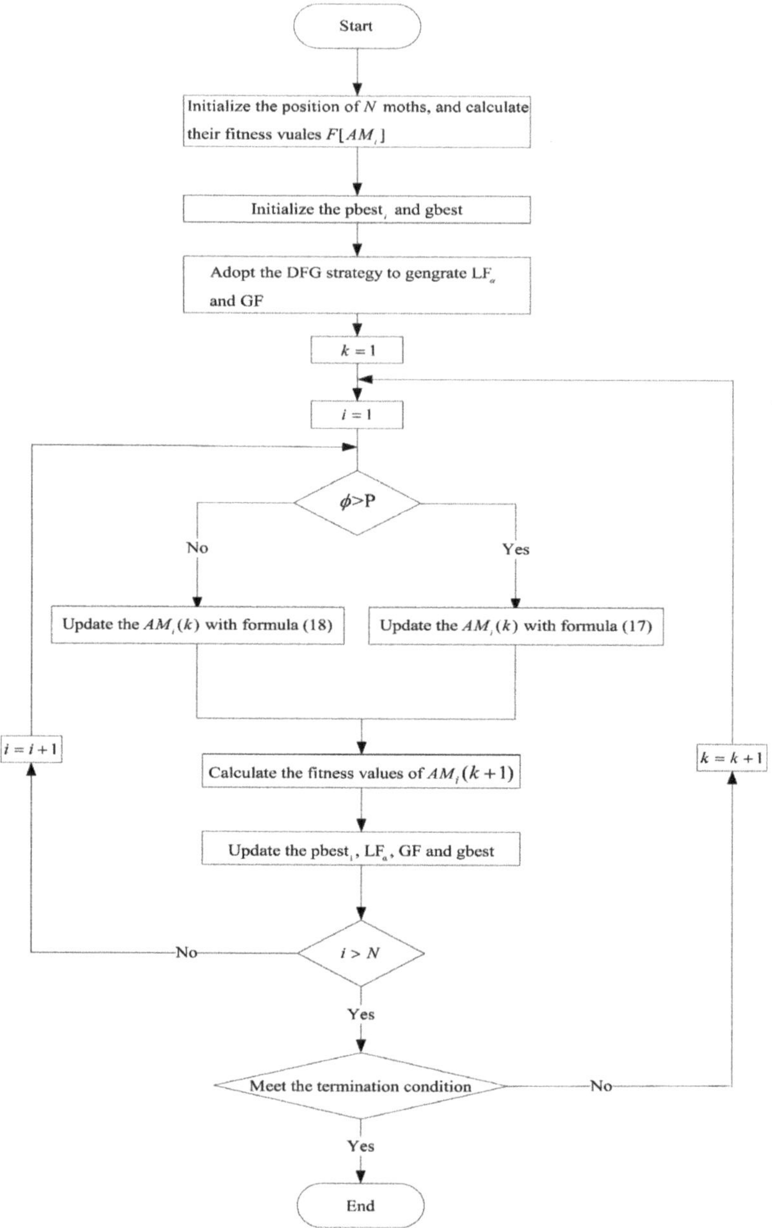

Figure 3. Flow chart of the improved moths-flames optimization (IMFO).

5. Results and Analysis

To show the validity of the proposed IMFO method, it is applied to identify the parameters of three different PV models involving the SDM, DDM, and PMM. For the SDM and DDM, reference [18] gives the measured current and voltage values from a 57 mm diameter commercial silicon R.T.C. France silicon solar cell working at a temperature of 33 °C under irradiance of 1000 W/m^2. Besides, as a PMM, the Photowatt-PWP201 is applied to evaluate the IMFO for identifying the corresponding parameters [18,38,39]. Particularly, the Photowatt-PWP201 contains 36 silicon cells with series conducting under 1000 W/m^2 at 45 °C. Considering fair comparison, we utilize the same search upper and lower boundaries [24,26,27] of each parameter of three PV models, as shown in Table 1.

Table 1. Parameters boundaries of the three different photovoltaic (PV) models.

Parameter	Single Diode/Double Diode		PV Module	
	Lower Bound	Upper Bound	Lower Bound	Upper Bound
$I_{ph}(A)$	0	1	0	2
$I_{sd}, I_{sd1}, I_{sd2}(\mu A)$	0	1	0	50
$R_S(\Omega)$	0	0.5	0	2
$R_{sh}(\Omega)$	0	100	0	2000
m, m_1, m_2	1	2	1	50

To indicate the excellent performance of the IMFO method, it has been compared with eight well-established and competitive methods. These compared methods were the basis MFO [30], opposition-based MFO method (OMFO) [33], hybrid water cycle–moth-flame optimization method (WCMFO) [34], brain storm optimization method (BSO) [40], comprehensive learning PSO (CLPSO) method [41], artificial bee colony (ABC) method [20], sine cosine algorithm (SCA) [42], and improved JAYA (IJAYA) method [26]. All the compared methods were independently conducted 30 times on each PV model. For each running, the maximum number of evaluations for the compared methods is set to 50,000. Table A1 of Appendix A shows the parameter settings of the comparison methods. Note that in the IMFO, the population size of moths is set as 100, and the number of sub-swarms and P are set as four and 0.4, respectively. Moreover, the accuracy of the nine compared methods were illustrated by comparing the best values in their RMSE. Also, their robustness and convergence speed were evaluated through the analysis of data results and convergence curves.

5.1. Results on the Single Diode Model

For the SDM, RMSE and the estimated five parameters values are shown in Table 2, where we highlight the overall best and the second best RMSE values in gray boldface and boldface, respectively. Interestingly, both the IMFO and WCMFO achieve the best RMSE value (9.8602 × 10^{-4}) among the nine algorithms; besides, the IJAYA acquires second best RMSE value (9.8606 × 10^{-4}), followed by ABC, CLPSO, MFO, OMFO, BSO, and SCA. RMSE is applied to denote the accuracy of experimental results because of the information of the accurate values of the parameters is not available. Although the gap between the best and second best RMSE values is extremely subtle, it is important to decrease the difference between the true and estimated parameter values in objective function. Because the smaller of objective value RMSE, the more accurate the estimated parameters. Moreover, the extracted best parameters from IMFO are employed to plot *I-V* and *P-V* curves. This further demonstrates the accuracy of the extracted parameters on SDM. From Figure A1 of Appendix A, we can see that the measured and simulated current and voltage values acquired by IMFO are extremely consistent over the entire voltage range. In addition, the individual absolute errors (IAE) between the simulated and measured current values are displayed in Table A2 of Appendix A. Note that compared with 2.5 × 10^{-3}, all the IAE values are smaller, demonstrating the preciseness of the extracted parameters from IMFO.

Table 2. Comparison among different algorithms on the single diode model.

Algorithm	$I_{ph}(A)$	$I_{sd}(\mu A)$	$R_s(\Omega)$	$R_{sh}(\Omega)$	m	RMSE
IMFO	0.76078	0.32296	0.03638	53.71456	1.48117	$\mathbf{9.8602 \times 10^{-4}}$
MFO	0.76092	0.30105	0.03596	51.81957	1.46935	9.9496×10^{-4}
OMFO	1.09171	0.69332	0.06003	93.11915	2.16144	1.1927×10^{-3}
WCMFO	0.76078	0.32314	0.03638	53.69502	1.48122	$\mathbf{9.8602 \times 10^{-4}}$
BSO	0.76090	0.99996	0.03137	97.35715	1.60455	2.4551×10^{-3}
CLPSO	0.76073	0.31629	0.03639	52.31786	1.47910	9.9455×10^{-4}
ABC	0.76085	0.33319	0.03623	53.72712	1.48431	9.9049×10^{-4}
SCA	0.76503	0.67937	0.03544	50.14796	1.56094	5.8058×10^{-3}
IJAYA	0.76077	0.32349	0.03637	53.89245	1.48133	$\mathbf{9.8606 \times 10^{-4}}$

Note: $\mathbf{9.8602 \times 10^{-4}}$: both the IMFO and WCMFO achieve the best RMSE value among the nine algorithms; $\mathbf{9.8606 \times 10^{-4}}$: the IJAYA acquires second best RMSE value (9.8606×10^{-4}), followed by ABC, CLPSO, MFO, OMFO, BSO, and SCA.

5.2. Results on the Double Diode Model

For the DDM, the values of seven extracted parameters with the comparison results of RMSE are presented in Table 3. It can be seen that the proposed IMFO obtain the best RMSE result (9.8252×10^{-4}) among the nine methods; the second best RMSE result (9.8371×10^{-4}) is achieved by WCMFO. Besides, the *I-V* and *P-V* curves are rebuilt in Figure A2 through using the best model that estimated by IMFO. Furthermore, via using IMFO, we can obtain IAE values listed in Table A3 of Appendix A, where each of all the IAE values is also smaller than 2.5×10^{-3}. In addition, we can see from the Figure A2 of Appendix A that the measured and simulated current and voltage values obtained by IMFO are highly in coincidence. Therefore, IMFO can extract the parameters of DDM with high accuracy.

Table 3. Comparison among different algorithms on the double diode model.

Algorithm	$I_{ph}(A)$	$I_{sd1}(\mu A)$	$R_s(\Omega)$	$R_{sh}(\Omega)$	m_1	$I_{sd2}(\mu A)$	m_2	RMSE
IMFO	0.76078	0.23350	0.03671	55.29970	1.45374	0.68372	2.00000	$\mathbf{9.8252 \times 10^{-4}}$
MFO	0.76087	0.15997	0.03699	52.76975	1.43092	0.55840	1.82735	1.0102×10^{-3}
OMFO	0.76078	0.01137	0.03644	53.27466	1.46548	0.30651	1.48014	9.8652×10^{-4}
WCMFO	0.76078	0.23956	0.03661	55.11475	1.45681	0.44291	1.90457	$\mathbf{9.8371 \times 10^{-4}}$
BSO	0.76100	0.00001	0.03134	92.63446	1.14459	0.99843	1.60458	2.4636×10^{-3}
CLPSO	0.76066	0.28752	0.03664	55.28951	1.95861	0.26868	1.46524	9.9224×10^{-4}
ABC	0.76051	0.24568	0.03641	58.63990	1.46175	0.26266	1.76608	1.0001×10^{-3}
SCA	0.76540	0.00000	0.02945	38.07191	1.02508	0.95693	1.60284	9.2482×10^{-3}
IJAYA	0.76077	0.64791	0.03675	54.86248	1.98667	0.22965	1.45222	9.8380×10^{-4}

Note: $\mathbf{9.8252 \times 10^{-4}}$: It can be seen that the proposed IMFO obtain the best RMSE result among the nine methods; the second best RMSE result ($\mathbf{9.8371 \times 10^{-4}}$) is achieved by WCMFO.

5.3. Results on the PV Module Model

For the PMM, it has five parameters that need to be estimated. Table 4 presents the best RMSE and the five extracted parameters values for each of the nine compared methods on 30 tests. From the Table 4, we can see that IMFO achieves the best RMSE value (2.4251×10^{-3}); the IJAYA has the second best RMSE value (2.2453×10^{-3}). Furthermore, Figure A3 of Appendix A shows that the *I-V* and *P-V* curves features of the estimated parameters by IMFO are good in coincidence with the measured values. In addition, Table A4 of Appendix A indicates that each of all the IAE values is smaller than 4.8×10^{-3}. From these results, IMFO can achieve the great accuracy parameters extraction.

Table 4. Comparison among different algorithms on the PV module model.

Algorithm	$I_{ph}(A)$	$I_{sd}(\mu A)$	$R_s(\Omega)$	$R_{sh}(\Omega)$	m	RMSE
IMFO	1.03052	3.47835	1.20139	980.46728	48.63854	$\mathbf{2.4251 \times 10^{-3}}$
MFO	1.02807	4.06598	1.19215	1185.52443	49.19573	2.4576×10^{-3}
OMFO	1.03062	3.41288	1.20337	961.83398	48.56592	2.4257×10^{-3}
WCMFO	1.03039	3.53468	1.19979	1003.12281	48.69988	2.4255×10^{-3}
BSO	1.03591	1.18207	1.31645	593.82890	44.80230	4.7604×10^{-3}
CLPSO	1.03013	3.62057	1.19763	1043.44361	48.79166	2.4282×10^{-3}
ABC	1.03095	3.79242	1.18896	946.21624	48.97719	2.4660×10^{-3}
SCA	1.04244	4.68166	1.20422	1204.05479	49.78753	1.0327×10^{-2}
IJAYA	1.03047	3.50872	1.20024	987.63650	48.67208	$\mathbf{2.4253 \times 10^{-3}}$

Note: IMFO achieves the best RMSE value $\mathbf{2.4251 \times 10^{-3}}$; the IJAYA has the second best RMSE value $\mathbf{2.4251 \times 10^{-3}}$.

5.4. Statistical Results and Convergence Curves

In this section, IMFO is compared with the other eight methods on the reliability and convergence speed by the convergence curves and statistical results. Table 5 displays the nine algorithms' statistical results over 30 independent tests on each of three PV models, where the minimum (min), mean and standard deviation (SD) values of RMSE represent the accuracy, average preciseness and reliability of the extracted parameters, respectively. Besides, we conduct the Wilcoxon signed-rank examination at 5% significant level [43] to verify the great differences between IMFO and other compared methods. Sign "+" in the Table 5 indicates that the IMFO is greatly superior to its competitors.

Table 5. Comparisons on the statistical results of different algorithms for three different PV models.

Model	Algorithm	RMSE				Sig.
		Min	Mean	Max	SD	
Single diode model	IMFO	$\mathbf{9.8602 \times 10^{-4}}$	$\mathbf{9.8767 \times 10^{-4}}$	$\mathbf{9.964110^{-4}}$	$\mathbf{2.181010^{-6}}$	
	MFO	9.9496×10^{-4}	1.9256×10^{-3}	2.584910^{-3}	4.925510^{-4}	+
	OMFO	1.1927×10^{-3}	2.1967×10^{-3}	5.836910^{-3}	8.52141^{-4}	+
	WCMFO	$\mathbf{9.8602 \times 10^{-4}}$	$\mathbf{1.0122 \times 10^{-3}}$	1.330510^{-3}	7.342310^{-5}	+
	BSO	2.4551×10^{-3}	2.6986×10^{-3}	3.330710^{-3}	1.96141^{-4}	+
	CLPSO	9.9455×10^{-4}	1.0507×10^{-3}	1.186510^{-3}	4.673010^{-5}	+
	ABC	9.9049×10^{-4}	1.2951×10^{-3}	1.927810^{-3}	2.548310^{-4}	+
	SCA	5.8058×10^{-3}	3.7413×10^{-2}	4.863610^{-2}	1.302310^{-2}	+
	IJAYA	$\mathbf{9.8606 \times 10^{-4}}$	1.0261×10^{-3}	$\mathbf{1.122310^{-3}}$	$\mathbf{4.160910^{-5}}$	+
Double diode model	IMFO	$\mathbf{9.8252 \times 10^{-4}}$	$\mathbf{9.9737 \times 10^{-4}}$	$\mathbf{1.140910^{-3}}$	$\mathbf{3.293910^{-5}}$	
	MFO	1.0102×10^{-3}	3.4150×10^{-3}	3.521510^{-2}	5.689710^{-3}	+
	OMFO	9.8652×10^{-4}	2.1677×10^{-3}	3.727710^{-3}	7.301810^{-4}	+
	WCMFO	$\mathbf{9.8371 \times 10^{-4}}$	1.1449×10^{-3}	2.264810^{-3}	3.080910^{-4}	+
	BSO	2.4636×10^{-3}	2.8700×10^{-3}	4.090610^{-3}	3.704510^{-4}	+
	CLPSO	$9.9224 \times 10{-4}$	1.0522×10^{-3}	$\mathbf{1.146210^{-3}}$	$\mathbf{4.314110^{-5}}$	+
	ABC	1.0001×10^{-3}	1.0936×10^{-3}	1.289610^{-3}	7.411110^{-5}	+
	SCA	9.2482×10^{-3}	4.2445×10^{-3}	2.228610^{-1}	3.624810^{-2}	+
	IJAYA	9.8380×10^{-4}	$\mathbf{1.0240 \times 10^{-3}}$	1.350710^{-3}	8.564710^{-5}	+
PV module model	IMFO	$\mathbf{2.4251 \times 10^{-3}}$	$\mathbf{2.4294 \times 10^{-3}}$	$\mathbf{2.500510^{-3}}$	$\mathbf{1.383110^{-5}}$	
	MFO	2.4576×10^{-3}	2.7962×10^{-3}	5.175410^{-3}	5.728310^{-4}	+
	OMFO	2.4257×10^{-3}	2.5901×10^{-3}	2.639110^{-3}	5.258610^{-5}	+
	WCMFO	2.4255×10^{-3}	1.1551×10^{-2}	2.742510^{-1}	4.961610^{-2}	+
	BSO	4.7604×10^{-3}	2.0011×10^{-1}	4.063510^{-1}	1.051610^{-1}	+
	CLPSO	2.4282×10^{-3}	2.4527×10^{-3}	2.527710^{-3}	2.382810^{-5}	+
	ABC	2.4660×10^{-3}	2.6049×10^{-3}	2.759510^{-3}	5.996210^{-5}	+
	SCA	1.0327×10^{-2}	1.2834×10^{-1}	2.743310^{-1}	1.074010^{-1}	+
	IJAYA	$\mathbf{2.4253 \times 10^{-3}}$	$\mathbf{2.4404 \times 10^{-3}}$	2.502610^{-3}	1.976710^{-5}	+

We can clearly see from the Table 5 that the IMFO algorithm achieves the more outstanding performance than other eight algorithms on the reliability and average accuracy of the three models. For the SDM, WCMFO and IJAYA obtain the second-best average accuracy and reliability, respectively. For the DDM, the second-best reliability and average accuracy are acquired by IJAYA and CLPSO, respectively. For the PMM, the IJAYA achieves the second-best result both on average accuracy and reliability. Moreover, the results of Wilcoxon signed-rank test from Table 5 demonstrate that IMFO presents the greatly superior to all the compared methods on all the three models. Figure A4 of Appendix A displays the box-plots, indicating the distribution of all the 30 independent runs' results acquired by the nine different methods. Note that the sign "+" in Figure A4 represents the abnormal value during the plotting. The span of the data distributions demonstrates that the superior performance is also obtained by IMFO algorithm.

Furthermore, Figure 4 gives the convergence curves of nine methods, which are plotted by using the average RMSE value on the 30 independent tests. Clearly, IMFO acquired the fastest convergence speed among the nine methods on each of the three PV models.

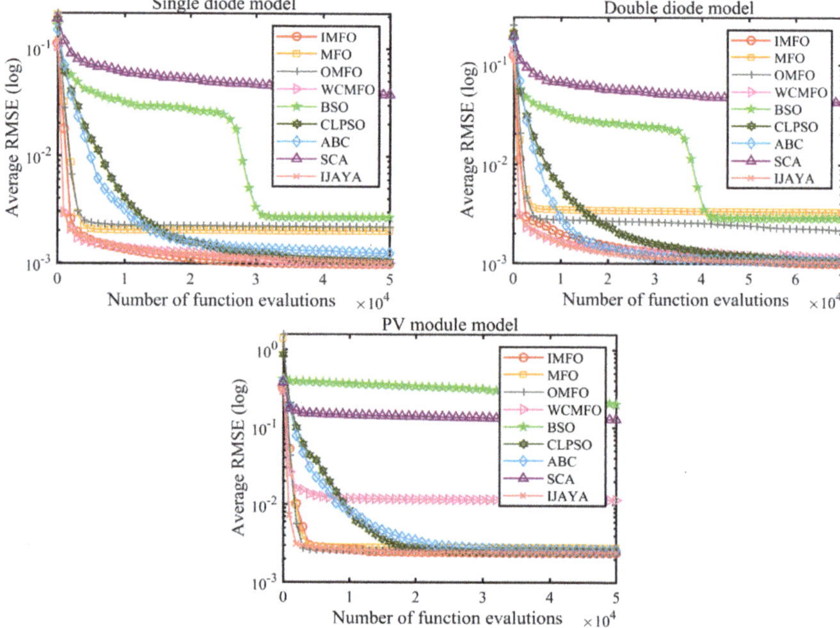

Figure 4. Convergence curves of different algorithms for three models.

In general, according to the above comparisons, we can indicate that the IMFO method expresses the superior performance on searching preciseness, reliability and faster convergence performance when extracting the three different PV models' parameters. Besides, IMFO performs the superior and competitive performance in contrast with eight mature heuristic methods.

6. Discussion

6.1. Discussion of Parameter

In this paper, the parameter P is used to rationally select one of the two update strategies for each moth at each iteration. Thus, it is significant to set a suitable P value for balancing the local exploitation and global exploration abilities. In this section, we test the influences of different P values on IMFO performance through comparing the preciseness and reliability when extracting the parameters of the three different PV models. The maximum number of evaluations is also set as 50,000 on each running; the population size of the moths and number of sub-swarms are set as 100 and 4, respectively.

We evaluate the IMFO with three different P values: 0.4, 0.5 and 0.6. In addition, the IMFO with each different P value is independently conducted 30 times on each model. Table 6 shows that $P = 0.4$ provides the best values on all the three different PV models according to all criteria. This means that $P = 0.4$ can obtain the best performance on accuracy and reliability when estimating the three models' parameters.

Table 6. Statistical results of root mean square error (RMSE) of IMFO with different settings of P on the three models.

Model	P	RMSE			
		Min	Mean	Max	SD
Single diode model	$P = 0.4$	**9.8602×10^{-4}**	**9.8767×10^{-4}**	**9.9641×10^{-4}**	**2.1810×10^{-6}**
	$P = 0.5$	9.8605×10^{-4}	**9.9166×10^{-4}**	**1.0558×10^{-3}**	**1.2596×10^{-5}**
	$P = 0.6$	**9.8604×10^{-4}**	1.0592×10^{-3}	1.6218×10^{-3}	1.5329×10^{-4}
Double diode model	$P = 0.4$	**9.8252×10^{-4}**	**9.9737×10^{-4}**	**1.1409×10^{-3}**	**3.2939×10^{-5}**
	$P = 0.5$	**9.8259×10^{-4}**	**1.0316×10^{-3}**	**1.3051×10^{-3}**	**9.3092×10^{-5}**
	$P = 0.6$	9.8463×10^{-4}	1.1169×10^{-3}	1.5621×10^{-3}	1.8730×10^{-4}
PV module model	$P = 0.4$	**2.4251×10^{-3}**	**2.4294×10^{-3}**	**2.5005×10^{-3}**	**1.3831×10^{-5}**
	$P = 0.5$	**2.4251×10^{-3}**	2.0553×10^{-2}	**2.7425×10^{-1}**	6.8963×10^{-2}
	$P = 0.6$	**2.4252×10^{-3}**	**1.1607×10^{-2}**	**2.7425×10^{-1}**	**4.9607×10^{-2}**

6.2. Discussion of IMFO

To demonstrate that IMFO can effectively escape the local optimum when solving multimodal problems, we further tested the performance of IMFO, MFO [30], OMFO [33], WCMFO [34], BSO [40], CLPSO [41], ABC [20], SCA [42], and IJAYA [26] on the CEC2017 benchmark suite [44]. Due to the space constraints, the 30 benchmark functions of CEC2017 were listed in Table A5 of the Appendix A. Moreover, according to the characteristics of the 30 benchmark functions, they can be classified into unimodal, multimodal, hybrid, and composition problems. In addition, each function has the same search boundary range (−100,100) on each dimension. Note that $F(x^*)$ is the optimal value of each function; x^* is the best solution corresponding to the optimal value. Here, owing to the limited space, we select F4, F7, and F8 from the multimodal functions, and F12 and F18 from the hybrid functions for evaluating the nine compared algorithms. Each evaluated algorithm was conducted 30 times independently on each of the five selected benchmarks. The mean values (Mean) of the corresponding errors are used as rating criteria. The comparison results of the nine algorithms with 30 dimensions are displayed in Table 7, where we highlight the best and second-best Mean results in gray boldface and boldface, respectively.

From the Table 7, we can see that IMFO achieves the best error means among the nine algorithms on all the five complex functions. Moreover, the indicators of IMFO on the five benchmark functions are all superior to the conventional MFO. For example, IMFO can provide the results of errors' mean values 5.06 and 7.30×10 on multimodal functions F4 and F7, respectively, which are both close to the optimal error value 0. However, MFO only can obtain the results of errors' mean values 8.91×10^2 and 3.42×10^2 on multimodal functions F4 and F7, respectively. The above analysis shows that IMFO can find a more promising solution than the MFO of the multimodal problem in the search space. This indicates that MFO is easily trapped into the local optimum when solving complex multimodal problems, however, IMFO can effectively escape the local optimum to find a more promising solution on the multimodal problems.

Table 7. Comparisons between IMFO and eight algorithms on five benchmark functions of CEC2017 with 30 dimensions.

Criteria	Algorithm	F4	F7	F8	F12	F18
	IMFO	**5.06**	**7.30×10**	**4.48×10**	**2.16×10^4**	**8.61×10^4**
	MFO	8.91×10^2	3.42×10^2	1.86×10^2	3.34×10^8	2.52×10^6
	OMFO	9.04×10	2.28×10^2	1.17×10^2	4.86×10^6	4.17×10^5
	WCMFO	7.74×10	6.63×10^2	2.09×10^2	7.12×10^5	**1.52×10^5**
Mean	BSO	8.86×10	3.66×10^2	1.25×10^2	1.63×10^6	1.72×10^5
	CLPSO	7.03×10	**8.84×10**	**5.25×10**	**4.27×10^5**	1.33×10^5
	ABC	**2.60×10**	1.06×10^2	9.02×10	9.48×10^5	2.74×10^5
	SCA	9.58×10^2	4.18×10^2	2.40×10^2	1.10×10^9	2.89×10^6
	IJAYA	1.38×10^2	2.30×10^2	1.18×10^2	6.54×10^6	3.06×10^5

7. Conclusions

The PV models' parameters are located in the multi-modal and nonlinear solution space with various local optima so that it is extremely tricky for various existing MFO algorithms to extract a set of optimal parameters of the PV models. This paper proposes an improved MFO (IMFO) algorithm for estimating the parameters of PV models. In IMFO, the DFG strategy has been first developed to generate two different types of high-performance target flames for effectively searching the global optimal solution. To effectively tackle the global exploration and local exploitation, two different update strategies are developed. In addition, we develop a probability to select one of above two update strategies for each moth during the iterative process, which ensures the balance between the global exploration and local exploitation. Due to effectively combining the DFG strategy with the two update strategies, IMFO can effectively accurately and reliably identify the different PV models' parameters.

In the future, IMFO will be adopted to settle multi-objective and constrained problems or economic dispatch issues in power systems. Furthermore, some other improvements will be proposed to expand the use of complex renewable energy algorithms by optimization algorithms.

Author Contributions: Conceptualization, H.S., C.L., and Z.Y.; Methodology, H.S., C.L., Y.X., and Q.K.; Software, H.S., C.L., H.W., and Z.C.; Writing—Original draft preparation, H.S. and C.L.; Writing—Review and editing, H.S. and C.L.

Funding: This work was supported by the National Natural Science Foundation of China under Grants 61503177, 81660299, and 61863028, by the China Scholarship Council under the State Scholarship Fund (CSC No. 201606825041), by the Science and Technology Department of Jiangxi Province of China under Grants 20161ACB21007, 20171BBE50071, and 20171BAB202033, and by the Education Department of Jiangxi province of China under Grants GJJ14228 and GJJ150197.

Conflicts of Interest: The authors declare no conflict of interest.

Appendix A

Table A1. Parameters settings of the compared algorithms.

Algorithm	Parameters
MFO	NP = 50
OMFO	NP = 40
WCMFO	NP = 40, $N_{sr} = 4$, $N_{stream} = N_{pop} - N_{sr}$, $\beta = 3/2$
BSO	NP = 50, $p_{r0} = 0.2$, $p_{r1} = 0.8$, $p_{r11} = 0.4$, $p_{r21} = 0.5$, c = 25, M = 5
CLPSO	NP = 40, w = 0.9-0.2, c = 1.49445, m = 5
ABC	NP = 50, $\alpha = 1$, limit = 100
SCA	NP = 50
IJAYA	NP = 50
IMFO	NP = 100, m = 4, P = 0.4

Table A2. Absolute error of IMFO for each measured on the single diode model.

Item	V_{measured}(V)	I_{measured}(A)	$I_{\text{simulated}}$(A)	IAE
1	−0.2057	0.7640	0.76408801	0.00008801
2	−0.1291	0.7620	0.76266328	0.00066328
3	−0.0588	0.7605	0.76135541	0.00085541
4	0.0057	0.7605	0.76015400	0.00034600
5	0.0646	0.7600	0.75905514	0.00094486
6	0.1185	0.7590	0.75804220	0.00095780
7	0.1678	0.7570	0.75709145	0.00009145
8	0.2132	0.7570	0.75614111	0.00085889
9	0.2545	0.7555	0.75508658	0.00041342
10	0.2924	0.7540	0.75366357	0.00033643
11	0.3269	0.7505	0.75139069	0.00089069
12	0.3585	0.7465	0.74735365	0.00085365
13	0.3873	0.7385	0.74011717	0.00161717
14	0.4137	0.7280	0.72738237	0.00061763
15	0.4373	0.7065	0.70697301	0.00047301
16	0.4590	0.6755	0.67528069	0.00021931
17	0.4784	0.6320	0.63075889	0.00124111
18	0.4960	0.5730	0.57192892	0.00107108
19	0.5119	0.4990	0.49960738	0.00060738
20	0.5265	0.4130	0.41364884	0.00064884
21	0.5398	0.3165	0.31750984	0.00100984
22	0.5521	0.2120	0.21215441	0.00015441
23	0.5633	0.1035	0.10225068	0.00124932
24	0.5736	−0.0100	−0.00871799	0.00128201
25	0.5833	−0.1230	−0.12550748	0.00250748
26	0.5900	−0.2100	−0.20847181	0.00152819

Table A3. Absolute error of IMFO for each measured on the double diode model.

Item	V_{measured}(V)	I_{measured}(A)	$I_{\text{simulated}}$(A)	IAE
1	−0.2057	0.7640	0.76399414	0.00000586
2	−0.1291	0.7620	0.76261019	0.00061019
3	−0.0588	0.7605	0.76133957	0.00083957
4	0.0057	0.7605	0.76017183	0.00032817
5	0.0646	0.7600	0.75910236	0.00089764
6	0.1185	0.7590	0.75811332	0.00088668
7	0.1678	0.7570	0.75717857	0.00017857
8	0.2132	0.7570	0.75623282	0.00076718
9	0.2545	0.7555	0.75516740	0.00033260
10	0.2924	0.7540	0.75371531	0.00028469
11	0.3269	0.7505	0.75139682	0.00089682
12	0.3585	0.7465	0.74730504	0.00080504
13	0.3873	0.7385	0.74001983	0.00151983
14	0.4137	0.7280	0.72725956	0.00074044
15	0.4373	0.7065	0.70686255	0.00036255
16	0.4590	0.6755	0.67521858	0.00028142
17	0.4784	0.6320	0.63076237	0.00123763
18	0.4960	0.5730	0.57199017	0.00100983
19	0.5119	0.4990	0.49969784	0.00069784
20	0.5265	0.4130	0.41372571	0.00072571
21	0.5398	0.3165	0.31754180	0.00104180
22	0.5521	0.2120	0.21212422	0.00012422
23	0.5633	0.1035	0.10216958	0.00133042
24	0.5736	−0.0100	−0.00878600	0.00121400
25	0.5833	−0.1230	−0.12553982	0.00253982
26	0.5900	−0.2100	−0.20837881	0.00162119

Table A4. Absolute error of IMFO for each measured on the PV module model.

Item	V_{measured} (V)	I_{measured} (A)	$I_{\text{simulated}}$ (A)	IAE
1	0.1248	1.0315	1.02912657	0.00237343
2	1.8093	1.0300	1.02738585	0.00261415
3	3.3511	1.0260	1.02574421	0.00025579
4	4.7622	1.0220	1.02410747	0.00210747
5	6.0538	1.0180	1.02229035	0.00429035
6	7.2364	1.0155	1.01992786	0.00442786
7	8.3189	1.0140	1.01635944	0.00235944
8	9.3097	1.0100	1.01049231	0.00049231
9	10.2163	1.0035	1.00062569	0.00287431
10	11.0449	0.9880	0.98454637	0.00345363
11	11.8018	0.9630	0.95952145	0.00347855
12	12.4929	0.9255	0.92284049	0.00265951
13	13.1231	0.8725	0.87260289	0.00010289
14	13.6983	0.8075	0.80727824	0.00022176
15	14.2221	0.7265	0.72834029	0.00184029
16	14.6995	0.6345	0.63714077	0.00264077
17	15.1346	0.5345	0.53621417	0.00171417
18	15.5311	0.4275	0.42951068	0.00201068
19	15.8929	0.3185	0.31877223	0.00027223
20	16.2229	0.2085	0.20738624	0.00111376
21	16.5241	0.1010	0.09616350	0.00483650
22	16.7987	−0.0080	−0.00832793	0.00032793
23	17.0499	−0.1110	−0.11093729	0.00006271
24	17.2793	−0.2090	−0.20924554	0.00024554
25	17.4885	−0.3030	−0.30085833	0.00214167

Table A5. CEC2017 benchmark functions with initialization and search range: $(-100,100)^{D}$.

Types	No.	Functions	$F[x^*]$
Unimodal	F1	Shifted and Rotated Bent Cigar Instance	100
	F2	Shifted and Rotated Sum of Different Power Instance	200
	F3	Shifted and Rotated Zakharov Instance	300
Multimodal	F4	Shifted and Rotated Rosenbrock's Instance	400
	F5	Shifted and Rotated Rastrigin's Instance	500
	F6	Shifted and Rotated Expanded Scaffer's F6 Instance	600
	F7	Shifted and Rotated Lunacek Bi_Rastrigin Instance	700
	F8	Shifted and Rotated Non-Continuous Rastrigin's Instance	800
	F9	Shifted and Rotated Levy Instance	900
	F10	Shifted and Rotated Schwefel's Instance	1000
Hybrid	F11	Hybrid Instance 1 (N = 3)	1100
	F12	Hybrid Instance 2 (N = 3)	1200
	F13	Hybrid Instance 3 (N = 3)	1300
	F14	Hybrid Instance 4 (N = 4)	1400
	F15	Hybrid Instance 5 (N = 4)	1500
	F16	Hybrid Instance 6 (N = 4)	1600
	F17	Hybrid Instance 6 (N = 5)	1700
	F18	Hybrid Instance 6 (N = 5)	1800
	F19	Hybrid Instance 6 (N = 5)	1900
	F20	Hybrid Instance 6 (N = 6)	2000
Composition	F21	Composition Instance 1 (N = 3)	2100
	F22	Composition Instance 2 (N = 3)	2200
	F23	Composition Instance 3 (N = 4)	2300
	F24	Composition Instance 4 (N = 4)	2400
	F25	Composition Instance 5 (N = 5)	2500
	F26	Composition Instance 6 (N = 5)	2600
	F27	Composition Instance 7 (N = 6)	2700
	F28	Composition Instance 8 (N = 6)	2800
	F29	Composition Instance 9 (N = 3)	2900
	F30	Composition Instance 10 (N = 3)	3000

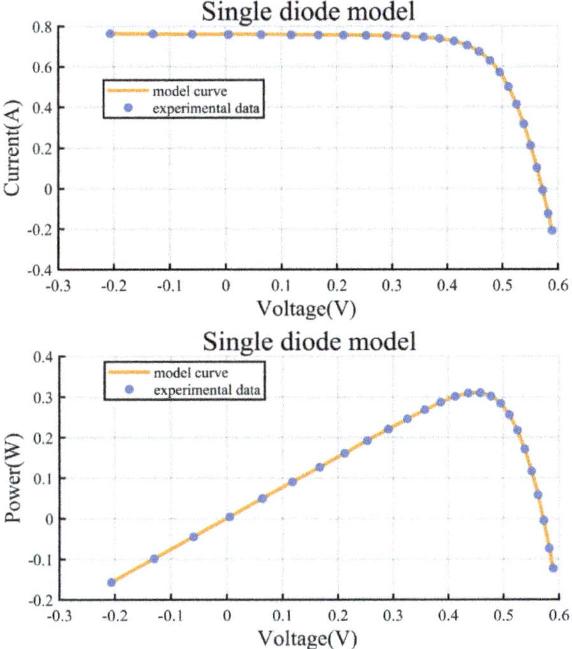

Figure A1. Comparisons between the experimental data and simulated data obtained by IMFO for single diode model: *I-V* and *P-V* characteristics.

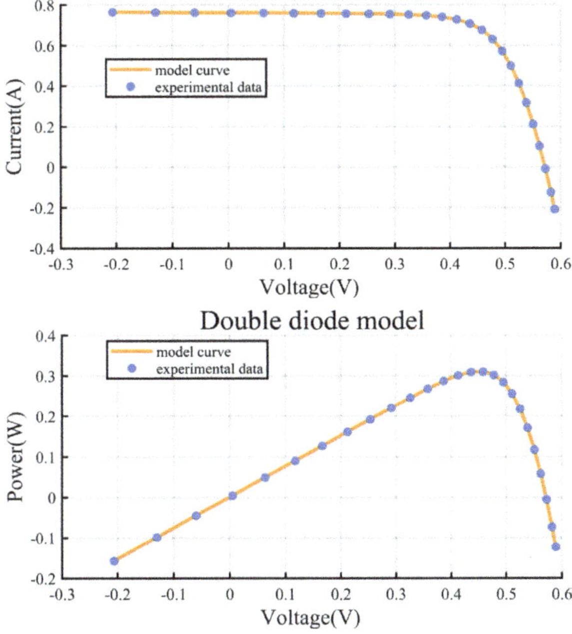

Figure A2. Comparisons between the experimental data and simulated data obtained by IMFO for double diode model: *I-V* and *P-V* characteristics.

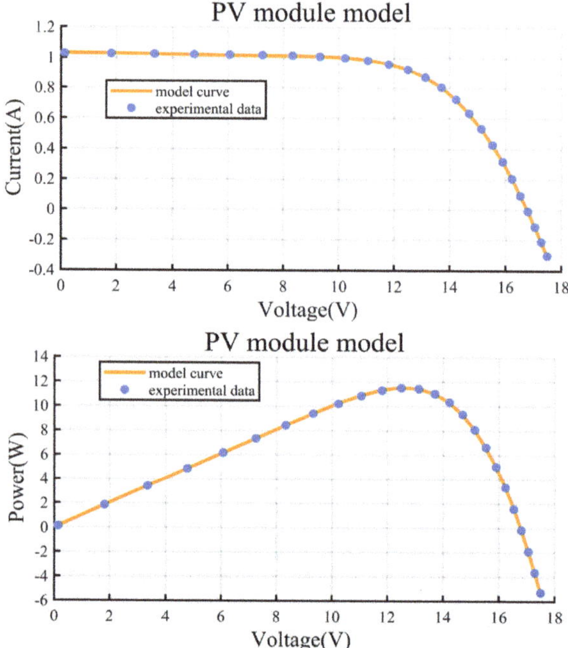

Figure A3. Comparisons between the experimental data and simulated data obtained by IMFO for PV module model: *I-V* and *P-V* characteristics.

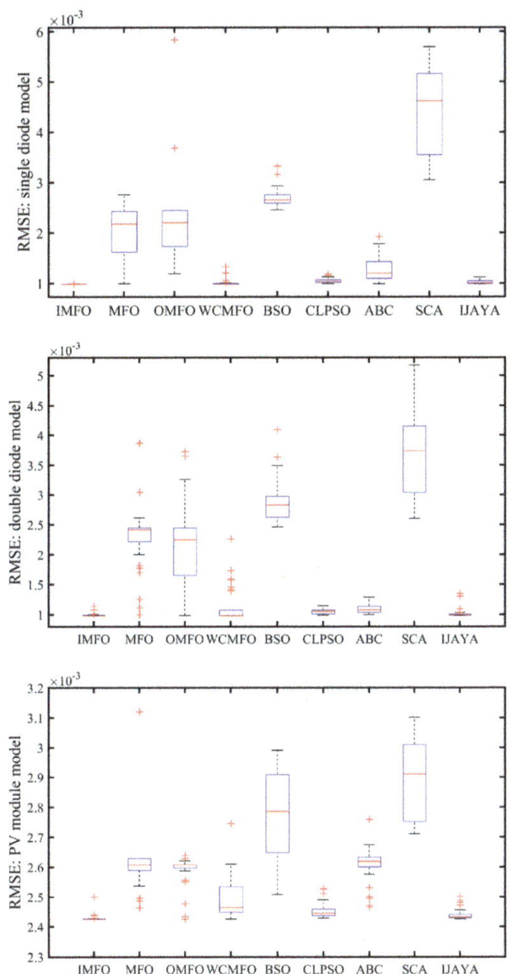

Figure A4. Best RMSE boxplot over 30 runs of different algorithms for three models.

References

1. Awan, A.B.; Khan, Z.A. Recent progress in renewable energy—Remedy of energy crisis in Pakistan. *Renew. Sustain. Energy Rev.* **2014**, *33*, 236–253. [CrossRef]
2. Irmak, E.; Ayaz, M.S.; Gok, S.G.; Sahin, A.B. A survey on public awareness towards renewable energy in Turkey. In Proceedings of the 2014 International Conference on Renewable Energy Research and Application (ICRERA), Milwaukee, WI, USA, 19–22 October 2014; pp. 932–937.
3. Jha, S.K.; Bilalovic, J.; Jha, A.; Patel, N.; Zhang, H. Renewable energy: Present research and future scope of Artifificial Intelligence. *Renew. Sustain. Energy Rev.* **2017**, *77*, 297–317. [CrossRef]
4. Lau, L.C.; Lee, K.T.; Mohamed, A.R. Global warming mitigation and renewable energy policy development from the Kyoto Protocol to the Copenhagen Accord—A comment. *Renew. Sustain. Energy Rev.* **2012**, *16*, 5280–5284. [CrossRef]
5. Mcelroy, M.B.; Chen, X. Wind and Solar Power in the United States: Status and Prospects. *CSEE J. Power Energy Syst.* **2017**, *3*, 1–6. [CrossRef]

6. Hanger, S.; Komendantova, N.; Schinke, B.; Zejli, D.; Ihlal, A.; Patt, A. Community acceptance of large-scale solar energy installations in developing countries: Evidence from Morocco. *Energy Res. Soc. Sci.* **2016**, *14*, 80–89. [CrossRef]

7. Abbassi, R.; Chebbi, S. Energy management strategy for a grid–connected wind-solar hybrid system with battery storage: Policy for optimizing conventional energy generation. *Int. Rev. Electr. Eng.* **2012**, *7*, 3979–3990.

8. Wu, Z.; Tazvinga, H.; Xia, X. Demand side management of photovoltaic-battery hybrid system. *Appl. Energy* **2015**, *148*, 294–304. [CrossRef]

9. Chen, Z.; Wu, L.; Lin, P.; Wu, Y.; Cheng, S.; Yan, J. Parameters identification of photovoltaic models using hybrid adaptive Nelder-Mead simplex algorithm based on eagle strategy. *Appl. Energy* **2016**, *182*, 47–57. [CrossRef]

10. Fathy, A.; Rezk, H. Parameter estimation of photovoltaic system using imperialist competitive algorithm. *Renew. Energy* **2017**, *111*, 307–320. [CrossRef]

11. Askarzadeh, A.; Rezazadeh, A. Parameter identification for solar cell models using harmony search-based algorithms. *Sol. Energy* **2012**, *86*, 3241–3249. [CrossRef]

12. Chan, D.S.H.; Phang, J.C.H. Analytical methods for the extraction of solar-cell single and double-diode model parameters from I–V characteristics. *IEEE Trans. Electron. Devices* **1987**, *34*, 286–293. [CrossRef]

13. Tong, N.T.; Kamolpattana, K.; Pora, W. A deterministic method for searching the maximum power point of a PV panel. In Proceedings of the International Conference on Electrical Engineering/Electronics, Computer, Telecommunications and Information Technology, Hua Hin, Thailand, 24–27 June 2015; pp. 1–6.

14. Bastidasrodriguez, J.D.; Petrone, G.; Ramospaja, C.A.; Spagnuolo, G. A genetic algorithm for identifying the single diode model parameters of a photovoltaic panel. *Math. Comput. Simul.* **2017**, *131*, 38–54. [CrossRef]

15. Ma, J.; Ting, T.O.; Man, K.L.; Zhang, N.; Guan, S.U.; Wong, P.W.H. Parameter estimation of photovoltaic models via cuckoo search. *J. Appl. Math.* **2013**, *2013*, 1–11. [CrossRef]

16. Ortiz-Conde, A.; Sïnchez, F.J.G.; Muci, J. New method to extract the model parameters of solar cells from the explicit analytic solutions of their illuminated characteristics. *Sol. Energy Mater. Sol. Cells* **2006**, *90*, 352–361. [CrossRef]

17. Saleem, H.; Karmalkar, S. An analytical method to extract the physical parameters of a solar cell from four points on the illuminated j-v curve. *IEEE Electron Device Lett.* **2009**, *30*, 349–352. [CrossRef]

18. Easwarakhanthan, T.; Bottin, J.; Bouhouch, I.; Boutrit, C. Nonlinear minimization algorithm for determining the solar cell parameters with microcomputers. *Int. J. Sol. Energy* **1986**, *4*, 1–12. [CrossRef]

19. Soon, J.J.; Low, K.S. Photovoltaic model identification using particle swarm optimization with inverse barrier constraint. *IEEE Trans. Power Electron.* **2012**, *27*, 3975–3983. [CrossRef]

20. Oliva, D.; Cuevas, E.; Pajares, G. Parameter identification of solar cells using artificial bee colony optimization. *Energy* **2014**, *72*, 93–102. [CrossRef]

21. Jiang, L.L.; Maskell, D.L.; Patra, J.C. Parameter estimation of solar cells and modules using an improved adaptive differential evolution algorithm. *Appl. Energy* **2013**, *112*, 185–193. [CrossRef]

22. Askarzadeh, A.; Rezazadeh, A. Extraction of maximum power point in solar cells using bird mating optimizer-based parameters identification approach. *Sol. Energy* **2013**, *90*, 123–133. [CrossRef]

23. Oliva, D.; El Aziz, M.A.; Hassanien, A.E. Parameter estimation of photovoltaic cells using an improved chaotic whale optimization algorithm. *Appl. Energy* **2017**, *200*, 141–154. [CrossRef]

24. Wu, Z.; Yu, D.; Kang, X. Parameter identification of photovoltaic cell model based on improved ant lion optimizer. *Energy Convers. Manag.* **2017**, *151*, 107–115. [CrossRef]

25. Rajasekar, N.; Kumar, N.K.; Venugopalan, R. Bacterial Foraging Algorithm based solar PV parameter estimation. *Sol. Energy* **2013**, *97*, 255–265. [CrossRef]

26. Yu, K.; Liang, J.; Qu, B.; Chen, X.; Wang, H. Parameters identification of photovoltaic models using an improved JAYA optimization algorithm. *Energy Convers. Manag.* **2017**, *150*, 742–753. [CrossRef]

27. Patel, S.J.; Panchal, A.K.; Kheraj, V. Extraction of solar cell parameters from a single current–voltage characteristic using teaching-learning-based optimization algorithm. *Appl. Energy* **2014**, *119*, 384–393. [CrossRef]

28. Alam, D.F.; Yousri, D.A.; Eteiba, M.B. Flower Pollination Algorithm based solar PV parameter estimation. *Energy Convers. Manag.* **2015**, *101*, 410–422. [CrossRef]

29. Mirjalili, S. Moth-flame optimization algorithm: A novel nature-inspired heuristic paradigm. *Knowl.-Based Syst.* **2015**, *89*, 228–249. [CrossRef]

30. Allam, D.; Yousri, D.A.; Eteiba, M.B. Parameters extraction of the three diode models for the multi-crystalline solar cell/module using Moth-Flame Optimization Algorithm. *Energy Convers. Manag.* **2016**, *123*, 535–548. [CrossRef]

31. Aghaie, R.; Farshad, M. Maximum Power Point Tracker for Photovoltaic Systems Based on Moth-Flame Optimization Considering Partial Shading Conditions. *J. Oper. Autom. Power Eng.* **2019**. [CrossRef]

32. Aouchiche, N.; Aitcheikh, M.S.; Becherif, M.; Ebrahim, M.A. AI-based global MPPT for partial shaded grid connected PV plant via MFO approach. *Sol. Energy* **2018**, *171*, 593–603. [CrossRef]

33. Apinantanakon, W.; Sunat, K. OMFO: A New Opposition-Based Moth-Flame Optimization Algorithm for Solving Unconstrained Optimization Problems. Recent Advances in Information and Communication Technology 2017, IC2IT 2017; In *Advances in Intelligent Systems and Computing*; Springer: Cham, Switzerland, 2017; Volume 566, pp. 22–31.

34. Khalilpourazari, S.; Khalilpourazary, S. An efficient hybrid algorithm based on Water Cycle and Moth-Flame Optimization algorithms for solving numerical and constrained engineering optimization problems. *Soft Comput.* **2019**, *2013*, 1–24. [CrossRef]

35. Wu, W.-M.; Li, Z.-X.; Lin, Z.-Y.; Wu, W.-Y.; Fang, D.-Y. Moth-flame optimization algorithm based on chaotic crisscross operator. *J. Comput. Eng. Appl.* **2018**, *54*, 136–141.

36. Wolf, M.; Noel, G.T.; Stirn, R.J. Investigation of the double exponential in the current–voltage characteristics of silicon solar cells. *Electron Devices IEEE Trans.* **1977**, *24*, 419–428. [CrossRef]

37. Alrashidi, M.R.; Alhajri, M.F.; Elnaggar, K.M.; Alothman, A.K. A new estimation approach for determining the i–v characteristics of solar cells. *Sol. Energy* **2011**, *85*, 1543–1550. [CrossRef]

38. Niu, Q.; Zhang, H.; Li, K. An improved TLBO with elite strategy for parameters identification of PEM fuel cell and solar cell models. *Int. J. Hydrogen Energy* **2014**, *39*, 3837–3854. [CrossRef]

39. Chen, X.; Yu, K.; Du, W.; Zhao, W.; Liu, G. Parameters identification of solar cell models using generalized oppositional teaching learning-based optimization. *Energy* **2016**, *99*, 170–180. [CrossRef]

40. Shi, Y. Brain storm optimization algorithm. In *International Conference on Swarm Intelligence*; Springer: Berlin/Heidelberg, Germany, 2011; pp. 303–309.

41. Liang, J.J.; Qin, A.K.; Suganthan, P.N.; Baskar, S. Comprehensive learning particle swarm optimizer for global optimization of multimodal functions. *IEEE Trans. Evol. Comput.* **2006**, *10*, 281–295. [CrossRef]

42. Mirjalili, S. SCA: A Sine Cosine Algorithm for solving optimization problems. *Knowl.-Based Syst.* **2016**, *96*, 120–133. [CrossRef]

43. Alcala-Fdez, J.; Sanchez, L.; Garcia, S.; del Jesus, M.J.; Ventura, S.; Garrell, J.; Otero, J.; Romero, C.; Bacardit, J.; Rivas, V.M.; et al. KEEL: A software tool to assess evolutionary algorithms for data mining problems. *Soft Comput.* **2009**, *13*, 307–318. [CrossRef]

44. Awad, N.; Ali, M.; Liang, J.; Qu, B.; Suganthan, P. *Problem Definitions and Evaluation Criteria for the CEC2017 Special Session and Competition on Single Objective Real-Parameter Numerical Optimization*; Technical Report; NTU: Singapore, 2016.

Article

Estimation of Single-Diode and Two-Diode Solar Cell Parameters by Using a Chaotic Optimization Approach

Martin Ćalasan [1,*]**, Dražen Jovanović** [2]**, Vesna Rubežić** [1]**, Saša Mujović** [1] **and Slobodan Đukanović** [1]

1   Faculty of Electrical Engineering, University of Montenegro, Džordža Vašingtona bb, 81000 Podgorica, Montenegro; vesnar@ucg.ac.me (V.R.); sasam@ucg.ac.me (S.M.); slobdj@ucg.ac.me (S.Đ.)
2   Montenegrin Electrical-energy distribution company—CEDIS, 81000 Podgorica, Montenegro; drazen1.dj@gmail.com
*   Correspondence: martinc@t-com.me; Tel.: +382-69615255

Received: 3 October 2019; Accepted: 29 October 2019; Published: 4 November 2019

Abstract: Estimation of single-diode and two-diode solar cell parameters by using chaotic optimization approach (COA) is addressed. The proposed approach is based on the use of experimentally determined current-voltage (*I-V*) characteristics. It outperforms a large number of other techniques in terms of average error between the measured and the estimated *I-V* values, as well as of time complexity. Implementation of the proposed approach on the *I-V* curves measured in laboratory environment for different values of solar irradiation and temperature prove its applicability in terms of accuracy, effectiveness and the ease of implementation for a wide range of practical environment conditions. The COA-based parameter estimation is, therefore, useful for PV power converter designers who require fast and accurate model for PV cell/module.

Keywords: Solar cell parameters; single-diode model; two-diode model; COA

---

1. Introduction

The contribution of solar energy in total electric energy production is growing constantly. As the price of solar inverters and solar panels constantly decreases, most countries are basing their energy policy on higher use of solar energy. Studies on energy networks, and especially testing of the integration of solar energy sources into power networks, requires accurate calculation of the solar output power, as well as accurate modeling of solar cells. For that reason, modeling of solar cells (corresponding equivalent circuit and accurate parameters value) represents a very popular research field.

In the literature, two basic models of the equivalent circuits of solar cell can be found, namely the single-diode model (SDM) [1] and the double-diode model (DDM) [2]. DDM considers the composite effect of the neutral region of the junction, and, therefore, models the solar cells more accurately [3]. However, it is characterized by seven unknown parameters. Because of the complexity of DDM, some authors reduce the number of unknown parameters [3,4], which can greatly affect the model accuracy [5]. In this work, we focus on both SDM and DDM, without neglecting any of the model parameters of the solar cell.

For the estimation of solar cell parameters, two main sets of "input data" and corresponding estimations can be found, namely

(a)   estimation based on datasheet information [6,7] and
(b)   estimation based on experimental data [8].

The former uses the datasheet information (open circuit voltage, short circuit current, voltage and current value at maximum power point characteristics) provided by photovoltaic (PV) manufacturers under standard test conditions. However, recent research [7] on the usage of datasheet values for solar cell parameter estimation shows that current-voltage characteristic is not unique when designers focus on three datasheet points (open circuit, short circuit and maximum power). It is shown that by observing only three points, we can have multiple *I-V* characteristics, although in reality a solar cell has a single defined *I-V* characteristic corresponding to a specific set of cell parameters. To obtain unique and accurate *I-V* characteristic of PV cell, experimental data on more than three major points are necessary [7]. Research [7] has also implied that only approaches based on experimental data generate accurate models.

Evaluating the performance of solar cells (or PV panels) requires as accurate an estimation of the equivalent circuit solar cell parameters as possible. The approaches used for this purpose can be categorized as follows:

(a)   analytical techniques [9–13],

(b)   numerical extraction [13] and

(c)   meta-heuristic techniques [14–59].

Analytical techniques provide mathematical expressions for solving equivalent circuit parameters based on some input data (manufacturer data or/and data obtained from measurements). A review and comparative assessment of non-iterative methods for the extraction of the single-diode model parameters of photovoltaic modules is given in [11]. In general, analytical techniques provide rapid solution. On the other hand, these techniques give erroneous results when the estimated and measure solar cell output characteristics are compared [12].

Numerical techniques are based on curve fitting, usually via iterative methods. However, the application of curve fitting to nonlinear diode equations is quite complex, making numerical determination of solar cell parameters unpopular [14].

Recently, meta-heuristic algorithms for solar cell parameter estimation have been proposed [14–58]. They impose no restrictions on the problem formulation, they are excellent in dealing with nonlinear equations, and they can be applied for different numbers of unknown parameters.

Of all the proposed techniques, none excels in terms of accuracy and efficiency with respect to others. This was our main incentive for doing research in this field. We propose both accurate and efficient parameter optimization of solar cell SDM and DDM through chaotic optimization approach (COA).

Recently, COA has been used in solving various optimization problems: parameter identification of Jiles-Atherton hysteresis model [59], single-phase transformer parameter estimation [60], design of PID parameters for automatic voltage regulation of synchronous machine [61], antenna array radiation pattern synthesis [62,63]. The main advantages of COA over other optimization techniques are easy implementation and short execution time [64]. It should be noted that different versions of chaotic algorithm have also been used in solar cell parameters estimation, namely chaotic heterogeneous comprehensive learning particle swarm optimizer variants [16], chaotic asexual reproduction optimization [33], mutative-scale parallel chaos optimization algorithm [41], chaos-embedded gravitational search algorithm [65], chaotic improved artificial bee colony algorithm [66], improved chaotic whale optimization algorithm [67], etc. Unlike methods proposed in [16,33,41,65–67], this paper will use COA based on Logistic map for solar cell parameter estimation. Logistic map has a simple form with one variable and one control parameter, and it can produce chaotic behavior similar to more complex chaotic systems. [68]. The power of this optimization approach is demonstrated in [60], where it was used for the estimations of the transformer's parameters.

The effectiveness of the proposed approach will be evaluated on different solar cells (different with respect to solar cell voltage and current level) found in the literature and the laboratory environment. Furthermore, COA-based parameter estimation will be compared with 50 various literature techniques for SDM, and with 12 various techniques for DDM. Also, we will compare parameters obtained by using the proposed method on the measured data with analytically and numerically obtained parameters. Finally, we will apply COA on the measured *I-V* characteristics using the laboratory environment.

The paper is organized as follows. SDM and DDM for solar cells are described in Section 2. In Section 3, COA and its implementation for solar cell parameters estimation are described. The comparison of solar cell parameters estimation accuracy obtained by using the COA-based and other methods, for one solar cell and one solar module, is presented in Section 4. The experimental setup for measuring *I-V* curves is presented in Section 5, along with the COA-based parameter estimation results. The concluding remarks are given in Section 6.

2. Mathematical Modeling of Single and Double Solar Cells

SDM is commonly used model for solar cell representation [1], and the equivalent circuit is shown in Figure 1a. The *I-V* relationship for this model can be described by the following equation:

$$I = I_{pv} - I_0\left(e^{\frac{V+IR_s}{n \cdot V_{th}}} - 1\right) - \frac{V + IR_s}{R_p} \tag{1}$$

where I_{pv} is the photo-generated current, R_s the series parasitic resistance, R_p the parallel parasitic resistance, I_0 the saturation current, n is ideality factor and $V_{th} = k_B T/q$ is the thermal voltage (k_B is Boltzmann constant equal to 1.38×10^{-23} J/K, T the temperature and q the electron charge equal to 1.602×10^{-19} C).

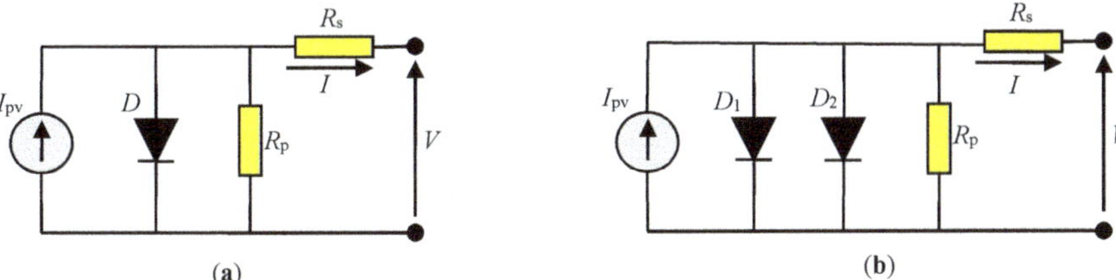

Figure 1. (a) SDM, and (b) DDM of a solar cell.

The equivalent circuit with DDM for the solar cell is shown in Figure 1b. Therefore, unlike the SDM model, the DDM model of the solar cell, in addition to the rectifying diode, includes one more diode to consider the space charge recombination current [48]. The *I-V* characteristic of DDM is given as

$$I = I_{pv} - I_{01}\left(e^{\frac{V+IR_s}{n_1 \cdot V_{th}}} - 1\right) - I_{02}\left(e^{\frac{V+IR_s}{n_2 \cdot V_{th}}} - 1\right) - \frac{V + IR_s}{R_p} \tag{2}$$

where I_{01} and I_{02} are the diffusion and saturation currents, whereas n_1 and n_2 are the diffusion and recombination diode ideality factors [48]. The ideality factor is discussed in [69,70], whereas [71] presents a method for ideality factor calculation.

3. COA and Objective Function

COA is a very powerful optimization technique that has found numerous scientific applications [59–63]. This approach is based on the theory of chaos, which is, in a mathematical sense, described by ordinary differential equations or by an iterative map [64].

Different chaotic systems, including the logistic map, lozi map, tent map and Lorenz system, can be found in the literature. In this paper, we will base COA on the logistic map [59–64].

The task of COA is to estimate a set of unknown parameters X which minimizes the objective function (*OF*). In our case, for SDM, X = $[R_s, R_p, I_{pv}, I_o, n]$, and for DDM, X = $[R_s, R_p, I_{pv}, I_{o1}, I_{o2}, n_1, n_2]$.

Therefore, in general, vector $X = [x_1, x_2, \ldots x_n]$ contains variables limited to the lower (LV) and upper (UV) permitted value, i.e, $x_i \in [L_i, U_i]$. On the other side, the OF for SDM is

$$OF = \sum_{t=1}^{P} \left(I_{pv} - I_0 \left(e^{\frac{V_t + I_t R_s}{n \cdot V_{th}}} - 1 \right) - \frac{V_t + I_t R_s}{R_p} - I_t \right) \tag{3}$$

whereas for DDM it reads

$$OF = \sum_{t=1}^{P} \left(I_{pv} - I_{o1} \left(e^{\frac{V_t + I_t R_s}{n_1 \cdot V_{th}}} - 1 \right) - I_{o2} \left(e^{\frac{V_t + I_t R_s}{n_2 \cdot V_{th}}} - 1 \right) - \frac{V_t + I_t R_s}{R_p} - I_t \right) \tag{4}$$

where P is the number of measured *I-V* pairs from the *I-V* characteristics, and V_t and I_t represent the voltage and current value of pair t.

Figure 2 presents the search procedure, i.e., the COA flowchart. The detailed description of COA flowchart can be found in [59].

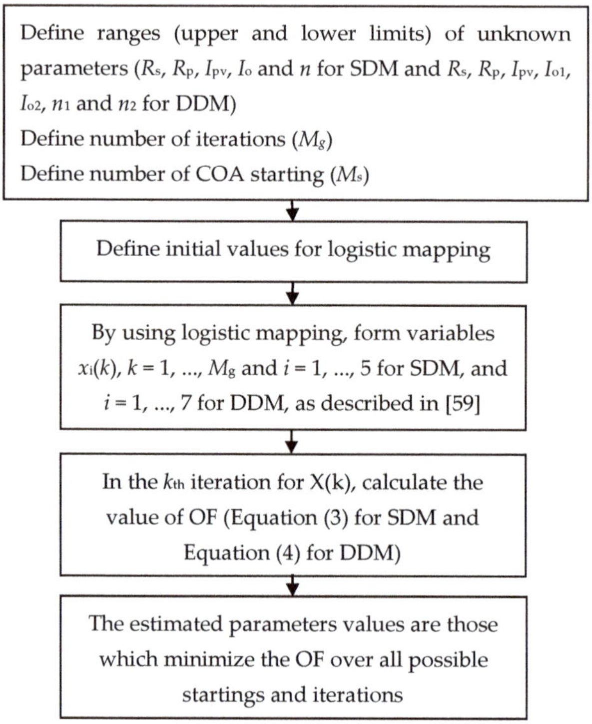

Define ranges (upper and lower limits) of unknown parameters (R_s, R_p, I_{pv}, I_o and n for SDM and R_s, R_p, I_{pv}, I_{o1}, I_{o2}, n_1 and n_2 for DDM)
Define number of iterations (M_g)
Define number of COA starting (M_s)

Define initial values for logistic mapping

By using logistic mapping, form variables $x_i(k)$, $k = 1, \ldots, M_g$ and $i = 1, \ldots, 5$ for SDM, and $i = 1, \ldots, 7$ for DDM, as described in [59]

In the k_{th} iteration for X(k), calculate the value of OF (Equation (3) for SDM and Equation (4) for DDM)

The estimated parameters values are those which minimize the OF over all possible startings and iterations

Figure 2. COA flowchart.

In this paper, the following COA parameters were used: $M = 1000$, $N = 50{,}000$. The COA-based estimation is compared with other approaches through the root mean square error ($RMSE$), defined as follows:

$$RMSE = \sqrt{\frac{\sum_{k=1}^{P} \left(I_{est,k} - I_{meas,k} \right)}{P}} \tag{5}$$

where $I_{est,k}$ and $I_{meas,k}$ represent the estimated and the measured values of solar output current in point k, respectively.

4. Simulation Results

To evaluate COA for solar cell parameters estimation, we first applied the proposed method to an experimental current-voltage characteristic extracted from the manufacturer's datasheets of a well-known R.T.C. France solar cell operating under standard test conditions.

The values of parameters obtained by using COA for the R.T.C. France solar cell are summarized, by year of publication, in Table 1, for SDM and Table 2 for DDM. These values are compared with the values of parameters published in recent papers (column Reference) for the same experimental data. During the estimation process, the parameter ranges for SDM estimation were $R_s(\Omega) \in [0.02, 0.05]$, $I_{pv}(A) \in [0.74, 0.78]$, $I_o(\mu A) \in [0.2, 0.4]$, $R_p(\Omega) \in [50, 55]$ and $n \in [1.35, 1.6]$, whereas for DDM, they were $R_s(\Omega) \in [0.02, 0.04]$, $R_p(\Omega) \in [54, 58]$, $n_1 \in [1.4, 1.5]$, $n_2 \in [1.95, 2]$, $I_{pv}(A) \in [0.75, 0.77]$ $I_{o1}(\mu A) \in [0.2, 0.25]$ and $I_{o2}(\mu A) \in [0.7, 0.8]$.

Table 1. Calculated SDM parameters for the R.T.C France solar cell.

No.	Algorithm	Reference	First Author, Year	I_{pv} (A)	I_0 (μA)	n	R_s (Ω)	R_p (Ω)	RMSE
	Proposed Method—COA			0.7607745	0.3230018	1.4811774	0.0363775	53.73	9.860221×10^{-4}
1.	HISA *	[15]	Dhruv, 2019	0.7607078	0.3106845918	1.47726778	0.03654694	52.88979426	9.8911×10^{-4}
2.	HCLPSO *	[16]	Dalia, 2019	0.76079	0.31062	1.4771	0.036548	52.885	1.12009×10^{-3}
3.	OBWOA *	[17]	Abd, 2018	0.76077	0.3232	1.5208	0.0363	53.6836	1.1417×10^{-3}
4.	MPSO *	[18]	Manel, 2018	0.760787	0.310683	1.475262	0.036546	52.88971	7.33007×10^{-3}
5.	ER-WCA	[19]	Kler D, 2017	0.760776	0.322699	1.481080	0.036381	53.69100	9.8609×10^{-4}
6	MSSO	[20]	Lin P, 2017	0.760777	0.323564	1.481244	0.036370	53.742465	1.0599×10^{-3}
7	BPFPA *	[21]	Ram JP, 2017	0.7600	0.3106	1.4774	0.0366	57.7151	1.2536×10^{-3}
8	ICA	[22]	Fathy A, 2017	0.7603	0.14650	1.4421	0.0389	41.1577	1.1582×10^{-1}
9	GOTLBO	[23]	Chen X, 2016	0.760780	0.331552	1.483820	0.036265	54.115426	9.8744×10^{-4}
10	CSO	[24]	Guo L, 2016	0.76078	0.3230	1.48118	0.03638	53.7185	9.8612×10^{-4}
11.	NM-MPSO	[25]	Hamid N, 2016	0.76078	0.32306	1.48120	0.03638	53.7222	9.8620×10^{-4}
12.	PCE	[26]	Zhang Y, 2016	0.760776	0.323021	1.481074	0.036377	53.718525	1.0606×10^{-3}
13.	TONG	[27]	Tong NT, 2016	0.7610	0.3635	1.4935	0.03660	62.574	2.3859×10^{-3}
14.	MABC	[28]	Jamadi M, 2016	0.760779	0.321323	1.481385	0.036389	53.39999	2.7610×10^{-3}
15.	MVO	[29]	Ali EE, 2016	0.7616	0.32094	1.5252	0.0365	59.5884	1.2680×10^{-1}
16.	DET	[30]	Chellaswamy C, 2016	0.751	0.315	1.487	0.036	54.532	2.4481×10^{-2}
17.	WCA			0.760908	0.4135540	1.504381	0.035363	57.669488	7.6069×10^{-3}
18.	TLBO			0.760809	0.312244	1.47578	0.036551	52.8405	7.2723×10^{-3}
19.	GWO	[31]	Jordehi AR, 2016	0.760996	0.2430388	1.451219	0.037732	45.116309	7.2845×10^{-3}
20.	TVACPSO			0.760788	0.3106827	1.475258	0.036547	52.889644	7.3438×10^{-3}
21.	PPSO	[32]	Ma J, 2016	0.7608	0.3230	1.4812	0.0364	53.7185	9.9161×10^{-4}
22.	CARO	[33]	Yuan X, 2015	0.76079	0.31724	1.48168	0.03644	53.0893	8.1969×10^{-3}
23.	LI	[34]	Lim LHI, 2015	0.7609438	0.3456572	1.48799169	0.03614233	49.482205	1.3462×10^{-3}
24.	MBA	[35]	El-Fergany A. 2015	0.7604	0.2348	1.4890	0.0388	44.61	1.1672×10^{-1}
25.	FPA *	[36]	Alam DF, 2015	0.76079	0.310677	1.47707	0.0365466	52.8771	1.2121×10^{-3}
26.	LMSA	[37]	Dkhichi F, 2014	0.76078	0.31849	1.47976	0.03643	53.32644	9.8649×10^{-4}
27.	DE			0.76068	0.35515	1.49080	0.03598	56.5533	1.0035×10^{-3}
28.	BBO	[38]	Niu Q, 2014	0.76098	0.86100	1.58742	0.03214	78.8555	2.3929×10^{-3}
29.	BBO-M			0.76078	0.31874	1.47984	0.03642	53.36227	9.8656×10^{-4}
30.	STLBO			0.76078	0.32302	1.48114	0.03638	53.7187	9.9763×10^{-4}
31.	TLBO	[39]	Niu Q, 2014	0.76074	0.32378	1.48136	0.03641	54.4029	1.0016×10^{-3}
32.	ABC	[40]	Oliva D, 2014	0.7608	0.3251	1.4817	0.0364	53.6433	1.0967×10^{-3}
33.	HPEPD	[8]	Laudani A, 2014	0.7607884	0.3102482	1.4769641	0.03655304	52.859056	1.1487×10^{-3}
34.	MPCOA	[41]	Yuan X, 2014	0.76073	0.32655	1.48168	0.03635	54.6328	2.3131×10^{-3}
35.	TLBO	[42]	Patel SJ, 2014	0.7608	0.3223	1.4837	0.0364	53.76027	9.6960×10^{-3}
36.	BMO	[43]	Askarzadeh A, 2013	0.76077	0.32479	1.48173	0.03636	53.8716	9.8622×10^{-4}
37.	ABSO	[44]	Askarzadeh A, 2013	0.76080	0.30623	1.47583	0.03659	52.2903	9.9125×10^{-4}
38.	IADE	[45]	Jiang LL, 2013	0.7607	0.33613	1.4852	0.03621	54.7643	9.9076×10^{-4}
39.	CS	[46]	Ma J, 2013	0.7608	0.323	1.4812	0.0364	53.7185	9.9161×10^{-4}
40.	ABSO			0.76080	0.30623	1.47986	0.03659	52.2903	1.4169×10^{-2}
41.	ABCDE			0.76077	0.32302	1.47986	0.03637	53.7185	4.8548×10^{-3}
42.	DE	[47]	Hachana O, 2013	0.76077	0.32302	1.48059	0.03637	53.7185	2.3423×10^{-3}
43.	MPSO			0.76077	0.32302	1.47086	0.03637	53.7185	3.9022×10^{-2}
44.	GGHS			0.76092	0.32620	1.48217	0.03631	53.0647	9.9089×10^{-4}
45.	HS	[48]	Askarzadeh A, 2012	0.76070	0.30495	1.47538	0.03663	53.5946	9.9515×10^{-4}
46.	IGHS			0.76077	0.34351	1.48740	0.03613	53.2845	1.0335×10^{-3}
47.	PS	[49]	AlHajri MF, 2012	0.7617	0.9980	1.6000	0.0313	64.10256	1.4936×10^{-2}
48.	SA	[50]	El-Naggar KM, 2012	0.7620	0.4798	1.5172	0.0345	43.10345	1.8998×10^{-2}
49.	GA	[51]	AlRashidi MR, 2011	0.7619	0.8087	1.5751	0.0299	42.37288	1.9078×10^{-2}
50.	PSO	[52]	Ye M, 2009	0.760798	0.322721	1.48382	0.0363940	53.7965	9.6545×10^{-3}

* for this method, a real RMSE are given [56].

Table 2. Calculated DDM parameters for the R.T.C France solar cell.

No.	Algorithm	Ref.	First Author, Year	I_{pv}(A)	I_{o1}(μA)	I_{o2}(μA)	R_s(Ω)	R_p(Ω)	n_1	n_2	RMSE
	Proposed Method—COA			0.76078105	0.2259742	0.749346	0.03674043	55.4854236	1.45101673	2	$9.82484852 \times 10^{-4}$
1.	GOFPANM	[53]	X Shuhui, 2017	0.7607811	0.7493476	0.2259743	0.0367404	55.485449	2	1.4510168	9.82485×10^{-4}
2.	SATLBO	[54]	Y Kunjie, 2017	0.76078	0.25093	0.545418	0.03663	55.117	1.45982	1.99941	9.82941×10^{-4}
3.	MSSO	[20]	P Lin, 2017	0.760748	0.234925	0.671593	0.036688	55.714662	1.454255	1.995305	1.059101×10^{-3}
4.	WDO	[55]	M Derick, 2017	0.7606	0.2531	0.0482	0.037433	52.6608	151.162	1.38434	1.095213×10^{-3}
5.	CSO	[24]	L Guo, 2016	0.76078	0.22732	0.72785	0.036737	55.3813	1.45151	1.99769	9.82532×10^{-4}
6.	GOTLBO	[23]	X Chen, 2016	0.760752	0.800195	0.220462	0.036783	56.0753	1.999973	1.448974	9.83152×10^{-4}
7.	PCE	[26]	Y Zhang, 2016	0.760781	0.226015	0.749340	0.03674	55.483160	1.450.923	2	9.8248×10^{-4}
8.	MABC	[28]	M Jamadi, 2016	0.7607821	0.24102992	0.6306922	0.03671215	54.7550094	1.4568573	2.0000.538	9.8276×10^{-4}
9.	FPA	[36]	DF Alam, 2015	0.760795	0.300088	0.166159	0.0363342	52.3475	1.47477	2	1.24239×10^{-3}
10.	BMO	[43]	A. Askarzadeh, 2013	0.76078	0.2111	0.87688	0.03682	558.081	1.44533	1.99.997	9.82661×10^{-4}
11.	ABSO	[44]	A. Askarzadeh, 2013	0.73078	0.26713	0.38191	0.03657	54.6219	1.46512	1.98152	9.8359×10^{-04}
12.	IGHS	[48]	A. Askarzadeh, 2012	0.76079	0.97310	0.16791	0.03690	56.8368	1.92126	1.42814	9.86572×10^{-4}

Tables 1 and 2 report parameters as they appear in the cited papers with no modification. However, in some papers in the Energy Conversion and Management journal (in Table 1 marked by *), inaccuracies occurred in parameter estimation of the PV cell using metaheuristic techniques. Namely, the results proposed in [15–18,21] do not correspond to the objective function [56].

The presented results, especially the value of RMSE, show that COA offers solar characteristics closer to the measured characteristics than the other existing methods, i.e., it outperforms other methods in terms of accuracy. In addition, by observing Tables 1 and 2, it is also evident that DDM characterizes solar cells more accurately than SDM, which supports the conclusion regarding DDM accuracy noted in [3].

It can be seen that COA outperforms several other techniques, such as evaporation rate-based water cycle algorithm (ER-WPA) [19] and cat swarm optimization (CSO) [24] for SDM, and with the generalized opposition-flower pollination algorithm-nelder-mead simplex method (GOFPANM) [53], by a small margin. However, the implementation of COA is simpler than implementation of ER-WPA, CSO and GOFPANM. Furthermore, COA is computationally less demanding than CSO since CSO requires changing the operation mode during the estimation process [24]. On the other side, GOFPANM is a hybrid algorithm which combines local and global search as well as different algorithms during estimation [53]. In general, most evolutionary algorithms have the complexity of $O((np + C_{of}\,p)N_i)$, where O is the big O notation, n is the dimension of the parameter space, p is the population size, N_i is the number of iterations and C_{of} is the complexity of the *OF*. The complexity of COA is $O(QC_{of})$, where Q is the number of points in the parameter space in which the OF is calculated. Therefore, the proposed COA-based estimation has significantly lower computational complexity than evolutionary algorithms.

To show the additional advantage of COA over other techniques, we conducted a comparison in terms of required time for one iteration. In that sense, in MATLAB 2015 (MathWorks, Natick, MA, USA) we have implemented the following algorithms for solar cell parameter estimation: evaporation rate-based water cycle algorithm (ER-WCA) [19], cuckoo search (CS) [46] and harmony search (HS) [48]. ER-WCA algorithm has a very good accuracy, very close to that obtained by the proposed method (see Table 1). On the other hand, HS and CS also have a good accuracy ($\sim 10^{-4}$). All computer simulations were carried out on a PC with Intel(R) Core (TM) i3-7020U CPU @ 2.30 GHz and 4 GB RAM. The obtained results, i.e., the mean, maximal and minimal required time per one iteration, obtained over 20 runs, are presented in Table 3. Clearly, the COA-based algorithm is the most efficient method, as it is characterized by the lowest value of required time per iteration. Note, in order to draw a fair comparison between the considered algorithms, MATLAB implementation follows the same rules for each algorithm (e.g., avoiding loops and using array operations such as dot product and matrix product whenever possible).

Table 3. Time per iteration comparison.

Algorithm	Mean Value of Requested Time (s)	Maximal Value of Requested Time (s)	Minimal Value of Requested Time (s)
COA	0.016416	0.017023	0.015871
ER-WCA [19]	0.021063	0.024145	0.019492
CS [46]	0.029179	0.037177	0.027130
HS [48]	0.021103	0.023264	0.020393

The measured *I-V* and *P-V* characteristics and the corresponding simulated characteristics, for parameters obtained by using COA, are shown in Figure 3. Very good agreement can be seen between the measured and estimated curves. Also, the difference between the DDM and SDM simulated curves is small but consistent and always in favor of DDM. In addition, in Table 4, we presented the estimated value of the unknown DDM parameters of the BPSolar MSX-60 module. These parameters are obtained by using COA as well as by using analytical, numerical, iteration and Newton methods presented in the literature. In the COA-based estimation, the ranges of parameters were $R_s(\Omega) \in [0.2\ 0.4]$, $R_p(\Omega) \in [150,\ 300]$, $n_1 \in [0.5,\ 1.5]$, $I_{pv}(A) \in [3.5,\ 4]$, $I_{o1}(A) \in \left[10^{-10},\ 10^{-6}\right]$, $I_{o2}(A) \in \left[10^{-10},\ 10^{-6}\right]$, and $n_2 \in [1.5,\ 2]$. From the presented results, it is clear that COA outperforms the considered non-metaheuristic methods for solar cell parameters determination in terms of accuracy.

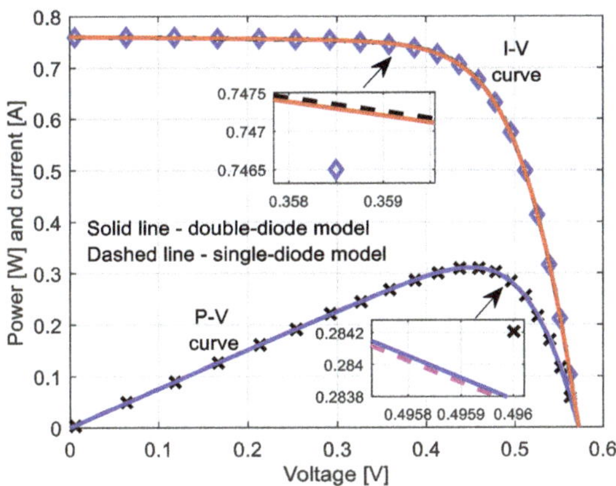

Figure 3. *I-V* and *P-V* characteristics of R.T.C. France solar cell.

Table 4. Calculated DDM parameters for the BPSolar MSX-60 module.

Parameter	Analytical Method [13]	Numerical Method [13]	Iteration Method [57]	Newton Method [58]	COA
I_{pv} (A)	3.8752	3.8046	3.8	3.8084	3.8418
I_{o1} (A)	3.6129×10^{-10}	3.9901×10^{-10}	4.704×10^{-10}	4.8723×10^{-10}	4.95821×10^{-8}
I_{o2} (A)	9.3773×10^{-6}	4.033×10^{-6}	4.704×10^{-10}	6.1528×10^{-10}	9.54961×10^{-9}
R_s (Ω)	0.3084	0.3397	0.35	0.3692	0.2495
R_p (Ω)	280.6449	280.2171	176.4	169.0471	267.57
n_1	1	0.99859	1	1.0003	1.2569
n_2	2	2.0014	1.2	1.9997	1.9345
RMSE	0.0358	0.0517	0.1211	0.1636	0.0194

The measured and corresponding simulated *I-V* and *P-V* characteristics, for parameters obtained by using COA and other methods, are shown in Figures 4 and 5, respectively. It is evident that COA outperforms the other methods in terms of approaching the measured characteristics.

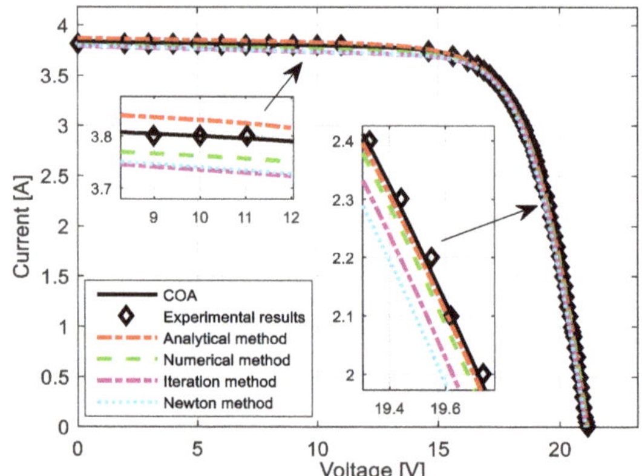

Figure 4. *I-V* characteristics of BPSolar MSX-60 module.

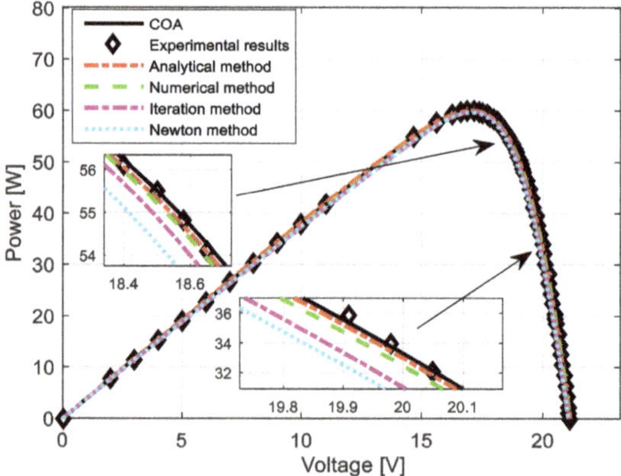

Figure 5. *P-V* characteristics of BPSolar MSX-60 module.

Based on all of the presented results, it can be concluded that COA can precisely estimate the solar cell/module circuit parameters, outperforming the other metaheuristics as well as analytical or numerical methods in terms of estimation accuracy.

5. Experimental Results and Analysis

To check the applicability and efficiency of COA for solar cell parameter estimation, we also observed solar cells from the Clean Energy Trainer setup. The main motivation to use these solar modules is that this setup enables adjustable solar insolation, USB data monitoring for PC-supported data acquisition and analysis, as well as highly advanced didactic software for system control and real-time data plotting.

The observed system contains of:

1. two solar modules and one module: 4 solar cells, 400 mW, 2 V, 0.5 A,
2. TES 1333R data logging Solar power meter—instrument with range of 2000 W/m^2, high resolution (0.1 W/m^2), and wide spectral resolution (400–1100 nm), etc.
3. lamp—special double spotlight lamp that simulates sunlight. It provides the optimal light spectrum for the solar module.
4. USB Data Monitor—used for data acquisition. Also, it is connected to the computer and software through the USB port.

5.　load—simulates electric consumer load.
6.　software—designed to facilitate system control, parameter monitoring, data acquisition and graphical representation of the collected data.

The experimental setup, installed in Laboratory for Automatics, at the Faculty of Electrical engineering, University of Montenegro, is presented in Figure 6.

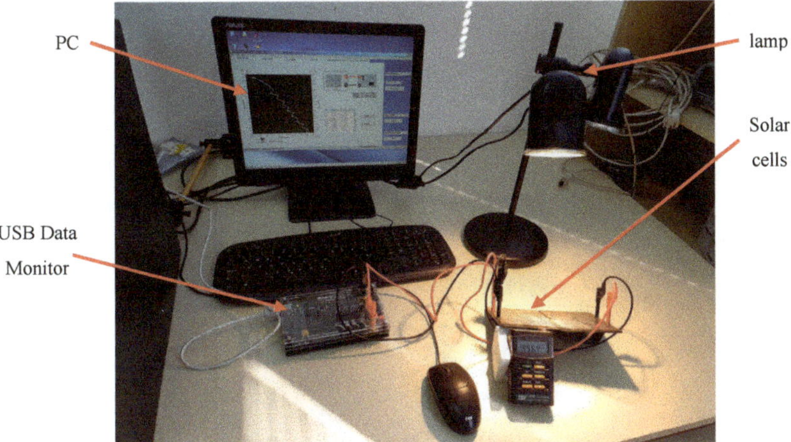

Figure 6. Experimental setup.

Firstly, we measured the *I-V* characteristics for insolation of 1285 W/m^2 and temperature of 42 °C. For the measured *I-V* pairs, we determined single and double diode solar cell parameters (see Table 5). The parameter ranges for SDM estimation were $R_s(\Omega) \in [0.1, 0.4]$, $I_{pv}(A) \in [0.2, 0.4]$, $I_o(A) \in [5 \times 10^{-8}, 15 \times 10^{-8}]$, $R_p(\Omega) \in [200, 600]$ and $n \in [0.2, 1]$, whereas for DDM were $R_s(\Omega) \in [0.1, 0.4]$, $R_p(\Omega) \in [600, 900]$, $n_1 \in [0.2, 1]$, $n_2 \in [1.95, 2]$, $I_{pv}(A) \in [0.2, 0.4]$, $I_{o1}(A) \in [5 \times 10^{-8}, 15 \times 10^{-8}]$, and $I_{o2}(A) \in [5 \times 10^{-8}, 15 \times 10^{-8}]$. Then we measured the *I-V* and *P-V* characteristics for different values of insolation and temperature. The corresponding simulated characteristics were determined by taking into account the change of parameters with insolation and temperature (see [13]). The measured and estimated *I-V* and *P-V* characteristics for different values of insolation and temperature are presented in Figures 7–10. The agreement between the measured and estimated characteristics is evident (see zoomed parts in these figures). Finally, we repeated the estimation procedure on all measured *I-V* characteristics. The estimated values of parameters were in range of ±4% of the initially observed, which confirms that we can use any of the measured characteristics for parameter estimation. On the other hand, by observing the data provided in Table 5, even for this module, it is evident that DDM is more accurate than SDM.

Table 5. Estimated value of experimentally tested solar module parameters.

SDM		DDM	
R_s (Ω)	0.2283	R_s (Ω)	0.2513
R_{sh} (Ω)	439.55	R_{sh} (Ω)	782.9911
I_o (A)	10.56×10^{-8}	I_{o1} (A)	6.8452×10^{-8}
I_{pv} (A)	0.2987	n_1	0.3342
n	0.3441	I_{pv} (A)	0.2972
		I_{o2} (A)	6.0643×10^{-8}
RMSE	4.3418×10^{-4}	n_2	1.9906
		RMSE	4.146×10^{-4}

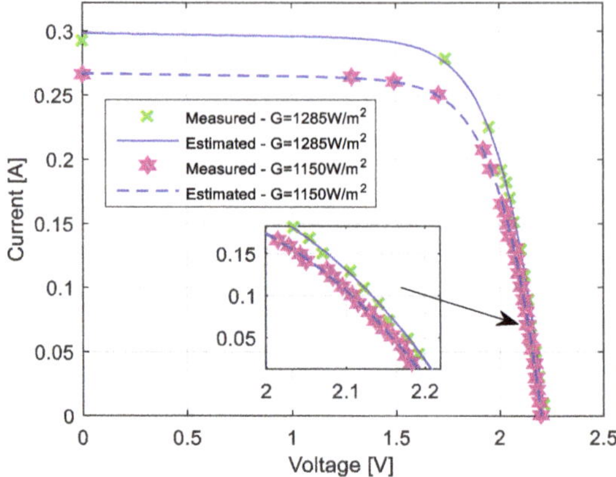

Figure 7. Current-voltage characteristics for two different insolation values and for temperature T = 42 °C.

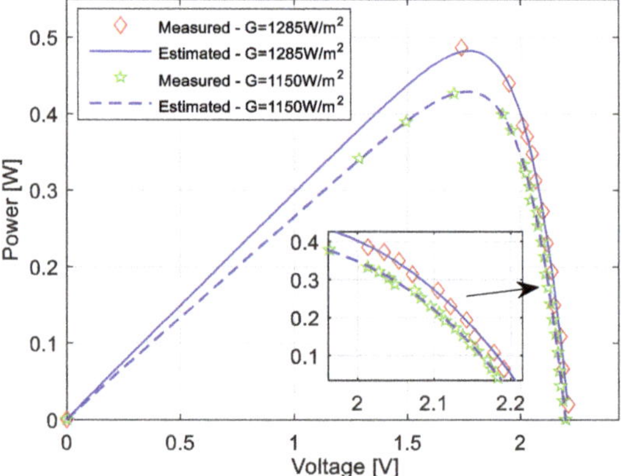

Figure 8. Power-voltage characteristics for two different insolation values and for temperature T = 42 °C.

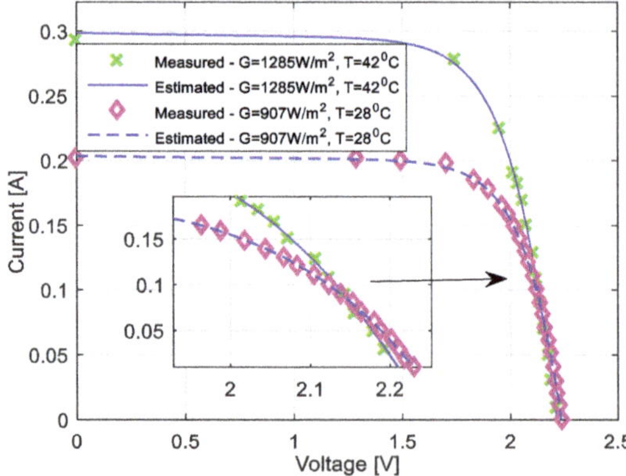

Figure 9. Current-voltage characteristics for two different temperatures and insolation values.

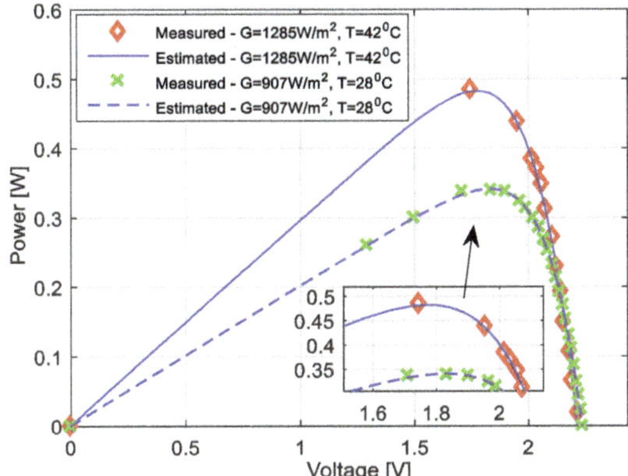

Figure 10. Power-voltage characteristics for two different temperatures and insolation values.

6. Conclusions

Modeling of solar cells is a very popular research direction, which is supported by numerous recent contributions in the literature. This paper proposes COA as a very successful approach for this purpose.

The proposed method is verified using practical data from various manufacturers. Its accuracy is confirmed by comparing its RMSE with numerous metaheuristics and non-metaheuristics methods for different solar cells. Experimental testing of COA applicability for parameter estimation is also implemented in laboratory environment. In all considered scenarios, a high level of accuracy is demonstrated. Apart from this, excellent matching of the simulated *I-V* and *P-V* curves with the measured characteristics additionally confirms the COA accuracy and its applicability for parameter estimation.

In future work, our attention will be focused on the usage of COA for estimation of solar cell parameters when solar cell output current is represented through the Lambert W function.

Author Contributions: Conceptualization, M.Ć. and S.Đ.; Methodology, M.Ć.; software, D.J. and V.R.; validation, V.R and M.Ć., formal analysis, S.M.; investigation, M.Ć., V.R and S.Đ.; resources, S.M.; writing—original draft preparation, M.Ć. and S.Đ.; writing—review and editing, S.Đ.; visualization, D.J.; supervision, M.Ć.

Funding: This work has been supported through European Union's Horizon 2020 research and innovation program under project CROSSBOW-CROSS BOrder management of variable renewable energies and storage units enabling a transnational Wholesale market (Grant No. 773430).

Conflicts of Interest: The authors declare no conflict of interest.

References

1. Moshksar, E.; Ghanbari, T. Adaptive Estimation Approach for Parameter Identification of Photovoltaic Modules. *IEEE J. Photovolt.* **2017**, *7*, 614–623. [CrossRef]
2. Ali, H.; Mojgan, H.; Saad, M.; Hussein, H. Solar cell parameters extraction based on single and double-diode models: A review. *Renew. Sustain. Energy Rev.* **2016**, *56*, 494–509.
3. Kashif, I.; Zainal, S.; Hamed, T. Simple, fast and accurate two-diode model for photovoltaic modules. *Sol. Energy Mater. Sol. Cells* **2011**, *95*, 586–594.
4. Chitti, B.B.; Suresh, G. A novel simplified two-diode model of photovoltaic (PV) module. *IEEE J. Photovolt.* **2014**, *4*, 1156–1161.
5. Chen, Y.; Sun, Y.; Meng, Z. An improved explicit double-diode model of solar cells: Fitness verification and parameter extraction. *Energy Convers. Manag.* **2018**, *169*, 345–358. [CrossRef]
6. Laudani, A.; Fulginei, F.R.; Salvini, A. Identification of the one-diode model for photovoltaic modules from datasheet values. *Sol. Energy* **2014**, *108*, 432–446. [CrossRef]

7. Biswas, P.P.; Suganthan, P.N.; Wu, G.; Amaratunga, G.A.J. Parameter estimation of solar cells using datasheet information with the application of an adaptive differential evolution algorithm. *Renew. Energy* **2019**, *132*, 425–438. [CrossRef]

8. Laudani, A.; Riganti, F.F.; Salvini, A. High performing extraction procedure for the one-diode model of a photovoltaic panel from experimental I-V curves by using reduced forms. *Sol. Energy* **2014**, *103*, 316–326. [CrossRef]

9. Chatterjee, A.; Keyhani, A.; Kapoor, D. Identification of photovoltaic source models. *IEEE Trans. Energy Convers.* **2011**, *24*, 883–889. [CrossRef]

10. Shongwe, S.; Hanif, M. Comparative analysis of different single-diode PV modeling methods. *IEEE J. Photovolt.* **2015**, *5*, 938–946. [CrossRef]

11. Batzelis, E. Non-Iterative Methods for the Extraction of the Single-Diode Model Parameters of Photovoltaic Modules: A Review and Comparative Assessment. *Energies* **2019**, *12*, 358. [CrossRef]

12. Jordehi, A.R. Parameter estimation of solar photovoltaic (PV) cells: A. review. *Renew. Sustain. Energy Rev.* **2016**, *61*, 354–371. [CrossRef]

13. Kumar, M.; Kumar, A. An efficient parameters extraction technique of photovoltaic models for performance assessment. *Sol. Energy* **2017**, *158*, 192–206. [CrossRef]

14. Kashif, I.; Salam, Z.; Mekhilef, S.; Shamsudin, A. Parameter extraction of solar photo voltaic modules using penalty based differential evolution. *Appl. Energy* **2012**, *99*, 297–308.

15. Dhruv, K.; Goswami, Y.; Kumar, R.V. A novel approach to parameter estimation of photovoltaic systems using hybridized optimizer. *Energy Convers. Manag.* **2019**, *187*, 486–511.

16. Dalia, Y.; Dalia, A.; Eteiba, M.B.; Ponnuthurai, N.S. Static and dynamic photovoltaic models' parameters identification using Chaotic Heterogeneous Comprehensive Learning Particle Swarm Optimizer variants. *Energy Convers. Manag.* **2019**, *182*, 546–563.

17. Abd, E.M.; Diego, O. Parameter estimation of solar cells diode models by an improved opposition-based whale optimization algorithm. *Energy Convers. Manag.* **2018**, *171*, 1843–1859.

18. Manel, M.; Anis, S.; Faouzi, M.M. Particle swarm optimisation with adaptive mutation strategy for photovoltaic solar cell/module parameter extraction. *Energy Convers. Manag.* **2018**, *175*, 151–163.

19. Kler, D.; Sharma, P.; Banerjee, A.; Rana, V.; Kumar, K.P.S. PV cell and module efficient parameters estimation using evaporation rate-based water cycle algorithm. *Swarm Evolut. Comput.* **2017**, *35*, 93–110. [CrossRef]

20. Lin, P.; Cheng, S.; Yeh, W.; Chen, Y.; Wu, L. Parameters extraction of solar cell models using a modified simplified swarm optimization algorithm. *Sol. Energy* **2017**, *144*, 594–603. [CrossRef]

21. Ram, J.P.; Babu, T.S.; Dragicevic, T.; Rajasekar, N. A new hybrid bee pollinator flower pollination algorithm for solar PV parameter estimation. *Energy Convers. Manag.* **2017**, *135*, 463–476. [CrossRef]

22. Fathy, A.; Rezk, H. Parameter estimation of photovoltaic system using imperialist competitive algorithm. *Renew. Energy* **2017**, *111*, 307–320. [CrossRef]

23. Chen, X.; Yu, K.; Du, W.; Zhao, W.; Liu, G. Parameters identification of solar cell models using generalized oppositional teaching learning-based optimization. *Energy* **2016**, *99*, 170–180. [CrossRef]

24. Guo, L.; Meng, Z.; Sun, Y.; Wang, L. Parameter identification and sensitivity analysis of solar cell models with cat swarm optimization algorithm. *Energy Convers. Manag.* **2016**, *108*, 520–528. [CrossRef]

25. Hamid, N.F.A.; Rahim, N.A.; Selvaraj, J. Solar cell parameters identification using hybrid Nelder-Mead and modified particle swarm optimization. *J. Renew. Sustain. Energy* **2016**, *8*, 1–10. [CrossRef]

26. Zhang, Y.; Lin, P.; Chen, Z.; Cheng, S. A population classification evolution algorithm for the parameter extraction of solar cell models. *Int. J. Photoenergy* **2016**, *2016*, 2174573. [CrossRef]

27. Tong, N.T.; Pora, W. A parameter extraction technique exploiting intrinsic properties of solar cells. *Appl. Energy* **2016**, *176*, 104–115. [CrossRef]

28. Jamadi, M.; Merrikh-Bayat, F.; Bigdeli, M. Very accurate parameter estimation of single- and double-diode solar cell models using a modified artificial bee colony algorithm. *Int. J. Energy Environ. Eng.* **2016**, *7*, 13–25. [CrossRef]

29. Ali, E.E.; El-Hameed, M.A.; El-Fergany, A.A.; El-Arini, M.M. Parameter extraction of photovoltaic generating units using multi-verse optimizer. *Sustain. Energy Technol. Assess.* **2016**, *17*, 68–76. [CrossRef]

30. Chellaswamy, C.; Ramesh, R. Parameter extraction of solar cell models based on adaptive differential evolution algorithm. *Renew. Energy* **2016**, *97*, 823–837. [CrossRef]

31. Jordehi, A.R. Time varying acceleration coefficients particle swarm optimisation (TVACPSO): A new optimisation algorithm for estimating parameters of PV cells and modules. *Energy Convers. Manag.* **2016**, *129*, 262–274. [CrossRef]
32. Ma, J.; Man, K.L.; Guan, S.U.; Ting, T.O.; Wong, P.W. Parameter estimation of photovoltaic model via parallel particle swarm optimization algorithm. *Int. J. Energy Res.* **2016**, *40*, 343–352. [CrossRef]
33. Yuan, X.; He, Y.; Liu, L. Parameter extraction of solar cell models using chaotic asexual reproduction optimization. *Neural Comput. Appl.* **2015**, *26*, 227–239. [CrossRef]
34. Lim, L.H.I.; Ye, Z.; Ye, J.; Yang, D.; Du, H. A linear identification of diode models from single I-V characteristics of PV panels. *IEEE Trans. Ind. Electron.* **2015**, *62*, 4181–4193. [CrossRef]
35. El-Fergany, A. Efficient tool to characterize photovoltaic generating systems using mine blast algorithm. *Electr. Power Compon. Syst.* **2015**, *43*, 890–901. [CrossRef]
36. Alam, D.F.; Yousri, D.A.; Eteiba, M.B. Flower pollination algorithm based solar PV parameter estimation. *Energy Convers. Manag.* **2015**, *101*, 410–422. [CrossRef]
37. Dkhichi, F.; Oukarfi, B.; Fakkar, A.; Belbounaguia, N. Parameter identification of solar cell model using Levenberg–Marquardt algorithm combined with simulated annealing. *Sol. Energy* **2014**, *110*, 781–788. [CrossRef]
38. Niu, Q.; Zhang, L.; Li, K. A biogeography-based optimization algorithm with mutation strategies for model parameter estimation of solar and fuel cells. *Energy Convers. Manag.* **2014**, *86*, 1173–1185. [CrossRef]
39. Niu, Q.; Zhang, H.; Li, K. An improved TLBO with elite strategy for parameters identification of PEM fuel cell and solar cell models. *Int. J. Hydrogen Energy* **2014**, *39*, 3837–3854. [CrossRef]
40. Oliva, D.; Cuevas, E.; Pajares, G. Parameter identification of solar cells using artificial bee colony optimization. *Energy* **2014**, *72*, 93–102. [CrossRef]
41. Yuan, X.; Xiang, Y.; He, Y. Parameter extraction of solar cell models using mutative-scale parallel chaos optimization algorithm. *Sol. Energy* **2014**, *108*, 238–251. [CrossRef]
42. Patel, S.J.; Panchal, A.K.; Kheraj, V. Extraction of solar cell parameters from a single current–voltage characteristic using teaching learning-based optimization algorithm. *Appl. Energy* **2014**, *119*, 384–393. [CrossRef]
43. Askarzadeh, A.; Rezazadeh, A. Extraction of maximum power point in solar cells using bird mating optimizer-based parameters identification approach. *Sol. Energy* **2013**, *90*, 123–133. [CrossRef]
44. Askarzadeh, A.; Rezazadeh, A. Artificial bee swarm optimization algorithm for parameters identification of solar cell models. *Appl. Energy* **2013**, *102*, 943–949. [CrossRef]
45. Jiang, L.L.; Maskell, D.L.; Patra, J.C. Parameter estimation of solar cells and modules using an improved adaptive differential evolution algorithm. *Appl. Energy* **2013**, *112*, 185–193. [CrossRef]
46. Ma, J.; Ting, T.O.; Man, K.L.; Zhang, N.; Guan, S.-U.; Wong, P.W.H. Parameter estimation of photovoltaic models via cuckoo search. *J. Appl. Math.* **2013**, *2013*, 362619. [CrossRef]
47. Hachana, O.; Hemsas, K.E.; Tina, G.M.; Ventura, C. Comparison of different metaheuristic algorithms for parameter identification of photovoltaic cell/module. *J. Renew. Sustain. Energy* **2013**, *5*, 053112. [CrossRef]
48. Askarzadeh, A.; Rezazadeh, A. Parameter identification for solar cell models using harmony search-based algorithms. *Sol. Energy* **2012**, *86*, 3241–3249. [CrossRef]
49. AlHajri, M.F.; El-Naggar, K.M.; AlRashidi, M.R.; Al-Othman, A.K. Optimal extraction of solar cell parameters using pattern search. *Renew. Energy* **2012**, *44*, 238–245. [CrossRef]
50. El-Naggar, K.M.; AlRashidi, M.R.; AlHajri, M.F.; Al-Othman, A.K. Simulated annealing algorithm for photovoltaic parameters identification. *Sol. Energy* **2012**, *86*, 266–274. [CrossRef]
51. AlRashidi, M.R.; AlHajri, M.F.; El-Naggar, K.M.; Al-Othman, A.K. A new estimation approach for determining the I-V characteristics of solar cells. *Sol. Energy* **2011**, *85*, 1543–1550. [CrossRef]
52. Ye, M.; Wang, X.; Xu, Y. Parameter extraction of solar cells using particle swarm optimization. *J. Appl. Phys.* **2009**, *105*, 1–8. [CrossRef]
53. Shuhui, X.; Yong, W. Parameter estimation of photovoltaic modules using a hybrid flower pollination algorithm. *Energy Convers. Manag.* **2017**, *144*, 53–68.
54. Kunjie, Y.; Xu, C.; Xin, W.; Zhenlei, W. Parameters identification of photovoltaic models using self-adaptive teaching-learning-based optimization. *Energy Convers. Manag.* **2017**, *145*, 233–246.
55. Derick, M.; Rani, C.; Rajesh, M.; Farrag, M.E.; Wang, Y.; Busawon, K. An improved optimization technique for estimation of solar photovoltaic parameters. *Sol. Energy* **2017**, *157*, 116–124. [CrossRef]

56. Gnetchejo, P.J.; Essiane, S.N.; Ele, P.; Wamkeue, R.; Wapet, D.M.; Ngoffe, S.P. Important notes on parameter estimation of solar photovoltaic cell. *Energy Convers. Manag.* **2019**, *197*, 111870. [CrossRef]
57. Ishaque, K.; Salam, Z.; Taheri, H. Accurate MATLAB Simulink PV System Simulator Based on a Two-Diode Model. *J. Power Electron.* **2011**, *11*, 179–187. [CrossRef]
58. Elbaset, A.A.; Ali, H.; Abd-El Sattar, M. Novel seven-parameter model for photovoltaic modules. *Sol. Energy Mater. Sol. Cells* **2014**, *130*, 442–455. [CrossRef]
59. Rubežić, V.; Lazović, L.; Jovanović, A. Parameter identification of Jiles-Atherton model using the chaotic optimization method. *Int. J. Comput. Math. Electr. Electron. Eng.* **2018**, *37*, 2067–2080. [CrossRef]
60. Ćalasan, M.; Mujičić, D.; Rubežić, V.; Radulović, M. Estimation of Equivalent Circuit Parameters of Single-Phase Transformer by Using Chaotic Optimization Approach. *Energies* **2019**, *12*, 1697. [CrossRef]
61. Coelho, L.S. Tuning of PID controller for an automatic regulator voltage system using chaotic optimization approach. *Chaos Solitons Fractals* **2009**, *39*, 1504–1514. [CrossRef]
62. Jovanović, A.; Lazović, L.; Rubežić, V. Adaptive Array Beamforming Using a Chaotic Beamforming Algorithm. *Int. J. Antennas Propag.* **2016**, *2016*, 8354204. [CrossRef]
63. Jovanović, A.; Lazović, L.; Rubežić, V. Radiation pattern synthesis using a Chaotic beamforming algorithm. *COMPEL Int. J. Comput. Math. Electr. Electron. Eng.* **2016**, *35*, 1814–1829. [CrossRef]
64. Yang, D.; Li, G.; Cheng, G. On the efficiency of chaos optimization algorithms for global optimization. *Chaos Solitons Fractals* **2007**, *34*, 1366–1375. [CrossRef]
65. Valdivia-González, A.; Zaldívar, D.; Cuevas, E.; Pérez-Cisneros, M.; Fausto, F.; González, A. A Chaos-Embedded Gravitational Search Algorithm for the Identification of Electrical Parameters of Photovoltaic Cells. *Energies* **2017**, *10*, 1052. [CrossRef]
66. Oliva, D.; Ewees, A.A.; Aziz, M.A.E.; Hassanien, A.E.; Peréz-Cisneros, M. A Chaotic Improved Artificial Bee Colony for Parameter Estimation of Photovoltaic Cells. *Energies* **2017**, *10*, 865. [CrossRef]
67. Oliva, D.; Aziz, M.A.E.; Hassanien, A.E. Parameter estimation of photovoltaic cells using an improved chaotic whale optimization algorithm. *Appl. Energy* **2017**, *200*, 141–154. [CrossRef]
68. Bertuglia, C.S.; Via, F. *Nonlinearity, Chaos, and Complexity*; Oxford University Press: Oxford, UK, 2005.
69. Tress, W. Interpretation and evolution of open-circuit voltage, recombination, ideality factor and subgap defect states during reversible light-soaking and irreversible degradation of perovskite solar cells. *Energy Environ. Sci.* **2017**. [CrossRef]
70. Wetzelaer, G.A.H.; Kuik, M.; Lenes, M.; Blom, P.W.M. Origin of the dark-current ideality factor in polymer:fullerene bulk heterojunction solar cells. *AIP Appl. Phys. Lett.* **2011**, *99*, 153506. [CrossRef]
71. Perovich, S.M.; Djukanovic, M.; Dlabac, T.; Nikolic, D.; Calasan, M. Concerning a novel mathematical approach to the solar cell junction ideality factor estimation. *Appl. Math. Model.* **2015**, *39*, 3248–3264. [CrossRef]

Article

A Solution of Implicit Model of Series-Parallel Photovoltaic Arrays by Using Deterministic and Metaheuristic Global Optimization Algorithms

Luis Miguel Pérez Archila [1], **Juan David Bastidas-Rodríguez** [2], **Rodrigo Correa** [1], **Luz Adriana Trejos Grisales** [3,*] **and Daniel Gonzalez-Montoya** [4]

[1] Escuela de Ingenierías Eléctrica, Electrónica y de Telecomunicaciones, Universidad Industrial de Santander, Bucaramanga 680003, Colombia; luis2188752@correo.uis.edu.co (L.M.P.A.); crcorrea@saber.uis.edu.co (R.C.)
[2] Departamento de Ingeniería Eléctrica, Electrónica y Computación, Universidad Nacional de Colombia, Manizales 170003, Colombia; jubastidasr@unal.edu.co
[3] Departamento de Electromecánica y Mecatrónica, Instituto Tecnológico Metropolitano, Medellín 050013, Colombia
[4] Departamento de Electrónica y Telecomunicaciones, Instituto Tecnológico Metropolitano, Medellín 050013, Colombia; danielgonzalez@itm.edu.co
* Correspondence: adrianatrejos@itm.edu.co; Tel.: +57-4-460-0727

Received: 19 December 2019; Accepted: 30 January 2020; Published: 12 February 2020

Abstract: The implicit model of photovoltaic (PV) arrays in series-parallel (SP) configuration does not require the LambertW function, since it uses the single-diode model, to represent each submodule, and the implicit current-voltage relationship to construct systems of nonlinear equations that describe the electrical behavior of a PV generator. However, the implicit model does not analyze different solution methods to reduce computation time. This paper formulates the solution of the implicit model of SP arrays as an optimization problem with restrictions for all the variables, i.e., submodules voltages, blocking diode voltage, and strings currents. Such an optimization problem is solved by using two deterministic (Trust-Region Dogleg and Levenberg Marquard) and two metaheuristics (Weighted Differential Evolution and Symbiotic Organism Search) optimization algorithms to reproduce the current–voltage (I–V) curves of small, medium, and large generators operating under homogeneous and non-homogeneous conditions. The performance of all optimization algorithms is evaluated with simulations and experiments. Simulation results indicate that both deterministic optimization algorithms correctly reproduce I–V curves in all the cases; nevertheless, the two metaheuristic optimization methods only reproduce the I–V curves for small generators, but not for medium and large generators. Finally, experimental results confirm the simulation results for small arrays and validate the reference model used in the simulations.

Keywords: implicit model solution; photovoltaic array; series–parallel; global optimization; partial shading; deterministic optimization algorithm; metaheuristic optimization algorithm

---

1. Introduction

The continuous effects of climate change and environmental contamination have motivated a growing interest in the use of renewable energies to replace conventional energy sources based on fossil fuels. Within renewable energies, photovoltaic (PV) energy offers interesting advantages: long life cycle, lack of mobile parts, low maintenance costs, modularity, among others [1]; that is why the installed PV capacity has been growing in the last years [2]. In a PV system, one of the most important elements is the generator; therefore, it is important to have models of this element to reproduce its electrical behavior when they operate under homogeneous and non-homogeneous conditions [3]. These models are used in different applications like the sizing of PV generators, the power, and energy production estimation, the analysis, and evaluation of maximum power point tracking techniques, model-based diagnosis, and other applications [4]. A PV generator can be connected in different configurations, nonetheless, series-parallel (SP) is one of the most used configurations. In an SP generator, two or more modules are connected in series to form a string and two or more strings are connected in parallel to form the generator. As a consequence, all the strings have the same voltage and can be analyzed independently [5]. It is worth noting that each module is formed by one or more submodules connected in series and each submodule has a bypass diode connected in antiparallel. When the irradiance and temperature of all the submodules in the generator are the same, it can be said that the generator operates under homogeneous conditions [6]. In these conditions, the whole generator can be represented by the single-diode model (SDM) scaled in voltage, according to the number of modules in the string, and scaled in current, according to the number of strings in parallel [7]. In this case, the current vs. voltage (I–V) curve has a single knee and the power vs. voltage (P-V) curve has a single maximum power point (MPP), which can be easily tracked. In real applications, the PV generator is commonly shaded, or partially shaded, by surrounding objects, passing clouds, or other objects. Hence, the operating conditions (i.e., irradiance and temperature) of the shaded submodules are different than the operating conditions of the rest of the submodules [8]. In this case, it is said that the generator operates in nonhomogeneous conditions and the generator's I–V curve may present multiple knees, which is translated into multiple MPPs in the P-V curve [9]. To model a PV generator operating in nonhomogeneous conditions, it is necessary to consider that each submodule in a module is represented by the SDM. In this way, it is possible to include the effects of the nonhomogeneous conditions in the electrical behavior of the generator [10]. Therefore, the generator can be analyzed as an equivalent circuit that is obtained by connecting strings in parallel, where each string is formed by a set of submodules connected in series [11,12].

In the literature, most of the models use the SDM to represent each submodule and they use the LambertW function [13] to obtain an explicit equation of the submodule's current as a function of its voltage [7,14–16]. Then, each string formed by N submodules and a blocking diode is modeled independently by a system of N + 1 nonlinear equations, where the unknowns are the voltages of the blocking diode and the N submodules [7,14–16]. Solving the system of nonlinear equations of each string, it is simple to calculate the string current and finally the generator's current. However, the evaluation of the LambertW function implies a high computational cost for the solution to the system of nonlinear equations associated with each string. For example, in [7] the modeling of a PV array based on the use of the Lambert W-function to obtain an explicit relationship between the voltage and current of any PV module is presented. In addition, the non-linear equations that describe the PV array are solved by means of explicit symbolic calculation of the inverse of the Jacobian matrix. A similar solution was published in [14], where a new formulation of the one-diode model of the PV module is solved using the LambertW function with the terminal voltage expressed as an explicit function of the current, this approximation resulting in a reduction of simulation calculation time. Another solution was presented in [16], where a model of the PV field is presented by means of a set non-linear equations characterized by a sparse Jacobian matrix, which requires a moderate computational burden, both in terms of memory use and processor speed. Optimization methods for estimating the PV module parameters are presented as a suitable option to overcome the drawbacks of deterministic and iterative methods. In addition, optimization methods are not only used in PV area for estimating parameters, since the reconfiguration of PV modules to mitigate the effect of partial shading conditions in PV arrays is currently one of the most studied topics [17]. The model proposed in [18] uses the implicit equation that describes the current-voltage relationship in the SDM. From such an equation, is it possible to construct a system of N + 2 implicit equations to model each string, where the unknowns are the string current and the N + 1 voltages of the N submodules and the blocking diode. Then, solving each string it is possible to calculate the generator's current as in the other models. In [18] the system of N + 2 implicit equations is solved by using the Trust-Region Dogleg (TRD) algorithm, which is a deterministic optimization method. However, the authors in [18] do not formulate the solution of the system of N + 2 implicit equations as an optimization problem with restrictions; moreover, the authors do not provide any justification for the selection of TRD for solving the optimization problem nor evaluate other deterministic and metaheuristic optimization algorithms. Finally, those works [7,14–16,18] are focused on the model, i.e., the system of nonlinear equations, and not in the optimization algorithms used to solve such models. Therefore, from those papers it is not clear which type of optimization methods should be used to reduce the calculation time.

This paper formulates the solution of the system of N + 2 implicit equations, associated with each string, as an optimization problem with restriction for the N + 2 unknowns, i.e., string current and the voltages of the N modules and the blocking diode. Moreover, the paper evaluates four optimization algorithms, two deterministic and two metaheuristics, to solve the problem and generate the I–V and P-V curves of generators with different sizes. The two deterministic algorithms are TRD and Levenberg-Marquardt (LMA), which are widely used to solve optimization problems in different areas; while the two metaheuristic optimization algorithms are Weighted Differential Evolution (WDE) and Symbiotic Organism Search (SOS), which have been recently used for PV applications. The evaluation of the optimization algorithms is performed with simulations of small, medium, and large generators in homogeneous and non-homogeneous conditions. Finally, experimental results are used to evaluate the different optimization algorithms for a small PV generator. In light of the previous analysis, the main advantages of the proposed procedure are listed as follows:

- A solution of the system of N + 2 implicit equation as an optimization problem with restrictions, which can be used to explore other optimization methods to reduce the model's solution time.
- Performance comparison of metaheuristic and deterministic optimization algorithms for solving the problem provides guidelines to continue exploring other optimization methods to reduce the computation time.
- A novel application of the algorithms of weighted differential evolution and search for symbiotic organisms to solve the implicit model of series-parallel photovoltaic arrays is presented.

The rest of the paper is organized as follows: Section 2 describes the implicit model used in this paper, Section 3 introduces the optimization problem and the operating principle of the optimization algorithms, Sections 4 and 5 show the simulation and experimental results, respectively, and Section 6 closes the paper with the conclusions.

2. Implicit Model of a Series-Parallel Generator

This section shows a brief description of the implicit model of SP generators proposed in [18], which includes the SDM used to represent each submodule, the system of N + 2 implicit equation of each string, and the calculation of the array current.

2.1. Submodule Model

The SDM (see Figure 1) has been widely used in the literature to represent the electrical behavior of PV submodules [10,19,20], which can be defined as a set of N_s cells connected in series. Such a model is able to represent mono-crystalline and poly-crystalline PV cells technologies [20]; in this paper PV modules formed by poly-crystalline cells have been used in simulation and experimental validations. In the SDM, I_{ph} represents the current generated by the photovoltaic effect, the diode D models the nonlinearities of the PN junction, and the resistors R_p and R_s represent the leakage currents and ohmic losses, respectively. The bypass diode (BD) is an additional connected in antiparallel to the submodule (see Figure 1) to limit the power dissipated in the cells in mismatching conditions. The calculation of the PV module parameters has been widely reported in literature, not only for the SDM model [21] but also for the two-diode [22] and three-diode model [23]. Nevertheless, the estimating parameter techniques have been more widely studied for the SDM. Such an issue has been addressed by means of different approaches which can be grouped into: analytical, deterministic and metaheuristic [24,25]. The first two groups present drawbacks concerning accuracy and convergence. Therefore, metaheuristic algorithms have been introduced as an attractive alternative for estimating the PV cells and modules parameters [25]. Such approaches may include particle swarm optimization [26], genetic algorithms [27], differential evolution [28], cuckoo search [29], among others.

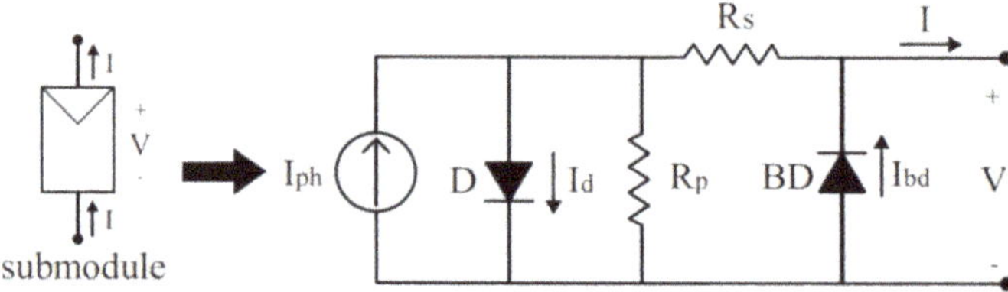

Figure 1. Submodule model including the bypass diode.

From the circuit shown in Figure 1, it is possible to write an implicit equation that describes the relationship between the submodule's output voltage (V) and current (I), as shown in (1), where I_{sat} and η are the inverse saturation current, and ideality factor of D, respectively, $V_{t,d}$ is the diode thermal voltage and I_{bd} is the current through BD. The thermal voltage is defined as $V_{t,d} = k \cdot T/q$, where k is the Boltzmann constant, q is the electron charge and T is the cell's temperature in K. Moreover, I_{db} is defined in (2), where $I_{sat,bd}$ and η_{bd} are the inverse saturation current and ideality factor of BD. It is worth noting that $V_{t,d}$ is the same for diodes D and BD since it can be assumed that the cell's temperature and the bypass diode temperature are the same if there isn't a fast change in the irradiance [30].

$$f(V,I) = -I + I_{ph} - I_{sat}\left[exp\left(\frac{V + (I - I_{bd}) \cdot R_s}{N_s \cdot \eta \cdot V_{t,d}}\right) - 1\right] - \frac{V + (I - I_{bd}) \cdot R_s}{R_p} + I_{bd} = 0. \quad (1)$$

$$I_{bd} = I_{sat,bd} \cdot \left[exp\left(\frac{-V}{\eta_{bd} \cdot V_{t,d}}\right) - 1\right]. \quad (2)$$

Although the SDM parameters (i.e., $I_{ph}, I_{sat}, V_{t,d}, R_s$ and R_h) may vary with the irradiance and temperature [31–33], I_{ph} is directly proportional to the irradiance; therefore, it can be used as an indication of the incident irradiance on the submodule.

2.2. Generator's Model

In general, an SP generator is formed by M parallel-connected strings (see Figure 2), where each string is formed by a set of PV modules (N_m) and a blocking diode connected in series. In turn, each module is formed by one or more submodules connected in series (N_{sm}); then, each string is formed by N submodules, where $N = N_m \cdot N_{sm}$. The generator is connected to a power converter, which sets the generator voltage (V_{gen}) to perform the MPP tracking; therefore, V_{gen} is known for the model and the objective is to find: the generator current (I_{gen}), the current of each string, the voltage of each submodule, and the voltage of each blocking diode.

Considering the V_{gen} is common for all the strings, each string can be analyzed and solved independently. Then, a string formed by N submodules and a blocking diode connected in series can be represented by a system of N + 2 nonlinear and implicit equations, as shown in (3) [18]; where $\vec{V}$ is a vector formed by the voltages of the N submodules and the blocking diode ($\vec{V} = [V_1, V_2, \cdots , V_N, V_{blk}]$) and I_{str} is the string current.

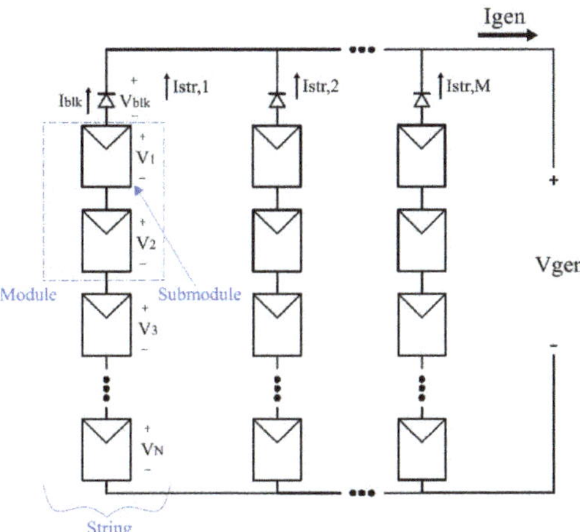

Figure 2. Series–parallel photovoltaic generator with M strings, of N submodules each, connected in parallel, where each string includes a blocking diode and each module has two submodules.

$$F\left(\vec{V}, I_{str}\right) = \begin{bmatrix} f_1\left(V_1, I_{str}\right) \\ f_2\left(V_2, I_{str}\right) \\ \vdots \\ f_N\left(V_N, I_{str}\right) \\ f_{N+1}\left(V_{blk}, I_{str}\right) \\ f_{N+2}\left(\vec{V}\right) = \sum_{i=1}^{N+1} V_i - V_{gen} \end{bmatrix} = \vec{0}. \tag{3}$$

The first N equations of (3), correspond the voltage-current relationship in each submodule given in (1), while equation N + 1 describes the model of the blocking diode and it is described in (4), where $I_{sat,blk}$, and η_{blk} are the inverse saturation current and the ideality factor of the blocking diode, respectively, and $V_{t,d}$ is the diode thermal voltage defined in Section 2.1, since the blocking diode temperature is assumed the same of the submodules. Finally, equation N + 2 in (3) is obtained from the Kirchhoff voltage law in the loop formed with the string and V_{gen}.

$$f_{N+1}\left(V_{blk}, I_{str}\right) = -I_{str} + I_{sat,blk} \cdot \left(exp\left(\frac{-V_{blk}}{\eta_{blk} \cdot V_{t,d}}\right) - 1\right) = 0. \tag{4}$$

All the electrical variables of the string are obtained by solving (3), i.e., I_{str} and $\vec{V}$; hence, after solving all the strings, Igen can be calculated using (5).

$$I_{gen} = \sum_{j=1}^{M} I_{str,j}. \tag{5}$$

3. Optimization Problem Formulation and Solution Methods

This Section defines the solution of (3) as an optimization problem with its restrictions and presents a brief explanation of the optimization algorithms used in the paper to solve the optimization problem.

3.1. Optimization Problem and Restrictions

Let $\mathbb{R}$ be the set of real numbers and $\mathbf{L}$ a non-empty subset of $\mathbb{R}^n$, i.e., $\mathbf{L} \subset \mathbb{R}^n$; then, the system of nonlinear equations given in (3) can be expressed as shown in (6), where $\vec{x} \in \mathbb{R}^n$

and $f_k : \mathbf{L} \rightarrow \mathbb{R}, \forall\, k = 1, 2, \ldots, N + 2$. The solution of (6) is given by all vectors $\vec{a} \in \mathbb{R}^n$, such that $f_k(\vec{a}) = 0, \forall\, k = 1, 2, \ldots, N + 2$.

$$\begin{cases} f_1(\vec{x}) = 0 \\ f_2(\vec{x}) = 0 \\ \quad\vdots \\ f_{N+2}(\vec{x}) = 0 \end{cases}. \tag{6}$$

Now, let's consider the function $F_{obj} : \mathbf{L} \rightarrow \mathbb{R}$ a function defined as shown in (7).

$$F_{obj} = \sum_{i=1}^{N+2} f_i(\vec{x})^2, \; with \; \vec{x} \in \mathbf{L}. \tag{7}$$

Note that F_{obj} is a well-defined function that guarantees the existence of a minimum, since $F_{obj}(\vec{x}) \geq 0$ for all $\vec{x} \in \mathbf{L}$ is a totally ordered set where exists the infimum of $F_{obj}(\mathbf{L}) \in \mathbb{R}$ and this infimum is non-negative, i.e., $inf\left\{ F_{obj}(\mathbf{L}) \right\} \geq 0$. Therefore, the minimum of F_{obj} over $\mathbf{L}$ exists and it is a non-negative minimum.

Theorem 1. *Suppose that the system of equations given in (3) has a solution $\mathbf{L}$ and $(\vec{a} \in \mathbf{L})$; hence, is a solution of (3) if and only if, $\vec{a}$ minimizes F_{obj} given in (7).*

Proof. If $\vec{a}$ is a solution of the system of equations given in (3), then $f_k(\vec{a}) = 0, \forall\, k = 1, 2, \ldots, N + 2$, therefore, $F_{obj}(\vec{a}) = 0$. Moreover, considering that $F_{obj}(\vec{x}) \geq 0$ for all $\vec{x} \in \mathbf{L}$, then $\vec{a}$ is the minimum point of F_{obj}. Now, if $\vec{a}$ minimizes F_{obj} but it is not a solution of (3), then $F_{obj}(\vec{a})$ must be positive, since $F_{obj}(\vec{x}) \geq 0$ for all $\vec{x} \in \mathbf{L}$. Taking into account that (3) has a solution in $\mathbf{L}$, then there is a solution $\vec{x}^* \in \mathbf{L}$ such that $F_{obj}(\vec{x}^*) = 0$ and $\vec{x}^* \neq \vec{a}$. Hence, $F_{obj}(\vec{x}^*) < F_{obj}(\vec{a})$, which contradicts that $\vec{a}$ is a minimum point for F_{obj}. $\square$

The last paragraph proves that the minimum point of the objective function given in (7) is the set of values that solve the system of equations shown in (3). Such a solution contains the unknown voltages in the string $(\vec{V})$ and the string current (I_{str}). However, to minimize (7) using numerical methods, it is necessary to establish the limits of I_{str} and each unknown in $\vec{V}$ as shown in Table 1.

Table 1. Upper and lower limits of the electrical variables in a string.

Variable	Lower Limit	Upper Limit
$V_1 \ldots V_N$	$-V_{bd,max}$	$V_{OC,STC}$
V_{blk}	$-V_{blk,max}$	$0\ V$
I_{str}	$0\ A$	$I_{SC,STC}$

The lower limit of the string current is 0 A; while the upper limit is $I_{SC,STC}$, which corresponds to the submodule short-circuit current in Standard Test Conditions (STC), i.e., irradiance of 1 kW/m^2 and the submodule temperature is 25 °C (i.e., $T = 298.15$ K). Moreover, the upper limit of the submodules voltages $(V_1 \ldots V_N)$ is open-circuit voltage in STC $(V_{OC,STC})$, which corresponds to the submodule voltage when $I = 0$ A. The lower limit of $V_1 \ldots V_N$ is the negative value of the maximum bypass diode voltage $(V_{bd,max})$, defined in (8), which corresponds to the bypass diode voltage when $I_{bd} = I_{SC,STC}$. This current corresponds to the worst-case scenario where the submodule is completely shaded, and the string current is maximum. Finally, the upper limit of the blocking diode voltage is 0 V, which corresponds to $I_{str} = 0$ A; while the lower limit corresponds to the negative value of the maximum blocking diode voltage $(V_{blk,max})$ defined in (9).

$$V_{bd,max} = \eta_{bd} \cdot V_{t,d} \cdot ln\left(\frac{I_{SC,STC}}{I_{sat,bd}} + 1 \right). \tag{8}$$

$$V_{blk,max} = \eta_{blk} \cdot V_{t,d} \cdot ln\left(\frac{I_{SC,STC}}{I_{sat,blk}} + 1\right). \tag{9}$$

3.2. Optimization Methods

This section presents a brief description of the optimization algorithms used to minimize the objective function introduced in (7) and to find the solution to the system of nonlinear equations associated with the string (3).

3.2.1. Trust-Region Dogleg (TRD)

This is a deterministic algorithm with a single search agent. It starts from an initial point x_0, that belongs to the domain of $f(x)$, and searches the point x^* that minimizes the function f through the iterative process indicated in (10), where x_k is the actual position of the search agent and p_k is the step of the k-th iteration.

$$x_{k+1} = x_k + p_k. \tag{10}$$

To determine the magnitude and direction of the vector p_k that approaches x_{k+1} to x^* is posed as an optimization problem, as shown in (11), where $m_k(p)$ is an approximate function of $f(x)$ formed by the two first terms of the Taylor series in a region Q_k that contains x_k.

$$\min_{p} \{m_k(p), \, p \in Q_k\}. \tag{11}$$

If $m_k(p)$ is a good approximation of the objective function in a region Q_k, then this region is denominated thrust-interval and it is defined as $Q_k = \{\mathbf{p} \mid |\mathbf{p}| < \Delta_k\}$, where Δ_k is the radius of the thrust-region in the k-th iteration. Once the vector p_k^* is found, i.e., the solution of (11), it is necessary to verify the fulfillment of the condition expressed in (12).

$$f(x_k + p_k^*) < f(x_k). \tag{12}$$

If condition (12) is fulfilled, the actual position is updated as $x_{k+1} = x_k + p_k^*$; then, the next step is to calculate a new model m_{k+1} and a new trust-region Q_{k+1} from the updated point x_{k+1}. Otherwise, if condition (12) is not fulfilled, the algorithm keeps the actual position, i.e., $x_{k+1} = x_k$, and the trust-region is reduced by reducing its radius ($\Delta_{k+1} < \Delta_k$). The details of this algorithm can be found in [34] along with different variants of this method.

3.2.2. Levenberg-Marquardt (LMA)

This is a deterministic optimization algorithm that requires the calculation of the jacobian matrix (J) and it is especially effective for solving nonlinear problems, where methods that used linear models, like Gauss–Newton, are not effective for the entire solution region [35]. The essence of this method is to select the search direction of the next step between the one given by the Gauss–Newton method and a direction close to the one provided by the gradient descent method. That direction search is selected by a criterion denominated μ_k.

Let's consider an optimization problem that has the structure of a trust-interval, as shown in (13) and (14) [35].

$$\min \frac{1}{2} \|J(x_k)(x - x_k) + rx_k\|_2^2. \tag{13}$$

$$s.t. \ \|x - x_k\| \leq \Delta_k. \tag{14}$$

This is a least-squares problem, which is linear with restrictions because $r(x_k)$ is a linearized model valid in the trust-region defined by Δ_k. The solution to the problem described in (13) and (14) is obtained by solving (15), where $s = x - x_k$, and the position of the next iteration is given by (16) [35].

$$(J(x_k)^T J(x_k) + \mu_k I)s = -J(x_k)^T r(x_k). \tag{15}$$

$$x_{k+1} = x_k - (J(x_k)^T J(x_k) + \mu I)^{-1} \cdot J(x_k)^T r(x_k). \tag{16}$$

When $\|(J(x_k)^T J(x_k))^{-1} J(x_k)^T r(x_k)\| \leq \Delta_k$, then $\mu_k = 0$ and (16) is solved by s_k; in this case, the best search direction is the same as the Gauss–Newton method. Otherwise, $\mu_k > 0$ and the solution of (16) fulfills $s_k = \Delta_k$; with this condition, it is possible to find the solution of (16) and the value of μ_k. When $\mu_k \gg 0$, the search direction is close to the one of the gradient descent method and (15) approaches to (17). Further details of the Levenberg–Marquardt algorithm are provided in [35].

$$\mu_k I s = -J(x_k)^T r(x_k). \tag{17}$$

3.2.3. Weighted Differential Evolution (WDE)

This metaheuristic optimization method uses the evolution concept to solve numerical optimization problems and it is characterized by calculating the search direction and the search steps in an evolutionary way [36]. The general process of the WDE algorithm is introduced in Figure 3, where the first step is to generate two populations denominated initial population (P) and subpopulation (SubP), both randomly created with a normal distribution. Then, SubP is mutated to create TempP and vector F and binary map M are generated, by using random variables. Population SubP, F and M are used in a mutation and crossover proves to generate the new population (T), which requires verification that all the individuals fulfill the problem restrictions (boundaries control). The individuals that do not fulfill the restrictions are adjusted to be within the problem borders. Now, the objective function is evaluated for all the individuals of T and SubP and the best individuals in both populations are identified ($T_{(i)}$ and $SubP_{(i)}$). If $T_{(i)}$ is a better solution than $SubP_{(i)}$, then $SubP_{(i)}$ is updated ($SubP_{(i)} = T_{(i)}$), and population P is updated from SubP; otherwise, TempP, F and M are calculated to generate a new T. Finally, the best individual of P is evaluated to verify if it fulfills the stop criteria; else, a new algorithm iteration starts.

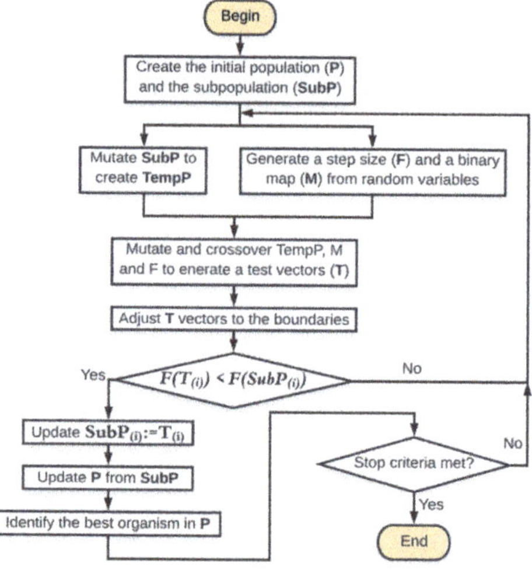

Figure 3. Flow chart of the weighted differential evolution algorithm.

3.2.4. Symbiotic Organism Search (SOS)

This is a metaheuristic optimization algorithm inspired by the different interactions between the living beings of different species that share an ecosystem, named symbiotic relationships. The first phase of the algorithm is to create an initial population of particles that represents an ecosystem, and each particle represents an organism. Then, it is necessary to define some parameters, like the solution boundaries and the stop criteria, before starting the iterative process, which is based on three types of relationships: mutualism, commensalism, and parasitism. In the mutualism phase, two organisms participate, X_i and X_j, where X_i is the i-th organism into the ecosystem and X_j is a randomly chosen organism within the ecosystem. In this phase, both organisms could be benefited from each other through the creation of a new vector (M_{vec}) described in (18)

$$M_{vec} = \frac{X_i + X_j}{2}.$$ (18)

Then, the algorithm calculates two new candidates to replace X_{inew} and X_{jnew} by using (19) and (20), which model the mutualist symbiotic relationship between two organisms. In (19) and (20) $rand(0,1)$ is a vector with random numbers between 0 and 1, and X_{best} is the organism with the lowest value of the cost function, also known as the organism with the best adaptation in the ecosystem. Moreover, in mutualist relationships some species benefit more than the others; this phenomenon is represented by the factors BF_1 and BF_2 in (19) and (20), which are randomly assigned with 1 or 2. Once X_{inew} and X_{jnew} are calculated, the algorithm evaluates the objective function for both organisms. If X_{inew} and X_{jnew} are better solutions than X_i and X_j, then X_{inew} and X_{jnew} replace X_i and X_j in the ecosystem.

$$X_{inew} = X_i + rand(0,1) \cdot (X_{best} - M_{vec} \cdot BF_1).$$ (19)

$$X_{jnew} = X_j + rand(0,1) \cdot (X_{best} - M_{vec} \cdot BF_2).$$ (20)

In the commensalism phase, X_i is the same organism used in the previous phase, while X_j is an organism randomly selected from the ecosystem, but it is different than X_j used in the mutualism stage. The commensalism relationship between X_i and X_j is represented by (21), where X_i benefits from X_j. If the objective function of the X_{inew} is less than the one of the objective function of X_i, then X_i is updated in the ecosystem.

$$X_{inew} = X_i + rand(-1,1) \cdot (X_{best} - X_j).$$ (21)

The last symbiotic relationship is parasitism, where the organism X_j is aleatory chosen from the ecosystem and X_i is the same organism used in the commensalism phase, but with a random modification of one of its components. The modified X_i is named parasite vector and if its objective function if better than X_j, then the parasite vector replaces X_j; otherwise, the parasite vector is discarded, and it is said that X_j developed immunity to the parasite. Once the three phases end for the i-th organism, the algorithm moves to the next organism to repeat the process until all the organisms in the ecosystem have been processed. At this point, one iteration if finished and the stop criteria are evaluated, if no stop condition is met, then the algorithm starts a new iteration. The detailed explanation of this algorithm is shown in [37]. Keep in mind that the evaluated algorithms (many of them) are sensitive to the assumed values for their multiple parameters. It is so sensitive that if they are not chosen with caution, they may not even converge to the expected solution even for PV arrays operating under uniform conditions. This aspect is a notable disadvantage of this metaheuristic solution strategy. Until now, there is no solid mathematical basis to predict the convergence and influence of these parameters. It is certainly possible that the results presented could have been obtained in shorter computation times, having selected other values for the algorithm parameters. However, it is worth noting that we tune the optimization algorithms in order to obtain the best results for all the algorithms. Particularly, the parameters of the metaheuristic methods selected for this work were defined randomly.

4. Simulations Results

This section introduces the simulation results for small, medium and large generators, i.e., PV generators formed by a single string with 6, 36 and 72 series-connected submodules, respectively. All the simulated generators are formed by Trina Solar TMS-PD05 de 270 W [38] modules, which are composed of three series-connected submodules. Moreover, the bypass diodes of the three submodules in this section are GF3045T [39]. It is worth noting that the generators considered in this section only have one string. This is because for generators with M strings each string is independently solved, which is equivalent to solve one string M times and then to sum the currents of all the strings to calculate the array current. The SDM parameters in STC are calculated, with the procedure proposed in [40], from the module datasheet information. Then, the parameters are adjusted for a module temperature of 44 °C ($T = 44$ °C) and an irradiance of 1 kW/m^2 ($G = 1$ kW/m^2), by following the procedure proposed in [31], obtaining the following values: $I_{ph} = 9.311$ A, $I_{sat} = 23.782$ ηA, $\eta = 1.097$, $R_s = 0.088$ Ω, and $R_p = 246.670$ Ω. Afterward, the bypass diode parameters are calculated from the datasheet information as follows: $I_{sat,bd} = 851,540$ µA and $\eta_{bd} = 1.634$. The irradiance condition of each submodule in the generator is represented by using the linear dependence on I_{ph} with the submodule irradiance. Therefore, the irradiance reduction in a submodule produced by some kind of shading is represented by a coefficient P_{irr} that multiplies I_{ph} and varies between 0 and 1. Thus the reduction of I_{ph} is proportional to the reduction of the submodule irradiance. The P_{irr} coefficients of all the string submodules are grouped in a vector ($\vec{P}_{irr}$) that represents the shading conditions of the string. The solutions obtained with the four optimization methods used in this paper are compared with the solution of the equivalent electrical circuit (EEC) of each generator. Those EECs for the small, medium, and large generators are implemented in Simulink of Matlab; hence, the errors calculated for the optimization algorithms are calculated concerning the EEC solution.

4.1. Small Generator

This generator is formed by two modules connected in series that correspond to six submodules. This low power generator may be used in grid-connected applications, by using a microinverter, or in stand-alone applications to charge a battery or to feed a set of lights. The simulation results of this generator operating under homogeneous (i.e., $\vec{P}_{irrh} = [1\,1\,1\,1\,1\,1]$) and partial shading (i.e., $\vec{P}_{irrnh} = [0.8\,0.8\,0.8\,0.8\,0.3\,0.3]$) conditions are introduced in Figures 4 and 5, respectively. Those figures show the I–V curves and the error in the current calculation for all the implemented optimization methods, which use the stop criteria shown in Table 2. Moreover, the evaluation criteria of the simulation results (see Table 3) are: the computation time, the root mean square error (RMSE), and the number of evaluations of the objective function (F_{eval}).

Table 2. Stop criteria for the optimization methods used for the small generator.

Stop Criteria	TRD	LMA	WDE	SOS
Number of particles	N/A	N/A	20	50
Maximun iteration number	1×10^3	1×10^3	1×10^4	5×10^4
Tolerance in the solution variation	1×10^{-3}	1×10^{-9}	N/A	N/A
Tolerance in the objective function	N/A	N/A	1×10^{-18}	1×10^{-18}

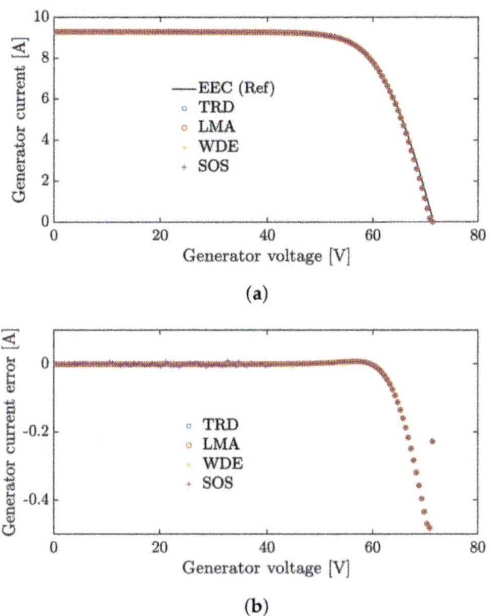

(a)

(b)

Figure 4. Simulation results of the small generator (six submodules) under homogeneous conditions. (**a**) Current vs. voltage (I–V) curve. (**b**) Current calculation error.

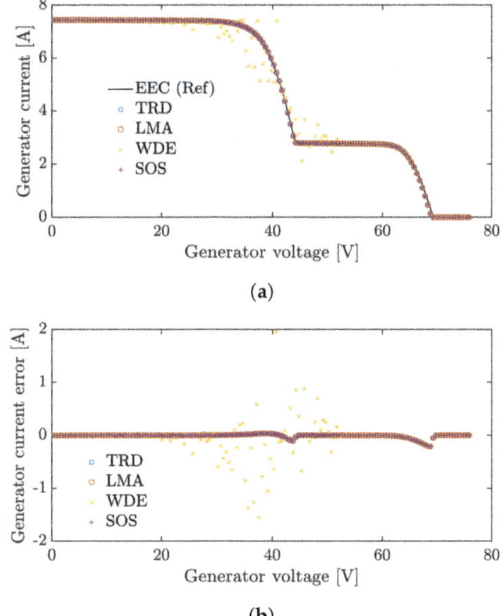

(a)

(b)

Figure 5. Simulation results of the small generator (six submodules) under partial shading conditions. (**a**) Current vs. voltage (I–V) curve. (**b**) Current calculation error.

In Figure 4a it can be observed that all the implemented optimization methods can reproduce the I–V curve of a small string (six submodules) under homogeneous conditions. Additionally, the errors in the string current calculation are similar for the different methods, as shown in Figure 4b and Table 3.

When the generator operates in partial shading conditions, three of the four algorithms (TRD, LMA, and SOS) can solve the system of implicit equations for all the voltages to reproduce the I–V with similar errors (see Figure 5 and Table 3). However, the WDE algorithm presents significant errors in the current calculation for voltages between 20 V and 55 V, as it is evidenced in Figure 5 and Table 3. Therefore, WDE is not a suitable algorithm to solve the model of a small generator in partial shading conditions. Note that under homogeneous and partial shading conditions the computation time and number of cost function evaluations (F_{eval}) of the deterministic optimization methods (TRD and LMA) are significantly less than the computation time and F_{eval} of the metaheuristic algorithms (WDE and SOS), as shown in Table 3. Moreover, the RMSE values of the methods able to solve the optimization problem provide similar errors in the I–V curve reproduction.

Table 3. Evaluation criteria for the small generator operating under homogeneous and partial shading conditions.

Algorithm	Computation Time (s)	RMSE (A)	F_{eval}
Homogeneous Conditions			
EEC(REF)	4.9	0	0
TRD	1.1	0.0957	2.03×10^4
LMA	1.2	0.0957	2.32×10^4
WDE	308.0	0.0957	1.36×10^7
SOS	164.0	0.0957	7.48×10^6
Partial Shading Conditions			
EEC(REF)	5.8	0	0
TRD	1.6	0.0387	2.69×10^4
LMA	1.6	0.0387	3.12×10^4
WDE	281.8	0.3488	1.24×10^7
SOS	161.3	0.0388	7.55×10^6

4.2. Medium Generator

In this case, the generator is formed by one string with 12 modules (36 submodules) connected in series, which is a typical configuration for a residential application with a string inverter (e.g., ABB UNO-7.6-TL-OUTD). For this generator, the stop criteria of the optimization algorithms are modified concerning the previous subsection, as shown in Table 4, to improve the performance of the algorithms.

Table 4. Stop criteria for the optimization methods used for the medium and large generators.

Stop Criteria	TRD	LMA	WDE	SOS
Number of particles	N/A	N/A	20	100
Maximun iteration number	4×10^3	4×10^3	1×10^4	1×10^5
Tolerance in the solution variation	1×10^{-5}	1×10^{-9}	N/A	N/A
Tolerance in the objective function	N/A	N/A	1×10^{-18}	1×10^{-18}

The medium generator I–V curves for homogeneous and partial shading conditions are presented in Figures 6 and 7, respectively; while the computation time, RMSE and F_{eval} are introduced in Table 5. Moreover, the vectors that define the homogeneous ($\vec{P}_{irrh}$) and partial shading conditions ($\vec{P}_{irrnh}$) have 36 elements and are defined as follows: all elements in $\vec{P}_{irrh}$ are 1 and $\vec{P}_{irrnh}$ has 24 elements equal to 0.8, 6 elements equal to 0.6, and 6 elements equal to 0.2.

The results in Figures 6 and 7 and Table 5 show that the deterministic methods (TRD and LMA) solve the system of equations of the string for homogeneous and partial shading conditions. Additionally, the RMSE values for the I–V curves are the same for both deterministic methods (less than 0.1 A); nevertheless, computation times of TRD are less than the computation times of LMA for both homogeneous and partial shading conditions. In general, metaheuristic methods present greater errors and computation times than the deterministic methods in the I–V curve's calculation. For homogeneous conditions, the current errors are concentrated in medium and high generator voltages (from 250 V to 400 V), as shown in Figure 6; those errors generate RMSE values between 13% and 60% greater than the ones of the deterministic methods. Whilst for partial shading conditions, the RMSE values of the metaheuristic methods between 15.3 and 21.3 times greater than the RMSE values of the deterministic methods, and the errors occur in the entire range of the generator voltage (see Figure 7). As observed, the computation times of the metaheuristic algorithms are two orders of magnitude greater than the computation times of the deterministic methods, which is also reflected in the number of cost function evaluations (F_{eval}). Finally, the results in Table 5 indicate that the implemented metaheuristic optimization methods are not suitable for solving the implicit model of a medium PV generator.

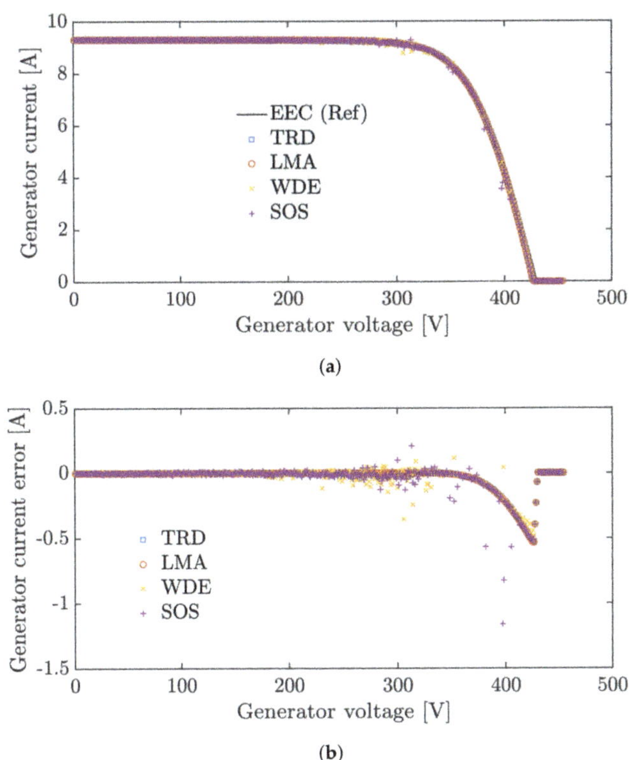

Figure 6. Simulation results of the medium generator (36 submodules) under homogeneous conditions. (a) Current vs. voltage (I–V) curve. (b) Current calculation error.

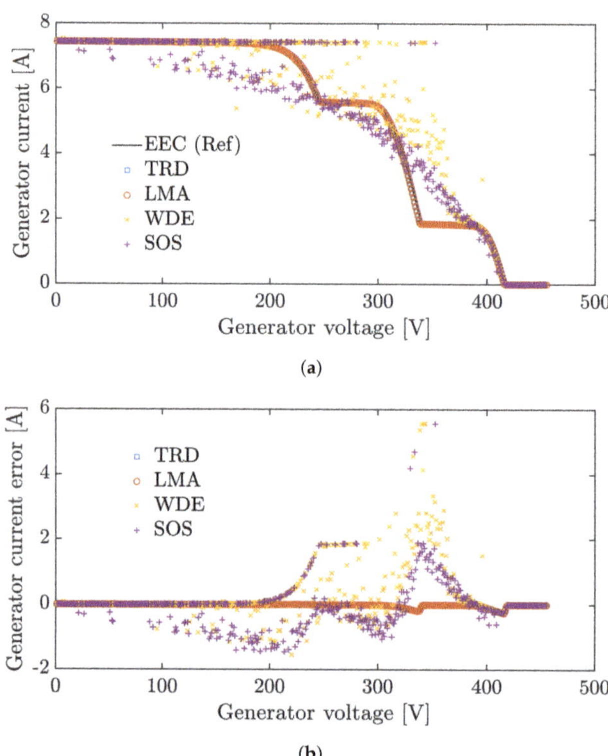

Figure 7. Simulation results of the medium generator (36 submodules) under partial shading conditions. (**a**) Current vs. voltage (I–V) curve. (**b**) Current calculation error.

Table 5. Evaluation criteria for the medium generator operating under homogeneous and partial shading conditions.

Algorithm	Computation Time (s)	RMSE (A)	F_{eval}
Homogeneous Conditions			
EEC(REF)	28.1	0	0
TRD	9.6	0.1034	2.78×10^5
LMA	10.0	0.1034	2.84×10^5
WDE	1898.8	0.1170	5.24×10^7
SOS	1271.3	0.1660	4.54×10^7
Partial Shading Conditions			
EEC(REF)	28.3	0	0
TRD	15.9	0.0460	4.56×10^5
LMA	20.3	0.0460	5.07×10^5
WDE	1784.4	1.0273	5.11×10^7
SOS	1296.9	0.7486	4.54×10^7

4.3. Large Generator

This generator is formed by a single string with 24 modules (72 submodules) connected in series, which can be found in industrial applications or high-power generators that use inverters whit maximum input voltages of 1 kV (e.g., Sunny Tripower 20000TL). The stop criteria are the same shown in Table 4 and the vectors that define homogeneous ($\vec{P}_{irrh}$) and partial shading ($\vec{P}_{irrnh}$) conditions have 72 elements defined as follows: all elements in ($\vec{P}_{irrh}$) are 1 and ($\vec{P}_{irrnh}$) has 30 elements equal to 0.8, 30 elements equal to 0.6, and 12 elements equal to 0.2. The generator I–V curves and the errors in the current calculation are shown in Figures 8 and 9 for homogeneous and partial shading conditions, respectively; while the computation time, RMSE and F_{eval} are presented in Table 6.

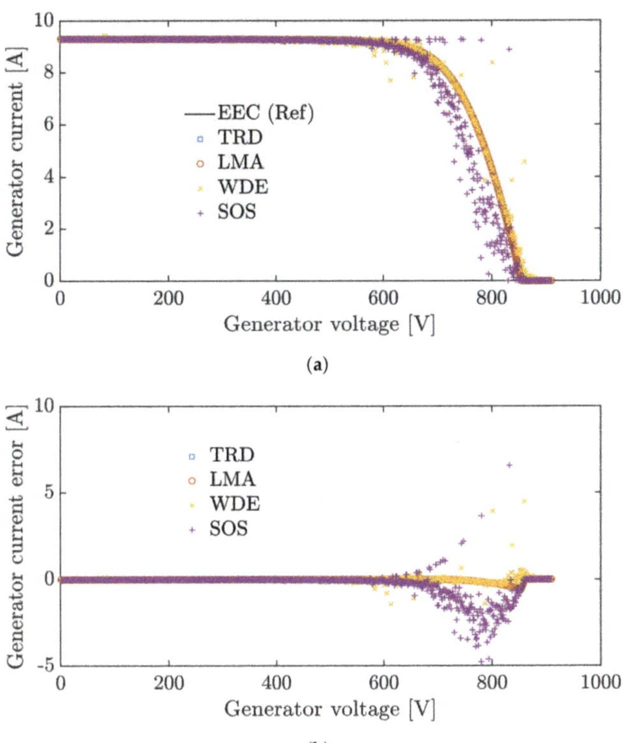

Figure 8. Simulation results of the large generator (72 submodules) under homogeneous conditions. (a) Current vs. voltage (I–V) curve. (b) Current calculation error.

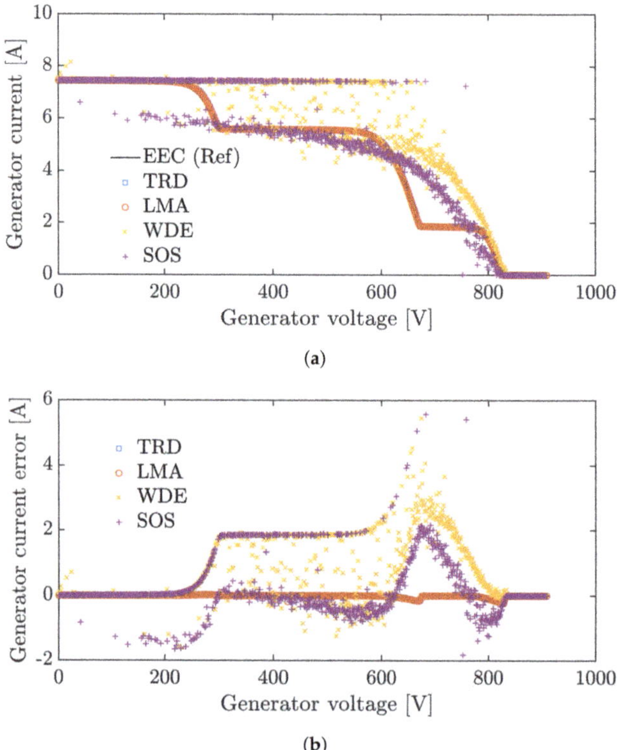

Figure 9. Simulation results of the large generator (72 submodules) under partial shading conditions. (**a**) Current vs. voltage (I–V) curve. (**b**) Current calculation error.

Table 6. Evaluation criteria for the large generator operating under homogeneous and partial shading conditions.

Homogeneous Conditions			
Algorithm	Computation Time (s)	RMSE (A)	F_{eval}
EEC(REF)	61.8	0	0
TRD	42.0	0.1041	1.13×10^6
LMA	36.6	0.1041	8.25×10^5
WDE	5410.8	0.2415	1.14×10^8
SOS	3047.3	0.8351	9.10×10^7
Partial Shading Conditions			
EEC(REF)	92.4	0	0
TRD	72.3	0.0404	1.81×10^6
LMA	85.6	0.0408	1.87×10^6
WDE	5586.4	1.3015	1.13×10^8
SOS	3124.7	0.9886	9.10×10^7

Once more, the results in Figures 8 and 9 and Table 6 indicate that the deterministic optimization methods (TRD and LMA) are capable to reproduce the I–V curves for the evaluated conditions with

RMSE less than 0.10 A. In homogeneous conditions, the computation time of LMA is 12.9% less than the one of TRD; while in partial shading conditions the computation time of TRD is 15.5% less than the computation time of LMA. Moreover, the values of F_{eval} for TRD and LMA have similar behavior to the computation time. The metaheuristic algorithms show errors and computation time greater than the deterministic algorithms in the reproduction of the I–V curves in the evaluated operating conditions. For homogeneous conditions (see Figure 8) the biggest errors are present in high voltages; as consequence, the RMSE values of the metaheuristic methods are significantly greater (between 132% and 700%) than the deterministic algorithms. In partial shading conditions (see Figure 9) the errors are in the whole voltage range, making the RMSE values of the metaheuristic methods between 24.2 and 31.9 times the RMSE values of the deterministic methods. Additionally, the computation time of the metaheuristic algorithms is two orders of magnitude greater than the one of the deterministic algorithms, which is similar to the differences in F_{eval} (see Table 6). After the evaluation of the four optimization algorithms for small, medium and large PV arrays, it seems that metaheuristic optimization methods do not worth their evaluations because the evaluated deterministic optimization methods outperform the evaluated metaheuristic methods. However, those results could not be obtained if the comparison is not performed; therefore, we consider that the evaluation is proposed in the paper is valid, since it is necessary to explore different optimization methods to reduce the computation time of the model. Such a reduction is important for applications like reconfiguration, energy production estimation, evaluation of MPPT techniques, among others.

5. Experimental Results

The performance of the optimization methods was validated with experimental data, which were obtained with the test bench shown in Figure 10 This test bench if formed by a PV with Matlab, a programmable electronic load BK Precision 8500, and a Trina Solar TMS-PD05 de 270 W [38] PV module. The PV module is the same one used in the simulations (formed by 3 submodules); however, the SDM parameters of the submodules ($I_{ph} = 9.316$ A, $I_{sat} = 1.549$ nA, $\eta = 0.972$, $R_s = 0.175\ \Omega$ and $R_p = 400.804\ \Omega$) and the bypass diodes ($I_{sat,bd} = 851{,}540$ µA and $\eta_{bd} = 0.2261$) are calculated from experimental I-V curves. The experimental validation considers two cables AWG 12 of 28.96 m each, which introduce losses that can be represented by resistance in series with the module of 313 mΩ. Therefore, such resistance was included in the R_s parameter of each submodule; thus, the ohmic losses introduced by the cables are lumped in the R_s of the submodules.

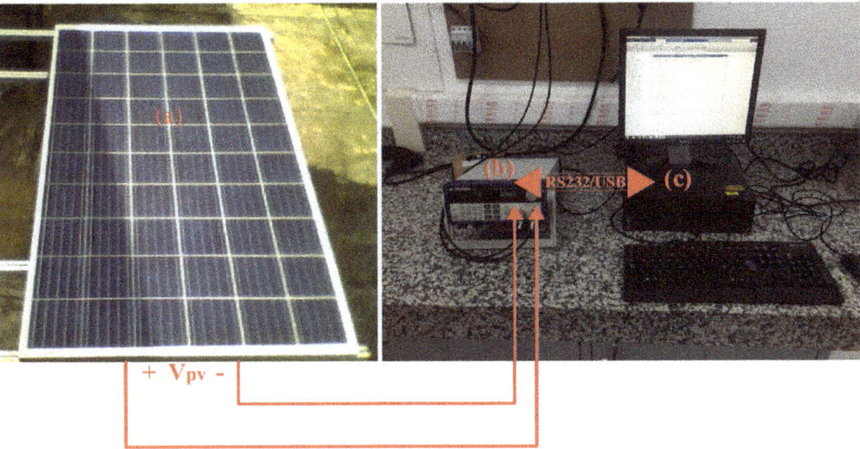

Figure 10. Test bench used to obtain the experimental data. (**a**) PV module with three submodules. (**b**) Electronic load. (**c**) PC with Matlab.

The small generator used for the experimental validation was formed by one string of three submodules, which was evaluated for homogeneous (C-1) and partial shading (C-2) defined by the vectors $(\vec{P}_{irr,1}) = [0.4\ 0.4\ 0.4]$ and $(\vec{P}_{irr,2}) = [0.620\ 0.598\ 0.210]$, respectively. Moreover, the stop criteria used for the optimization algorithms were the same ones defined in Table 2. The experimental results for conditions C-1 and C-2 are summarized in Figure 11 and Table 7. Figure 11 introduces the I–V curves obtained with the experimental test bench, the four optimization algorithms (TRD, LMA, WDE, and SOS), and the equivalent electrical circuit (EEC) for conditions C-1 and C-2; while Table 7 shows the computation time, RMSE values, and F_{eval} for the different methods.

In general, the experimental results agree with the simulation results obtained for small generators (Section 4.1) regarding computation time and current errors. On the one hand, in homogeneous conditions (C-1), the four optimization algorithms generate I–V curves that fit the experimental data with practically the same RMSE values. Moreover, the error in the current calculation for condition C-1 increases for high voltages (circles in Figure 11b), which agrees with the behavior shown in the simulation results for small generators (see Figure 4b). On the other hand, in partial shading conditions (C-2) three algorithms (TRD, LMA and SOS) reproduce the experimental I–V curves with the same RMSE values. However, the WDE presents significant errors in the current calculation for medium voltages (blue x in Figure 11b), which agree with the simulation results in Figure 5 of Section 4.1 (small generators). The behavior of the computation time in the experimental validation is similar to the one presented in simulation results. The computation time of the metaheuristic algorithms are, approximately, two orders of magnitude greater than the deterministic algorithms. Additionally, the computation time of the deterministic methods was less than the EEC in a proportion similar to the one presented above. The computation time of the four optimization algorithms is reflected in the F_{eval} values, which are similar to the values in the simulation results. Finally, Table 7 shows that the smallest errors were obtained with the EEC model, which verifies the usefulness of this model to be used as the reference in the simulation results.

Table 7. Evaluation criteria for a small generator operating under homogeneous and partial shading conditions from experimental results.

Scenario	Method	Computation Time (s)	RMSE (A)	F_{eval}
	1 EEC	3.04	0.0860	0
	2 TRD	0.67	0.0766	1.18×10^4
C-1	3 LMA	0.78	0.0766	1.33×10^4
	4 WDE	74.33	0.0766	3.57×10^6
	5 SOS	98.17	0.0766	4.75×10^6
	1 EEC	3.15	0.0898	0
	2 TRD	1.02	0.1012	1.82×10^4
C-2	3 LMA	1.23	0.1012	2.22×10^4
	4 WDE	83.40	0.3698	3.87×10^6
	5 SOS	104.33	0.1012	4.89×10^6

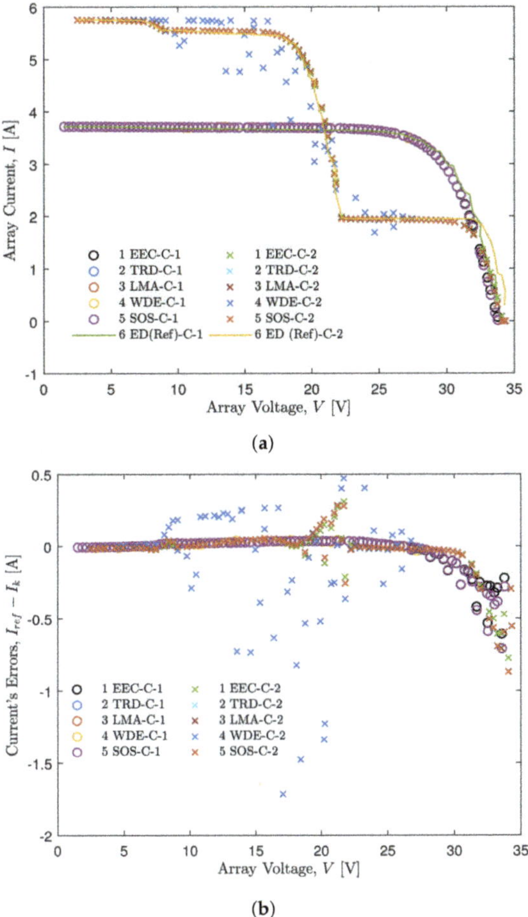

Figure 11. Experimental results in a small generator (3 submodules) under homogeneous and partial shading conditions. (**a**) Current vs. voltage (I–V) curve. (**b**) Current calculation error.

6. Conclusions

The solution of the implicit model of a PV generator in SP configuration as an optimization problem has been proposed and solved with deterministic and metaheuristic algorithms. The implicit model of an NxM SP generator (M strings with N submodules each) is formed by M systems of N + 2 implicit equations, where each system of equations corresponds to one string and it can be solved independently. Moreover, in a system of N + 2 implicit equations, the unknowns correspond to the N submodules voltages, the blocking diode voltage, and the string current. To solve a system of N + 2 implicit equations, the paper proposes an objective function and defines the upper and lower restrictions for the unknowns. Finally, the paper evaluates two deterministic optimization algorithms (TRD and LMA) and two metaheuristic algorithms (WDE and SOS) to solve the optimization problem and generate the I–V curves of small, medium, and large generators.

The proposed objective function and the performance of the four evaluated methods were validated with simulation and experimental results. The simulations show that the deterministic optimization methods can solve the optimization problem for small, medium and large generators operating in homogeneous and partial shading conditions; therefore, with these algorithms, it is possible to reproduce the generator I–V curves. Moreover, simulation results show that the metaheuristic algorithms are not able to solve the optimization problem in all cases. On the one hand, SOS correctly reproduces the I–V curves for small generators under homogeneous and non-homogeneous conditions and medium generators in homogeneous conditions. Nevertheless, for the other simulation scenarios, the errors were significant. On the other hand, WDE only solved the optimization problem for small generators under homogeneous conditions; while the errors were large for the other cases concerning the deterministic methods. We believe that metaheuristic algorithms performed worse because due to the random selection of parameter values instead of the nature of the selected algorithms (weighted differential evolution and search for symbiotic organisms). This could even be explained based on the No Free Lunch Theorems (NFL) for Optimization. Additionally, experimental results confirm simulation results for small generators in homogeneous and non-homogeneous conditions, where all the methods correctly reproduce the I–V curves except for WDE under partial shading conditions. Moreover, experimental results illustrate the usefulness of the EEC model as a reference for simulation results.

Finally, the results suggest that the solution of the implicit model of SP generators should be performed with deterministic algorithms, which suggests the necessity of evaluating other metaheuristic optimization methods to solve the proposed problem and reduce the computation time. Those presented results should be used to evaluate different solutions for energy yield calculations or economic analysis of a PV system; hence as future work, an addition to the presented algorithms with energy predictions or economic calculations will be considered.

Author Contributions: Conceptualization and methodology was performed by J.D.B.-R. and R.C.; software was developed by L.M.P.A. and J.D.B.-R.; simulation were performed by L.M.P.A.; experiments were developed by L.A.T.G. and D.G.-M.; analyses were performed by L.M.P.A., J.D.B.-R., R.C., L.A.T.G. and D.G.-M.; writing of the original paper draft was performed by L.M.P.A., J.D.B.-R., R.C.; writing of the reviewed draft was made by R.C., L.A.T.G. and D.G.-M. All authors have read and agree to the published version of the manuscript.

Funding: This work was funded by Universidad Industrial de Santander under the scholarship "Crédito Condonable para Estudiantes de Posgrado" and Universidad Nacional de Colombia Sede Manizales under de project "Desarrollo de un micro-inversor solar multietapa de inyección a red con capacidad de seguimiento del punto de máxima potencia" code HERMES 43049.

Conflicts of Interest: The authors declare no conflict of interest.

References

1. Lu, Q.; Yu, H.; Zhao, K.; Leng, Y.; Hou, J.; Xie, P. Residential demand response considering distributed PV consumption: A model based on China's PV policy. *Energy* **2019**, *172*, 443–456. [CrossRef]
2. International Energy Agency (IEA). *2018 Snapshot of Global Photovoltaic Markets*; IEA: Paris, France, 2018; pp. 1–16, ISBN 978-3-906042-72-5.
3. Tsafarakis, O.; Sinapis, K.; van Sark, W.G.J.H.M. A time-series data analysis methodology for effective monitoring of partially shaded photovoltaic systems. *Energies* **2019**, *12*, 1722. [CrossRef]
4. Manganiello, P.; Balato, M.; Vitelli, M. A survey on mismatching and aging of PV modules: The closed loop. *IEEE Trans. Ind. Electron.* **2015**, *62*, 7276–7286. [CrossRef]
5. Pendem, S.R. Performance Evaluation of Series, Series-Parallel and Honey-Comb PV Array Configurations under Partial Shading Conditions. In Proceedings of the 2017 7th International Conference on Power Systems (ICPS), Pune, India, 21–23 December 2017; pp. 749–754. [CrossRef]
6. Petrone, G.; Ramos-paja, C.A.; Spagnuolo, G. PV Simulation under Homogeneous Conditions. In *Photovoltaic Sources Modeling*; John Wiley & Sons, Ltd.: Chichester, UK, 2017; Chapter 3, pp. 45–80. [CrossRef]
7. Orozco-gutierrez, M.L.; Ramirez-scarpetta, J.M.; Spagnuolo, G.; Ramos-paja, C.A. A technique for mismatched PV array simulation. *Renew. Energy* **2013**, *55*, 417–427. [CrossRef]
8. Pandian, A.; Bansal, K.; Thiruvadigal, D.J.; Sakthivel, S. Fire hazards and overheating caused by shading faults on photo voltaic solar panel. *Fire Technol.* **2016**, *52*, 349–364. [CrossRef]
9. Sameeullah, M.; Swarup, A. Modeling of PV module to study the performance of MPPT controller under partial shading condition. In Proceedings of the India International Conference on Power Electronics, IICPE, Kurukshetra, India, 8–10 December 2014; pp. 1–6. [CrossRef]

10. Qing, X.; Sun, H.; Feng, X.; Chung, C.Y. Submodule-based modeling and simulation of a series-parallel photovoltaic array under mismatch conditions. *IEEE J. Photovolt.* **2017**, *7*, 1731–1739. [CrossRef]

11. Gallardo-saavedra, S.; Karlsson, B. Simulation, validation and analysis of shading e ff ects on a PV system. *Sol. Energy* **2018**, *170*, 828–839. [CrossRef]

12. Humada, A.M.; Hojabri, M.; Mekhilef, S.; Hamada, H.M. Solar cell parameters extraction based on single and double-diode models : A review *Renew. Sustain. Energy Rev.* **2016**, *56*, 494–509. [CrossRef]

13. Corless, R.M.; Gonnet, G.H.; Hare, D.E.G.; Jeffrey, D.J.; Knuth, D.E. On the Lambert W function. *Adv. Comput. Math.* **1996**, *5*, 329–359. [CrossRef]

14. Batzelis, E.I.; Routsolias, I.A.; Papathanassiou, S.A. An explicit pv string model based on the lambert w function and simplified mpp expressions for operation under partial shading. *IEEE Trans. Sustain. Energy* **2014**, *5*, 301–312. [CrossRef]

15. Petrone, G.; Ramos-paja, C.A. Modeling of photovoltaic fields in mismatched conditions for energy yield evaluations. *Electr. Power Syst. Res.* **2011**, *81*, 1003–1013. [CrossRef]

16. Petrone, G.; Spagnuolo, G.; Vitelli, M. Analytical model of mismatched photovoltaic fields by means of Lambert W-function. *Sol. Energy Mater. Sol. Cells* **2007**, *91*, 1652–1657. [CrossRef]

17. Sai Krishna, G.; Moger, T. Reconfiguration strategies for reducing partial shading effects in photovoltaic arrays: State of the art. *Sol. Energy* **2019**, *182*, 429–452. [CrossRef]

18. Bastidas-Rodriguez, J.D.; Cruz-Duarte, J.M.; Correa, R. Mismatched series–parallel photovoltaic generator modeling: An implicit current–voltage approach. *IEEE J. Photovolt.* **2019**, 1–7. [CrossRef]

19. Pendem, S.R.; Mikkili, S. Modelling and performance assessment of PV array topologies under partial shading conditions to mitigate the mismatching power losses. *Sol. Energy* **2018**, *160*, 303–321. [CrossRef]

20. Petrone, G.; Ramos-paja, C.A.; Spagnuolo, G. Single-diode Model Parameter Identification. In *Photovoltaic Sources Modeling*; John Wiley & Sons, Ltd.: Chichester, UK, 2017; Chapter 2, pp. 21–44. [CrossRef]

21. Hansen, C. *Parameter Estimation for Single Diode Models of Photovoltaics Modules*; Technical Report; Sandia National Laboratories: Albuquerque, NM, USA, 2015.

22. Chin, V.; Salam, Z.; Ishaque, K. An accurate and fast computational algorithm for the two-diode model of PV module based on a hybrid method. *IEEE Trans. Ind. Electron.* **2017**, *64*, 6212–6222. [CrossRef]

23. Elazab, O.S.; Hasanien, H.M.; Alsaidan, I.; Abdelaziz, A.Y.; Muyeen, S.M. Parameter estimation of three diode photovoltaic model using grasshopper optimization algorithm. *Energies* **2020**, *13*, 497. [CrossRef]

24. Khan, M.F.; Ali, G.; Khan, A.K. A Review of Estimating Solar Photovoltaic Cell Parameters. In Proceedings of the 2019 International Conference on Computing, Mathematics and Engineering Technologies—iCoMET 2019, Sukkur, Pakistan, 30–31 January 2019.

25. Sheng, H.; Li, C.; Wang, H.; Yan, Z.; Xiong, Y.; Cao, Z.; Kuang, Q. Parameters extraction of photovoltaic models using an improved Moth-Flame optimization. *Energies* **2019**, *12*, 3527. [CrossRef]

26. Yousri, D.; Allam, D.; Eteiba, M.B.; Suganthan, P.N. Static and dynamic photovoltaic models' parameters identification using Chaotic Heterogeneous Comprehensive Learning Particle Swarm Optimizer variants. *Energy Convers. Manag.* **2019**, *182*, 546–563. [CrossRef]

27. Bastidas-Rodriguez, J.D.; Petrone, G.; Ramos-Paja, C.A.; Spagnuolo, G. A genetic algorithm for identifying the single diode model parameters of a photovoltaic panel. *Math. Comput. Simul.* **2017**, *131*, 38–54. [CrossRef]

28. Jiang, L.L.; Maskell, D.L.; Patra, J.C. Parameter estimation of solar cells and modules using an improved adaptive differential evolution algorithm. *Appl. Energy* **2013**, *112*, 185–193. [CrossRef]

29. Kang, T.; Yao, J.; Jin, M.; Yang, S.; Duong, T. A novel improved cuckoo search algorithm for parameter estimation of photovoltaic (PV) models. *Energies* **2018**, *11*, 1060. [CrossRef]

30. Ko, S.W.; Ju, Y.C.; Hwang, H.M.; So, J.H.; Jung, Y.S.; Song, H.J.; Song, H.E.; Kim, S.H.; Kang, G.H. Electric and thermal characteristics of photovoltaic modules under partial shading and with a damaged bypass diode. *Energy* **2017**, *128*, 232–243. [CrossRef]

31. De Soto, W.; Klein, S.; Beckman, W. Improvement and validation of a model for photovoltaic array performance. *Sol. Energy* **2006**, *80*, 78–88. [CrossRef]

32. Villalva, M.; Gazoli, J.; Filho, E. Comprehensive approach to modeling and simulation of photovoltaic arrays. *IEEE Trans. Power Electron.* **2009**, *24*, 1198–1208. [CrossRef]

33. Bai, J.; Liu, S.; Hao, Y.; Zhang, Z.; Jiang, M.; Zhang, Y. Development of a new compound method to extract the five parameters of PV modules. *Energy Convers. Manag.* **2014**, *79*, 294–303. [CrossRef]

34. Nocedal, J.; Wright, S. *Numerical Optimization*; Springer Series in Operations Research and Financial Engineering; Springer: New York, NY, USA, 2006. [CrossRef]
35. Sun, W.; Yuan, Y.X. *Optimization Theory and Methods*; Springer Optimization and Its Applications; Kluwer Academic Publishers: Boston, MA, USA, 2006; Volume 1. [CrossRef]
36. Civicioglu, P.; Besdok, E.; Gunen, M.A.; Atasever, U.H. Weighted differential evolution algorithm for numerical function optimization: A comparative study with cuckoo search, artificial bee colony, adaptive differential evolution, and backtracking search optimization algorithms. *Neural Comput. Appl.* **2018**, *5*. [CrossRef]
37. Cheng, M.Y.; Prayogo, D. Symbiotic Organisms Search: A new metaheuristic optimization algorithm. *Comput. Struct.* **2014**, *139*, 98–112. [CrossRef]
38. Trina-Solar. *TSM-PD05 Datasheet*; Trina-Solar: Changzhou, China, 2017.
39. Huajing Microelectronics. *GF3045T Photovoltaic Bypass Diodes Datasheet*; Huajing Microelectronics: Zhuhai, China, 2018.
40. Accarino, J.; Petrone, G.; Ramos-Paja, C.A.; Spagnuolo, G. Symbolic algebra for the calculation of the series and parallel resistances in PV module model. In Proceedings of the 4th International Conference on Clean Electrical Power: Renewable Energy Resources Impact, ICCEP 2013, Alghero, Italy, 11–13 June 2013; pp. 62–66. [CrossRef]

Article

An ANFIS-Based Modeling Comparison Study for Photovoltaic Power at Different Geographical Places in Mexico

Nun Pitalúa-Díaz [1],*, Fernando Arellano-Valmaña [2], Jose A. Ruz-Hernandez [2], Yasuhiro Matsumoto [3], Hussain Alazki [2], Enrique J. Herrera-López [4], Jesús Fernando Hinojosa-Palafox [1], A. García-Juárez [1], Ricardo Arturo Pérez-Enciso [1] and Enrique Fernando Velázquez-Contreras [1]

[1] Departamento de Ingeniería Industrial, Departamento de Ingeniería Química y Metalurgia, Departamento de Investigación en Física, Departamento de Investigación en Polímeros y Materiales, Universidad de Sonora, Blvd. Luis Encinas y Rosales S/N, Col. Centro. Hermosillo 83000, Sonora C.P., Mexico

[2] Facultad de Ingeniería, Universidad Autónoma del Carmen, Calle 56 No. 4 Esq. Avenida Concordia Col. Benito Juárez C.P. 24180 Cd. Del Carmen, Campeche, Mexico

[3] Departamento de Ingeniería Eléctrica, Centro de Investigación y de Estudios Avanzados del IPN, Av. Instituto Politécnico Nacional 2508, La Laguna Ticoman, C.P. 07360 Ciudad de México, CDMX, Mexico

[4] Biotecnología Industrial, Sublínea Bioelectrónica, Centro de Investigación y Asistencia en Tecnología y Diseño del Estado de Jalisco A.C., Camino Arenero 1227, Col. El Bajío del Arenal, C.P. 45019 Zapopan Jalisco, Mexico

* Correspondence: nun.pitalua@unison.mx; Tel.: +52-1-662-259-2159

Received: 5 June 2019; Accepted: 9 July 2019; Published: 11 July 2019

Abstract: In this manuscript, distinct approaches were used in order to obtain the best electrical power estimation from photovoltaic systems located at different selected places in Mexico. Multiple Linear Regression (MLR) and Gradient Descent Optimization (GDO) were applied as statistical methods and they were compared against an Adaptive Neuro-Fuzzy Inference System (ANFIS) as an intelligent technique. The data gathered involved solar radiation, outside temperature, wind speed, daylight hour and photovoltaic power; collected from on-site real-time measurements at Mexico City and Hermosillo City, Sonora State. According to our results, all three methods achieved satisfactory performances, since low values were obtained for the convergence error. The GDO improved the MLR results, minimizing the overall error percentage value from 7.2% to 6.9% for Sonora and from 2.0% to 1.9% for Mexico City; nonetheless, ANFIS overcomes both statistical methods, achieving a 5.8% error percentage value for Sonora and 1.6% for Mexico City. The results demonstrated an improvement by applying intelligent systems against statistical techniques achieving a lesser mean average error.

Keywords: ANFIS; statistical method; gradient descent; photovoltaic system; sustainable development

---

1. Introduction

Considerable research has been developed internationally in the field of photovoltaic systems and power generation [1–3]. Mexico is a country that receives abundant solar energy, with the northwest region being the one with the highest annual incidence of solar radiation, achieving radiation indexes between 5.6 and 6.2 kWh/m^2 per day; nevertheless, its advances on the photovoltaic field are scarse nowadays, although there are many possibilities of research in this topic [4].

There are many scientific reports on statistical methods to estimate power generation [5–8]. Multiple Linear Regression (MLR) includes any study area, since it finds an approach to the relation among variables [9].

A satisfactory photovoltaic power estimation involving meteorological variables was carried out in [10] in which a Gradient Descent Optimization (GDO) minimizes the error value between the real and estimated variables after many iterations; even so, it is mentioned that other techniques may obtain better results despite their lower or greater complexity.

On the other hand, intelligent techniques have acquired a worldwide reputation as simple methods to represent and replicate the behavior of a process with a not-so-understood performance. These techniques have potential to model, precisely, linear and non-linear processes, the latter being their strongest application. Some studies have used intelligent techniques to analyze photovoltaic power behavior, e.g., in [11] an artificial neural network (ANN) was used to obtain a model to forecast the photovoltaic energy; however, solar radiation was the only meteorological variable analyzed. In [12], different ANN techniques were used to perform a comparative study of systems predicting photovoltaic thermal energy data; nevertheless, due to solar radiation measurement devices being expensive and requiring periodic maintenance, some results employed global solar radiation (GSR) estimations through ANNs. In [13], it was established that a photovoltaic system is affected by many meteorological conditions; a predictive model was proposed considering outside temperature, solar radiation and direction and speed of wind. In [14], an ANN-Adaptive Neuro-Fuzzy Inference System (ANFIS)-based forecast model for predicting the photovoltaic generation and wind energy generation is presented and considers the susceptibilities to which renewable energies are exposed due to nature's vagaries. In [15], it is proved that an adaptive neuro-fuzzy inference system technique provides a reliable tool to estimate temperature from photovoltaic systems. In [16], the solar still productivity prediction aims to be improved by using an ANFIS due to its simple maintenance and ready affordability.

The importance of achieving a satisfactory model design with a minimum error between estimated and real values is crucial in precision studies or management tasks in which certain differences among them may result in economic problems or loss of information.

As can be seen, diverse statistical and intelligent methods have been applied to estimate photovoltaic power generation. However, this topic still is an open field that requires better estimation methods, in particular, those related to intelligent systems.

Since it is relevant to develop new estimation strategies in photovoltaic systems, the aim of this work is to compare ANFIS methodology with MLR and GDO as statistical approaches by comparing electrical power estimation data from photovoltaic systems [10]. It was observed that all three methods achieve a satisfactory estimation performance, but ANFIS had a better estimation capacity.

2. Methodology

2.1. Statistical Methods

2.1.1. Multiple Linear Regression

To create a statistical representative model of the power generated by solar energy, the concept of multiple linear regression was applied. The relation that exists among all the input variables and the output is represented by a "linear regression". Depending on the number of input variables, the regression can be simple or multiple [6,17–19]. The purpose of multiple linear regression is to find an estimation of the real output through an equation involving all the gathered data, as shown in Equation (1).

$$y = \beta_0 + \beta_1 x_1 + \beta_2 x_2 + \ldots + \beta_k x_k + \varepsilon \tag{1}$$

where y is the estimated output, x_k is the kth input variable, β_k is the characteristic coefficient of every variable and ε is the error between the model and the real data.

To obtain Equation (1), a matrix "X" and a vector "y" are generated. The number of columns for "X" is $k+1$, with the condition that the first column is filled with ones. The dimension of the columns and the vector "y" depends on the total quantity of data "n". Consequently, Equation (2) is applied to find the respective coefficient values.

$$X = \begin{bmatrix} 1 & x_{11} & x_{12} & \cdots & x_{1k} \\ 1 & x_{21} & x_{22} & \cdots & x_{2k} \\ 1 & x_{31} & x_{32} & \cdots & x_{3k} \\ \vdots & \vdots & \vdots & & \vdots \\ 1 & x_{n1} & x_{n2} & \cdots & x_{nk} \end{bmatrix} \qquad y = \begin{bmatrix} y_1 \\ y_2 \\ \vdots \\ y_n \end{bmatrix} \qquad \beta = \left(X'X\right)^{-1}\left(X'y\right) \tag{2}$$

The data collected from Hermosillo, Sonora (located in the northwest region of Mexico) were obtained with the support of the University of Sonora (UNISON), while the data from Mexico City were issued by the Centro de Investigación y de Estudios Avanzados (CINVESTAV) campus Zacatenco. Solar radiation, outside temperature, wind speed and daylight hour (time) served as the input meteorological variables, while the electric power was the output for both of the PV systems. Each data sample was registered every 5 min. According to the above, the matrix "X" and the vector "y" from Equation (2) are shown in Equation (3).

$$X = \begin{bmatrix} 1 & Solar\,Radiation & Temperature & Wind\,Speed & Time \end{bmatrix}$$
$$y = [Electric\,Power] \tag{3}$$

2.1.2. Gradient Descent Optimization

The GDO has been used to estimate different behaviors in several studies [20,21]. As seen in [10], this method seeks the optimum β_k coefficients now symbolized by θ. To achieve the minimum possible error value in Equation (4), the cost function is used.

$$J[\theta] = \frac{1}{2m}\sum_{i=1}^{m}\left(h_\theta\left(x^{(i)}\right) - y^{(i)}\right)^2 \tag{4}$$

where $\theta = \beta$ is as mentioned in Equation (1), $x\,(i)$ represents the ith row of matrix "X" and $y\,(i)$ is the value of the ith row of vector "y", both described in Equation (2). It is important to notice that, in this case, the number of columns in matrix "X" is equal to the number of variables involved, omitting the column full of ones. The gradient descent, denoted by Equation (5), aims to converge to the cost function minimum by its partial derivative. The quickness of the convergence is given by α.

$$\theta_k := \theta_k - \alpha\frac{\partial}{\partial\theta_k}J[\theta] \tag{5}$$

Equation (6) represents the substitution of Equation (4) into Equation (5) and it is known as the gradient descent implementation on linear regression method, which has to be repeated n-times until the convergence is done.

$$\theta_k := \theta_k - \alpha\frac{1}{m}\sum_{i=1}^{m}\left(h_\theta\left(x^{(i)}\right) - y^{(i)}\right)\cdot x_k^{(i)} \tag{6}$$

Once all the coefficients have been computed, they are substituted into Equation (7)

$$h_\theta\left(x\right) = \theta_0 + \theta_1 x_1 + \theta_2 x_2 + ... + \theta_k x_k + \varepsilon \tag{7}$$

where $h_\theta\left(x\right)$ is the estimated output, x_k is the kth input variable, θ_k is the characteristic coefficient for every variable and ε is the error between the model and the real data.

2.2. Intelligent Technique

An intelligent technique consists of a dynamic learning process in order to generate an output as close as possible to the required one. Among the best intelligent systems are the artificial neural network, fuzzy systems and the adaptive neuro-fuzzy inference system [22,23]. For this contribution, an ANFIS is considered combining the strengths of a neural network and fuzzy systems, obtaining a better performance.

An artificial neural network (ANN) imitates the processing of a human brain and it has a great number of processing units (neurons or nodes) working in parallel. These nodes can be circular or squared in order to represent different adaptive capabilities. A circular node employs fixed operations that cannot be altered at any time (sum or product of inputs), unlike a squared node, which can be modified by the user (activation functions).

All the neurons are highly connected among each other through links (synapse) with weights. The network has a layer for all the inputs, a layer for one or more outputs and one or more hidden layers between the input and the output.

A neural network requires previous assumptions so it can learn from examples by adjusting the connection weights. The learning may be supervised if the right output is specified (being the case for this work) or non-supervised if it has to explore the relations between the patterns and learn to categorize the inputs [11,24].

A fuzzy system is able to apply conditional sentences as a human brain would. The objective of every system that uses fuzzy logic is to describe the degrees of the output sentences (given by a series of rules) according to the input ones. The "if–then" fuzzy rules are sentences with the form "if an event A happens then an event B will occur", where both events are known as the labels of fuzzy sets from their corresponding membership functions (MF). The strength of this system is to be able to capture all the imprecise modes of the reasoning, for example:

If the pressure is high, *then* the volume is small

where, analogous to the neural network, each and every variable has a specific type of membership function according to its behavior, which can be triangular, trapezoidal, gaussian, etc., as well as to its quantity [25].

Figure 1 shows three trapezoidal membership functions for the variable called "pressure". Each of them represents a range of values labeled as low (blue), medium (red) and high (yellow). This advantage allows to embrace all the possible data and to classify them as necessary.

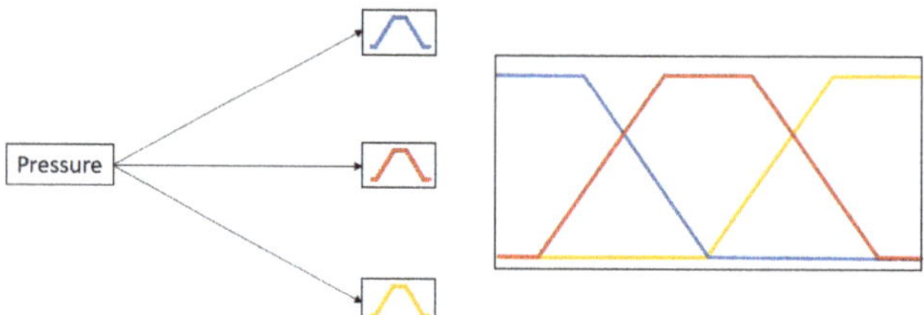

Figure 1. Example of three trapezoidal membership functions.

ANFIS is the union of both methods mentioned above, the neural network and fuzzy systems, in which the advantages of both cooperate in order to achieve easier and faster estimations [26].

The ANFIS consists of if–then rules and input–output pairs. Moreover, for the training, the learning algorithms of neural networks are used. In order to simplify the explanations, a basic fuzzy inference system consists of two inputs (x and y) and one output (z) [27]. Figure 2 shows that, if each input consists of three membership functions (represented as squared nodes, because they can be modified by the user), then the input layer would have six neurons (three for each input) and, in turn, the output layer would just have one neuron. As for the hidden layers, the first one of them will be formed by each of the if–then fuzzy rules mentioned before; by this, an ANFIS is created.

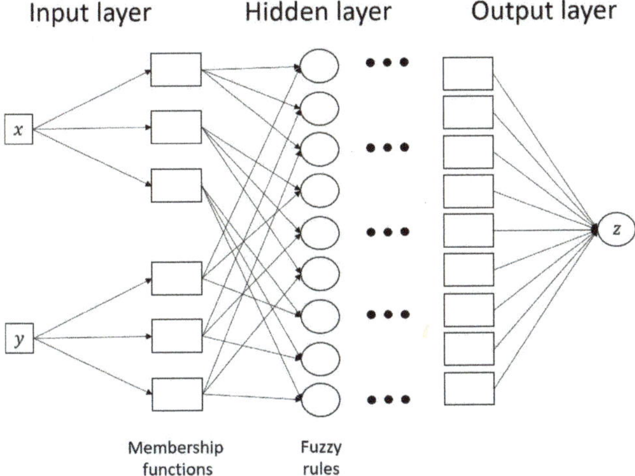

Figure 2. Adaptive Neuro-Fuzzy Inference System (ANFIS) structure with two inputs and three membership functions for each one.

The variables used to train the ANFIS are radiation, outside temperature, wind speed, daylight hour and electrical power, as mentioned in Section 2.1.1. Figure 3 presents the MF of input variables for the intelligent system. The behavior of each variable helps to identify the type of MF to be declared. Solar radiation, wind speed and daylight hour, due to their rapid change with respect to time, have 4, 3 and 3 triangular MF, respectively. As for temperature, which presents a slow variation with time, it has 3 gaussian MF.

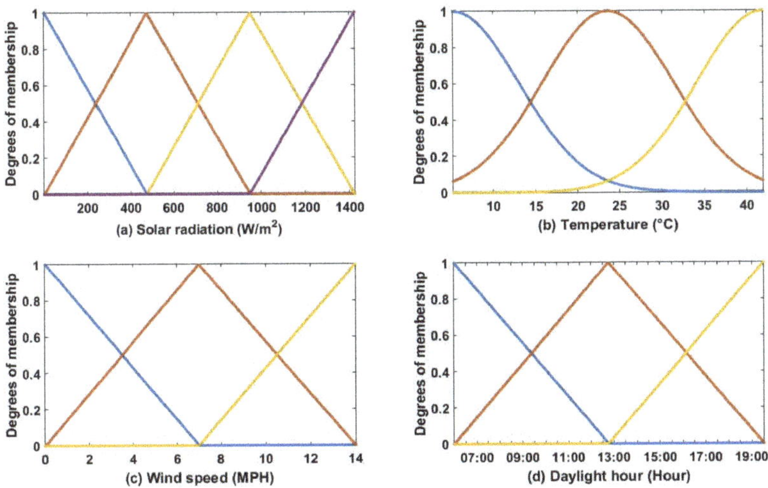

Figure 3. Membership functions for ANFIS: (**a**) Triangular for solar radiation; (**b**) gaussian for temperature; (**c**) triangular for wind speed; (**d**) triangular for daylight hour.

According to the above, the structure of the ANFIS is presented in Figure 4. The method applied to train the ANFIS is called hybrid training. This process consists of two steps; the first one computes the result of the next linked node according to the nodes behind, i.e., based on Figure 2 each circular node depends on two rectangular nodes (MF of variables x and y). Once all the nodes have been calculated and the output is obtained, the second step must identify the error in each node and minimize it during the training iterations in order to improve the estimation result. The first and second steps are based on least squares and gradient descent, respectively.

This ANFIS model was applied for both locations by using the MATLAB® software in order to prove its effectiveness against the statistical method, regardless of the place where the system is installed.

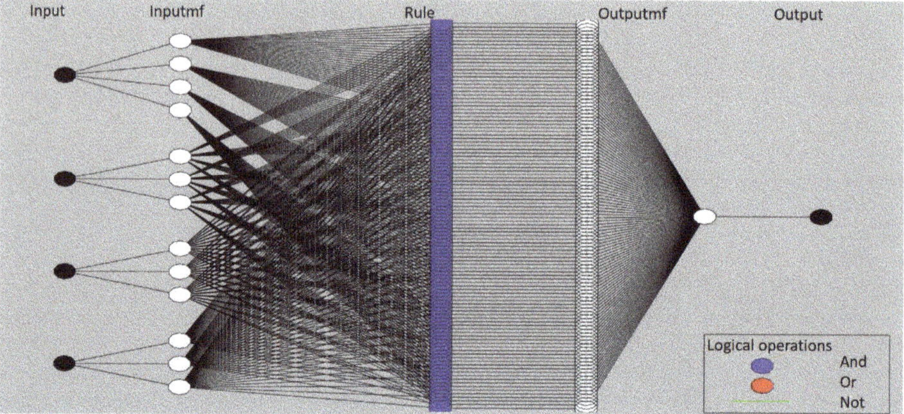

Figure 4. ANFIS structure.

3. Results

Seven different cases or sets of data were computed to analyze which estimation technique executes a better performance for the Hermosillo site (HS) and the Mexico City site (MCS). Every case was represented by data collected during a whole month; consequently, the first case corresponds to six months of collected data, while the seventh case stands for the accumulation of all data. According to Section 2.1.1, the time step of each estimation, regardless of the method, was five minutes. The former was considered to generate a greater amount of data allowing a better estimation result by the ANFIS. Figures 5 and 6 present the physical systems for each location, HS and MCS, respectively.

Figure 5. Physical system for the Hermosillo site (HS).

Figure 6. Physical system for the Mexico City site (MCS).

3.1. Hermosillo Site

Table 1 shows the resulting equations for the monthly and overall electrical power estimation by MLR and GDO with 1000 repetitions to minimize Equation (4), considering the data gathered from HS as mentioned in Sections 1 and 2, and by using the MATLAB® software. Figures 7 and 8 present the resulting estimation for each method considering a random period to achieve a reliable comparison and a better appreciation of the behavior (24th November to 24th December). The computational load for the overall case using MLR and GDO resulted in total durations of approximately 30 s and 10 min, respectively.

Table 1. Monthly and overall Multiple Linear Regression (MLR) and Gradient Descent Optimization (GDO) equations for HS.

Method	Month	Equations
MLR	Aug	$y = -2305.4 + 1.4182x_1 + 116.4097x_2 + 8.686x_3 - 2078.9x_4$
	Sep	$y = -498.9858 + 1.8082x_1 + 36.7926x_2 + 32.4585x_3 - 1042.6x_4$
	Oct	$y = -190.6163 + 1.6737x_1 + 38.1824x_2 + 9.8075x_3 - 1075.2x_4$
	Nov	$y = -120.4414 + 2.1624x_1 + 13.5544x_2 + 18.5136x_3 - 379.8749x_4$
	Dec	$y = 26.2042 + 1.9806x_1 + 12.1240x_2 + 12.0973x_3 - 405.2137x_4$
	Jan	$y = 142.9390 + 2.1120x_1 + 18.8379x_2 + 20.2070x_3 - 792.1223x_4$
	Total	$y = 84.9221 + 2.0035x_1 + 9.7233x_2 + 15.5707x_3 - 386.3517x_4$
GDO	Aug	$y = 0.0067 + 2.2334x_1 + 0.2415x_2 + 0.0186x_3 + 0.0036x_4$
	Sep	$y = 0.0050 + 2.2366x_1 + 0.1711x_2 + 0.0182x_3 + 0.0025x_4$
	Oct	$y = 0.0064 + 2.2069x_1 + 0.1931x_2 + 0.0251x_3 + 0.0032x_4$
	Nov	$y = 0.0029 + 2.2732x_1 + 0.0764x_2 + 0.0108x_3 + 0.0014x_4$
	Dec	$y = 0.0036 + 2.1411x_1 + 0.0814x_2 + 0.0114x_3 + 0.0018x_4$
	Jan	$y = 0.0044 + 2.3776x_1 + 0.0932x_2 + 0.0138x_3 + 0.0020x_4$
	Total	$y = 0.0044 + 2.2387x_1 + 0.1156x_2 + 0.0154x_3 + 0.0022x_4$

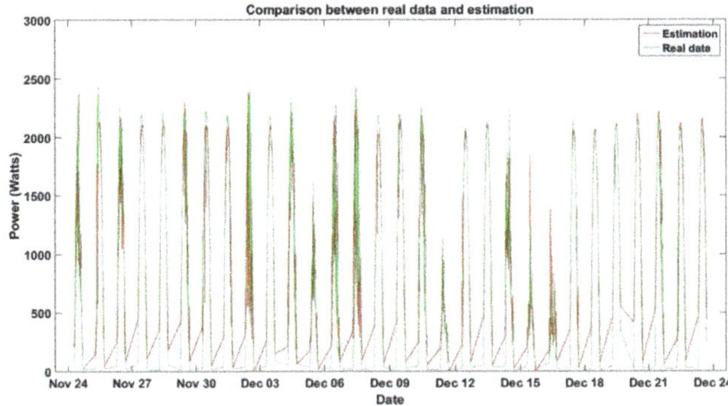

Figure 7. MLR results from HS.

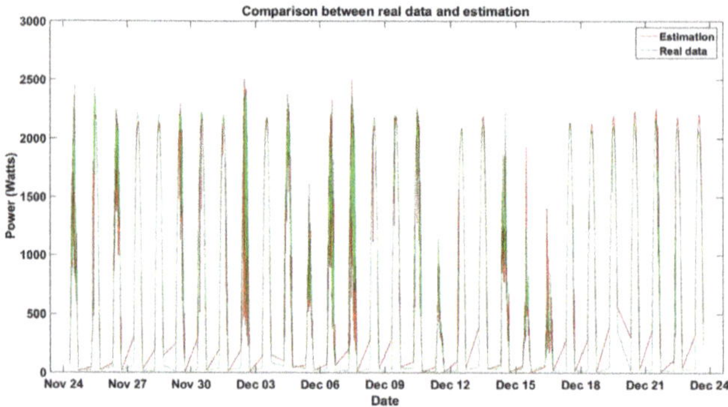

Figure 8. GDO results from HS.

Table 2 displays the monthly and overall trained parameters applying the ANFIS technique with the structure specified in Section 2.2 by using the MATLAB® software, with a hybrid mode and 1000 iterations. Finally, Figure 9 shows the resulting estimation within the same time range, i.e., total durations of approximately 30 s and 10 min in MLR and GDO, respectively. The computational load for the overall case using this method resulted in a total duration training of approximately 1.5 h.

Table 2. Monthly and overall trained membership function (MF) parameters for HS.

Month	Trained MF Parameters										
	Solar Radiation			Outside Temperature		Wind Speed			Daylight Hour		
	a	b	c	σ	c	a	b	c	a	b	c
Aug	−394.1	5.593	405.2	2.463	26.73	−3	0.9994	4.997	0.05206	0.3462	0.5417
	5.593	405.2	804.9	2.434	32.47	1.005	4.999	9	0.3214	0.5721	0.7766
	405.2	804.9	1205	2.434	38.23	5	9	13	0.5417	0.7538	1.01
	804.9	1205	1604								
Sep	−394.2	0.4883	395.2	4.142	22.39	−5.5	0.001136	5.497	0.02085	0.2968	0.5208
	0.4883	395.2	789.9	4.138	32.14	-1.51×10^{-11}	5.501	11	0.2645	0.5625	0.73
	395.2	789.9	1185	4.137	41.89	5.5	11	16.5	0.5	0.7479	1.021
	789.9	1185	1579								
Oct	−474	0.1488	474.3	4.193	16.56	−7	0.0005153	6.999	0.0396	0.3274	0.5382
	0.1489	474.3	948.5	4.184	26.42	0.04423	7.001	14	0.2215	0.5069	0.75
	474.3	948.5	1423	4.185	36.28	6.925	14	21	0.5021	0.7389	0.9879
	948.5	1423	1897								
Nov	−375.6	0.3992	376.4	4.399	12.4	−6	−0.00118	5.873	0.08051	0.336	0.5347
	0.3991	376.4	752.4	4.381	22.72	−0.005881	5.998	12	0.2085	0.5	0.7248
	376.4	752.4	1128	4.389	33.05	5.92	12	18	0.5	0.7117	0.9466
	752.4	1128	1504								
Dec	−390.7	1.12	393	5.623	5.238	−5.5	0.004238	5.418	0.08844	0.3785	0.5287
	1.12	393	784.8	5.599	18.41	−0.0002569	5.502	11	0.2034	0.5139	0.7245
	393	784.8	1177	5.606	31.6	5.552	11	16.5	0.5546	0.684	0.943
	784.8	1177	1568								
Jan	−408.1	0.9953	410.1	4.522	7.837	−4.5	−0.001668	4.493	0.08968	0.3261	0.5139
	0.9953	410.1	819.2	4.513	18.47	0.09047	4.497	9	0.3044	0.5216	0.7491
	410.1	819.2	1228	4.512	29.11	4.496	8.999	13.5	0.5417	0.7097	0.9521
	819.2	1228	1637								
Total	−474	0.1488	474.3	7.79	5.226	−7	0.001869	6.848	0.01973	0.3254	0.4923
	0.1488	474.3	948.5	7.783	23.55	−0.0008495	7.001	14	0.2726	0.5248	0.9075
	474.3	948.5	1423	7.781	41.89	6.917	14	21	0.5243	0.6577	0.9961
	948.5	1423	1897								

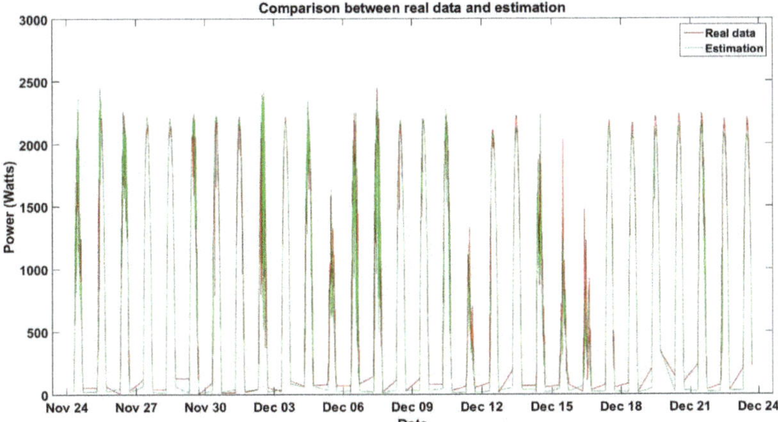

Figure 9. ANFIS results from HS.

Each case analyzed contemplates a different range of time (different month) and its respective ANFIS gets trained according to the data gathered within that period. Given that meteorological inputs vary in time, so do the ANFIS coefficients for every case; nevertheless, even if they are changing, the variation range is not too wide, as can be seen from solar radiation, outside temperature and daylight hour. As for wind speed, its variation is due to a smaller relationship with the output than with the other inputs, as well as to the different behaviors in each month.

3.2. Mexico City Site

Analogous to the HS case, the same procedure was applied to the data registered from the Mexico City site (MCS) mentioned in Sections 1 and 2, achieving approximately the same time duration to find the beta and theta parameters. For the MLR and GDO methods, Table 3 presents the monthly and overall estimation equations. Figures 10 and 11 depict the resulting MLR and GDO estimation in the period 24th November to 24th December (same as HS) for a better appreciation of the convergence of the method.

Table 3. Monthly and overall MLR and GDO equations for MCS.

Method	Month	Equations
MLR	Jul	$y = 727.5535 + 52.7725x_1 + 8.6209x_2 + 28.9473x_3 - 947.31789x_4$
	Ago	$y = 2527.4 + 55.7540x_1 - 110.5462x_2 + 19.4716x_3 + 849.8508x_4$
	Sep	$y = 1093.9 + 58.4495x_1 - 27.2714x_2 + 28.7514x_3 - 82.2214x_4$
	Oct	$y = -99.2276 + 57.0806x_1 - 20.0065x_2 + 52.4617x_3 + 1326.5x_4$
	Nov	$y = -1059.8 + 57.7075x_1 - 32.4315x_2 + 12.1205x_3 + 3016.3x_4$
	Dec	$y = -1430.5 + 53.7683x_1 - 15.8521x_2 + 109.1599x_3 + 2953.6x_4$
	Total	$y = 70.2756 + 55.1011x_1 + 33.8665x_2 + 90.3112x_3 - 689.5238x_4$
GDO	Jul	$y = 0.0927 + 53.4273x_1 + 2.2298x_2 + 0.1005x_3 + 0.0491x_4$
	Ago	$y = 0.1083 + 56.0178x_1 + 2.5973x_2 + 0.1441x_3 + 0.0580x_4$
	Sep	$y = 0.1159 + 59.0818x_1 + 2.7233x_2 + 0.1431x_3 + 0.0625x_4$
	Oct	$y = 0.1035 + 57.3927x_1 + 2.2950x_2 + 0.1899x_3 + 0.0560x_4$
	Nov	$y = 0.0908 + 57.2318x_1 + 1.9788x_2 + 0.1036x_3 + 0.0463x_4$
	Dec	$y = 0.0913 + 53.3033x_1 + 1.8920x_2 + 0.0441x_3 + 0.0477x_4$
	Total	$y = 0.1005 + 55.9815x_1 + 2.2910x_2 + 0.1290x_3 + 0.0533x_4$

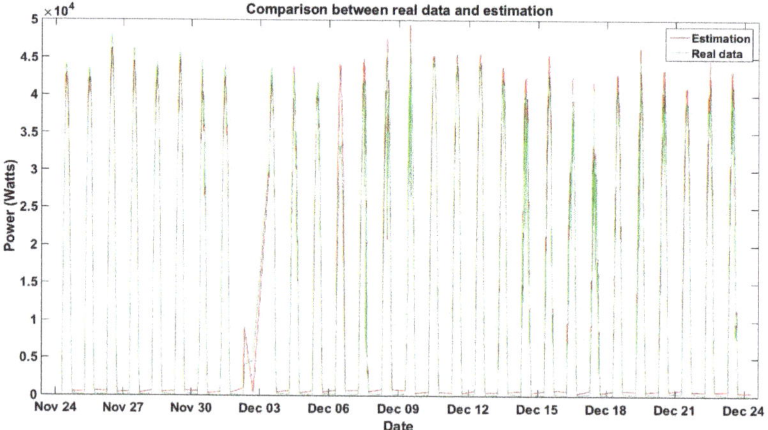

Figure 10. MLR results from MCS.

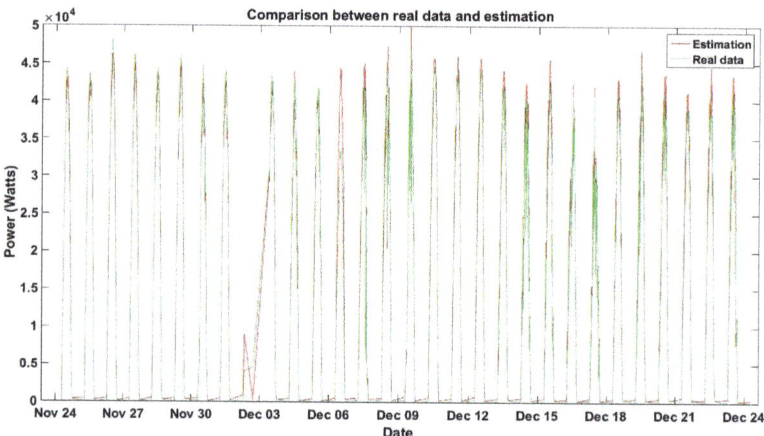

Figure 11. GDO results from MCS.

The same ANFIS structure with 1000 epoch hybrid training for the HS case was employed to obtain the estimation model of the photovoltaic system with the data collected from MCS. Table 4 shows the monthly and overall trained MF parameters. Figure 12 displays the comparison between the resulting estimation against real data with the same time range as for MLR and GDO. The time duration to train the neuro-fuzzy system was approximately the same as in the HS case.

Table 4. Monthly and overall trained MF parameters for MCS.

Month	Trained MF Parameters										
	Solar Radiation			Outside Temperature		Wind Speed			Daylight Hour		
	a	b	c	σ	c	a	b	c	a	b	c
Jul	−356	1	358	4.274	13.02	−8.5	−0.005509	8.509	0.01522	0.3264	0.5521
	1	358	715	4.238	23	0.1882	8.494	17	0.2917	0.5493	0.8816
	355	712	1070	4.227	33.01	8.492	17	25.5	0.5612	0.7535	1.102
	715	1072	1429								
	a	b	c	σ	c	a	b	c	a	b	c
Ago	−371.7	1	373.7	4.481	14.01	−8	−0.002784	7.967	0.02276	0.4896	0.4934
	1	373.7	746.3	4.452	24.5	0.1432	7.997	16	0.343	0.5799	0.7157
	373.7	746.3	1119	4.45	34.9	7.984	16	24	0.5584	0.8143	1.099
	746.3	1119	1492								
	a	b	c	σ	c	a	b	c	a	b	c
Sep	−377.7	1	379.7	5.096	10	−10.5	−0.0001171	10.5	0.04167	0.3015	0.5584
	1	379.7	758.3	5.096	22	0.01522	10.5	21	0.2981	0.5556	0.8166
	379.7	758.3	1137	5.096	34	10.5	21	31.5	0.566	0.8042	1.069
	758.3	1137	1516								
	a	b	c	σ	c	a	b	c	a	b	c
Oct	−353.7	1	355.7	4.463	9.003	−7.5	−0.001309	7.497	0.01042	0.2663	0.4574
	1	355.7	710.3	4.453	19.5	0.0376	7.499	15	0.3247	0.4896	0.8325
	355.7	710.3	1065	4.455	30	7.504	15	22.5	0.4612	0.7494	1.052
	710.3	1065	1420								
	a	b	c	σ	c	a	b	c	a	b	c
Nov	−314.7	1	316.7	5.124	6.014	−6	0.007692	5.994	0.03837	0.3368	0.5183
	1	316.7	632.3	5.067	17.99	−0.01071	6.005	12	0.307	0.4896	0.736
	316.7	632.3	948	5.065	30.02	5.907	12	18	0.4931	0.6581	0.9685
	632.3	948	1264								
	a	b	c	σ	c	a	b	c	a	b	c
Dec	−295.7	1	297.7	5.313	3.003	−5.5	−0.0008596	5.494	0.04514	0.2633	0.5084
	1	297.7	594.3	5.31	15.5	−2.43E-10	5.499	11	0.3472	0.4868	0.7411
	297.7	594.3	891	5.314	28	5.505	11	16.5	0.4805	0.7325	0.9757
	594.3	891	1188								
	a	b	c	σ	c	a	b	c	a	b	c
Total	−377.7	1	379.7	6.813	3.01	−10.5	−0.00304	10.5	−0.008574	0.3395	0.498
	0.9999	379.7	758.3	6.791	19	0.04036	10.5	21	0.2713	0.5833	0.8143
	379.7	758.3	1137	6.783	35.01	10.5	21	31.5	0.5576	0.8047	1.115
	758.3	1137	1516								

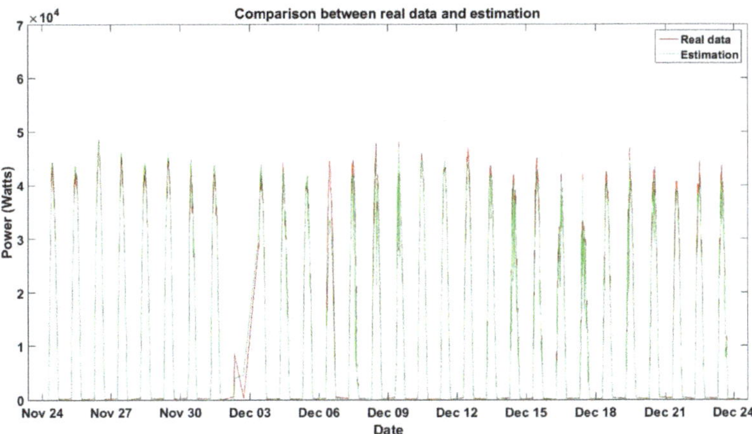

Figure 12. ANFIS results from MCS.

3.3. Error Analysis

In order to demonstrate the impact of the intelligent technique over statistical approaches, an error analysis was applied for every result obtained in Sections 3.1 and 3.2 for HS and MCS. Considering [10],

Mean Absolute Error (MAE) and Mean Absolute Percentage Error (MAPE) were used to achieve a solid comparison between MLR, GDO and ANFIS. MAE is one of the most implemented errors in estimation studies to measure and analyze precision and is described by Equation (8), and MAPE computes the percentage error value as in Equation (9).

$$MAE = \frac{\sum\limits_{s=1}^{N} |P_m - P_e|}{N} \tag{8}$$

$$MAPE_{(range)_\%} = \frac{\sum\limits_{s=1}^{N} \left| \frac{P_m - P_e}{\max(P_m) - \min(P_m)} \right|}{N} \cdot 100 \tag{9}$$

where "s" is the sample taken into consideration, "N" is the total amount of data samples gathered, "P_m" is the photovoltaic electrical power real measure and "P_e" is the estimated value.

Table 5 presents the monthly and overall MAE and MAPE for each applied technique in HS. By the analysis of these results, an improvement of GDO over MLR can be observed by achieving a lesses error value; nonetheless, the ANFIS overcomes both MLR and GDO, successfully obtaining lower error values for every time range. By increasing "N" in Equations (8) and (9) from the first to the last data sample, Figure 13 displays graphically the overall MAE and MAPE behavior of each implemented technique, showing and proving a clearly better performance by ANFIS against the statistical methods.

Table 5. Error comparison between MLR, GDO and ANFIS for HS.

	MLR		GDO		ANFIS	
Month	**MAE**	**MAPE**	**MAE**	**MAPE**	**MAE**	**MAPE**
	(W)	**(%)**	**(W)**	**(%)**	**(W)**	**(%)**
Aug	296.5	11.8	299.8	11.9	23.8	0.9
Sep	222.7	8.3	203.7	7.6	96.3	3.6
Oct	276.9	9.5	268.0	9.2	168.1	5.8
Nov	142.7	5.3	138.2	5.1	96.2	3.6
Dec	190.8	7.5	178.9	7.0	129.4	5.1
Jan	192.0	7.1	184.5	6.8	126.3	4.7
Total	209.3	7.2	200.1	6.9	169.2	5.8

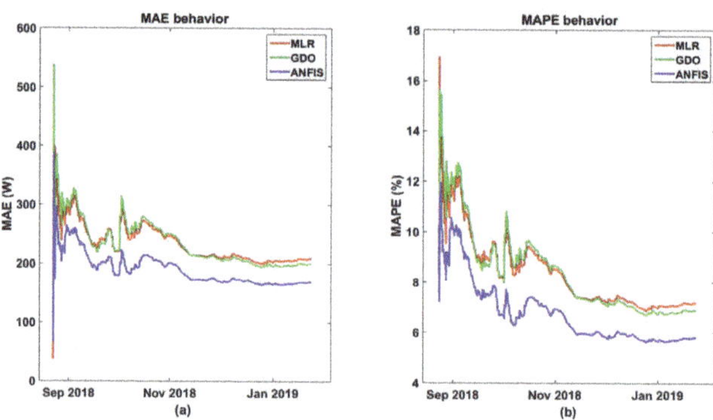

Figure 13. Error behavior comparison between MLR, GDO and ANFIS from HS: (**a**) MAE behavior; (**b**) MAPE behavior.

Table 6 presents the monthly and overall MAE and MAPE for each applied technique in MCS. As well as for the HS case; an improvement of GDO over MLR can be seen; nonetheless, the ANFIS overcomes again both MLR and GDO, successfully obtaining lower error values for every time range.

Analogous to Figure 13, Figure 14 displays graphically the overall MAE and MAPE behavior of each implemented technique for MCS, proving as well an improved performance by ANFIS compared to the statistical methods.

Table 6. Error comparison between MLR, GDO and ANFIS for MCS.

Month	MLR		GDO		ANFIS	
	MAE	MAPE	MAE	MAPE	MAE	MAPE
	(W)	(%)	(W)	(%)	(W)	(%)
Jul	588.8	1.0	590.6	1.0	334.1	0.6
Aug	1234.4	2.1	1192.9	2.0	732.9	1.3
Sep	727.6	1.3	726.3	1.3	345.8	0.6
Oct	733.6	1.3	741.8	1.3	441.4	0.8
Nov	715.5	1.3	741.4	1.4	477.6	0.9
Dec	794.1	1.7	812.8	1.7	638.3	1.4
Total	1142.2	2.0	1130.4	1.9	924.7	1.6

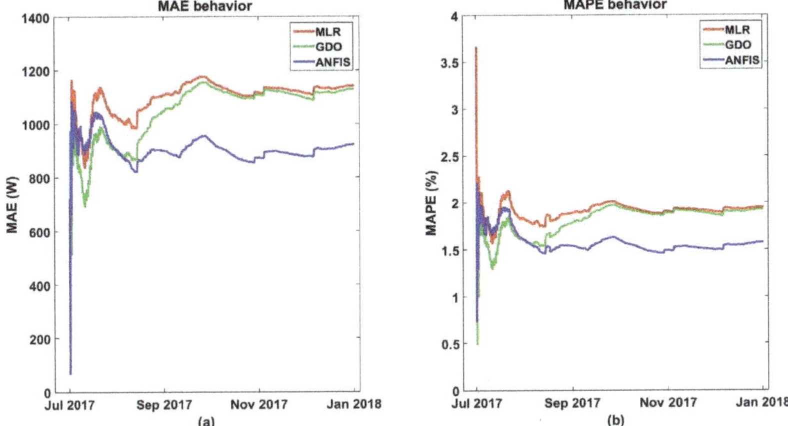

Figure 14. Error behavior comparison between MLR, GDO and ANFIS from MCS: (**a**) MAE behavior; (**b**) MAPE behavior.

4. Conclusions

An ANFIS was used to estimate the photovoltaic electrical power generated by solar energy at two different locations, Hermosillo and Mexico City. The intelligent technique demonstrated a better performance than the MLR and GDO as statistical methods. The reported results for MCS showed that all three methods achieved a satisfactory estimation performance by their comparison against real measured values; nonetheless, the ANFIS system clearly displayed an improvement among the methods employed. For the case of HS, the MAE overall values with respect to GDO and MLR were 200.1 W and 209.3 W, and the MAPE overall values with respect to GDO and MLR were 6.9% and 7.2%. For the MCS, the MAE overall values with respect to GDO and MLR were 1130.4 W and 1142.2 W, and the MAPE overall values with respect to GDO and MLR were 1.9% and 2.0%. Consequently, GDO produced better results than MLR overall and in almost every monthly case; however, ANFIS outperformed both MLR and GDO in every case, achieving overall results with respect to MAE and MAPE of 169.2 W and 5.8% for HS and 924.7 W and 1.6% for MCS.

As outlined in Section 3, even when the ANFIS computational time is considerably greater than the one involved in statistical methods, the neuro-fuzzy result displays a better performance. It should be mentioned that, although the intelligent method took approximately 90 min to be trained, it is quite an acceptable amount of time and can even be considered fast in terms of neuro-fuzzy models with months of data.

During a period of a year, the times of dusk and dawn are modified by minutes, resulting in different times to end and begin each day; as outlined in Section 3.1, for every transition day, dusk and dawn times may vary due to external variables, e.g., cloudiness. These issues generate discrepancies between the estimated and the real data, as seen in the results section, regardless the applied method; nonetheless, these may be minimized by gathering a greater amount of data to fulfil the range of changing time values.

The ANFIS proved to have a better performance than the conventional statistical methods, demonstrating that this kind of intelligent system is a potential tool to be considered for power estimation in Mexico.

Author Contributions: Research ideas and global design: N.P.-D., J.A.R.-H. and F.A.-V.; analysis of the results: Y.M., E.F.V.-C. and E.J.H.-L.; revision of the document: H.A., A.G.-J. and J.F.H.-P.; data gathering: R.A.P.-E.

Funding: This research received no external funding.

Acknowledgments: The authors would like to thank the University of Carmen (UNACAR), University of Sonora (UNISON), Centro de Investigación y de Estudios Avanzados (CINVESTAV) campus Zacatenco and Consejo Nacional de Ciencia y Tecnología (CONACYT), with the master scholarship program and the mobility scholarship 291249.

Conflicts of Interest: The authors declare no conflict of interest.

Abbreviations

The following abbreviations are used in this manuscript:

ANFIS	Adaptive Neuto-Fuzzy Inference System
ANN	Artificial Neural Network
°C	Celsius degree. It is a temperature scale used by the International Systems of Units
GDO	Gradient Descent Optimization
GSR	Global Solar Radiation
HS	Hermosillo Site
kWh/m^2	Quantity of kilowatts during an hour striking a squared meter area
MAE	Mean Absolute Error
MAPE	Mean Absolute Percentage Error
MCS	Mexico City Site
MF	Membership Function
MLR	Multiple Linear Regression
MPH	Miles Per Hour
PV	Photovoltaic
W	Watt. It is a unit of power defined as a derived unit of 1 joule per second (J/s)
W/m^2	Quantity of watts striking a squared meter area

References

1. Zahedi, A. Solar photovoltaic (PV) energy; latest developments in the building integrated and hybrid PV systems. *Renew. Energy* **2006**, *31*, 711–718. [CrossRef]
2. Kazem, H.A.; Yousif, J.H.; Chaichan, M.T. Modeling of daily solar energy system prediction using support vector machine for Oman. *Int. J. Appl. Eng. Res.* **2016**, *11*, 10166–10172. [CrossRef]
3. Awan, A.B.; Zubair, M.; P., P.R.; Abokhalil, A.G. Solar Energy Resource Analysis and Evaluation of Photovoltaic System Performance in Various Regions of Saudi Arabia. *Sustainability* **2018**, *10*, 1129. [CrossRef]
4. Perea-Moreno, A.J.; Hernandez-Escobedo, Q.; Garrido, J.; Verdugo-Diaz, J.; Perea-Moreno, A.J.; Hernandez-Escobedo, Q.; Garrido, J.; Verdugo-Diaz, J.D. Stand-Alone Photovoltaic System Assessment in Warmer Urban Areas in Mexico. *Energies* **2018**, *11*, 284. [CrossRef]
5. Gardiner, S.K.; Crabb, D.P. Examination of different pointwise linear regression methods for determining visual field progression. *Investig. Ophtalmol. Vis. Sci.* **2002**, *43*, 1400–1407. Available online: https://iovs.arvojournals.org/article.aspx?articleid=2123577 (accessed on 10 July 2019).

6. Verma, S.; Bartosova, A.; Markus, M.; Cooke, R.; Um, M.J.; Park, D.; Verma, S.; Bartosova, A.; Markus, M.; Cooke, R.; et al. Quantifying the Role of Large Floods in Riverine Nutrient Loadings Using Linear Regression and Analysis of Covariance. *Sustainability* **2018**, *10*, 2876. [CrossRef]

7. Grimaccia, F.; Mussetta, M.; Zich, R. Neuro-fuzzy predictive model for PV energy production based on weather forecast. In Proceedings of the 2011 IEEE International Conference on Fuzzy Systems (FUZZ-IEEE 2011), Taipei, Taiwan, 27–30 June 2011; IEEE: Piscataway, NJ, USA, 2011; pp. 2454–2457. [CrossRef]

8. Kashyap, Y.; Bansal, A.; Sao, A.K. Solar radiation forecasting with multiple parameters neural networks. *Renew. Sustain. Energy Rev.* **2015**, *49*, 825–835. [CrossRef]

9. He, J.; Zelikovsky, A. MLR-tagging: Informative SNP selection for unphased genotypes based on multiple linear regression. *Bioinformatics* **2006**, *22*, 2558–2561. [CrossRef] [PubMed]

10. Ruz-Hernandez, J.A.; Matsumoto, Y.; Arellano-Valmaña, F.; Pitalúa-Díaz, N.; Cabanillas-López, R.E.; Abril-García, J.H.; Herrera-López, E.J.; Velázquez-Contreras, E.F. Meteorological Variables' Influence on Electric Power Generation for Photovoltaic Systems Located at Different Geographical Zones in Mexico. *Appl. Sci.* **2019**, *9*, 1649. [CrossRef]

11. Ding, M.; Wang, L.; Bi, R. An ANN-based Approach for Forecasting the Power Output of Photovoltaic System. *Procedia Environ. Sci.* **2011**, *11*, 1308–1315. [CrossRef]

12. Yousif, J.H.; Kazem, H.A.; Alattar, N.N.; Elhassan, I.I. A comparison study based on artificial neural network for assessing PV/T solar energy production. *Case Stud. Therm. Eng.* **2019**, *13*, 100407. [CrossRef]

13. Wen-Tao, Z.; Shuai, W.; Xin-Hui, D. Research of power prediction about photovoltaic power system: Based on bp neural network. *Multi-Criteria Anal. Air Pollut. Urban Environ. Due Road Traffic* **2017**, *18*, 1614–1623. Available online: https://www.researchgate.net/profile/Claudia_Moisa/publication/323144140_SUSTAINABLE_DEVELOPMENT_THROUGH_CONVERSION_TO_ORGANIC_AGRICULTURE_-_IMPLICATIONS_ON_THE_FINANCIAL_INDICATORS_OF_FIRMS/links/5c25f7bfa6fdccfc706d4a64/SUSTAINABLE-DEVELOPMENT-THROUGH-CONVERSION-TO-ORGANIC-AGRICULTURE-IMPLICATIONS-ON-THE-FINANCIAL-INDICATORS-OF-FIRMS.pdf#page=313 (accessed on 10 July 2019).

14. Kuamr, K.R.; Kalavathi, M.S. ANN-ANFIS Based Forecast Model for Predicting PV and Wind Energy Generation. In Proceedings of the World Congress on Engineering 2016, WCE 2016, London, UK, 29 June–1 July 2016; Volume I. Available online: http://www.iaeng.org/publication/WCE2016/WCE2016_pp322-327.pdf (accessed on 10 July 2019).

15. Bassam, A.; May Tzuc, O.; Escalante Soberanis, M.; Ricalde, L.; Cruz, B.; Bassam, A.; May Tzuc, O.; Escalante Soberanis, M.; Ricalde, L.J.; Cruz, B. Temperature Estimation for Photovoltaic Array Using an Adaptive Neuro Fuzzy Inference System. *Sustainability* **2017**, *9*, 1399. [CrossRef]

16. Mashaly, A.F.; Alazba, A.A. ANFIS modeling and sensitivity analysis for estimating solar still productivity using measured operational and meteorological parameters. *Water Sci. Technol. Water Supply* **2018**, *18*, 1437–1448. [CrossRef]

17. Song, Y.; Park, M.; Song, Y.S.; Park, M.J. A Study on Estimation Equation for Damage and Recovery Costs Considering Human Losses Focused on Natural Disasters in the Republic of Korea. *Sustainability* **2018**, *10*, 3103. [CrossRef]

18. Clack, C.T.M. Modeling Solar Irradiance and Solar PV Power Output to Create a Resource Assessment Using Linear Multiple Multivariate Regression. *J. Appl. Meteorol. Climatol.* **2017**, *56*, 109–125. [CrossRef]

19. Montgomery, D.C. *Design and Analysis of Experiments*; John Wiley & Sons: Hoboken, NJ, USA, 2017; ISBN 978-1-119-11347-8. Available online: https://www.wiley.com/en-mx/Design+and+Analysis+of+Experiments2C+9th+Edition-p-9781119113478 (accessed on 10 July 2019).

20. Bairán, J.M.; Marí, A.; Cladera, A. Analysis of shear resisting actions by means of optimization of strut and tie models taking into account crack patterns. *Hormigón Y Acero* **2018**, *69*, 197–206. [CrossRef]

21. Arjoune, Y.; Mrabet, Z.; Kaabouch, N.; Arjoune, Y.; Mrabet, Z.E.; Kaabouch, N. Multi-Attributes, Utility-Based, Channel Quality Ranking Mechanism for Cognitive Radio Networks. *Appl. Sci.* **2018**, *8*, 628. [CrossRef]

22. Chen, S.H.; Jakeman, A.J.; Norton, J.P. Artificial Intelligence techniques: An introduction to their use for modelling environmental systems. *Math. Comput. Simul.* **2008**, *78*, 379–400. [CrossRef]

23. Matich, D.J. Redes Neuronales: Conceptos básicos y aplicaciones. *Universidad Tecnológica Nacional México.* Available online: ftp://decsai.ugr.es/pub/usuarios/castro/Material-Redes-Neuronales/Libros/matich-redesneuronales.pdf (accessed on 10 July 2019).
24. Aggarwal, S.; Saini, L. Solar energy prediction using linear and non-linear regularization models: A study on AMS (American Meteorological Society) 2013–14 Solar Energy Prediction Contest. *Energy* **2014**, *78*, 247–256. [CrossRef]
25. Pitalúa-Díaz, N.; Penaloza, U.C.; Ruz-Hernandez, J.A.; Jimenez, R.L. *Introducción a Los Sistemas Inteligentes*; Departamento de Editorial Universitaria (UABC): Mexicali, Baja California, Mexico, 2008; ISBN 978-607-7753-47-6. Available online: https://libreriauabc.com/products/introduccion-a-los-sistemas-inteligentes (accessed on 10 July 2019).
26. Jang, J.S. ANFIS: Adaptive-network-based fuzzy inference system. *IEEE Trans. Syst. Man Cybern.* **1993**, *23*, 665–685. [CrossRef]
27. Boyacioglu, M.A.; Avci, D. An Adaptive Network-Based Fuzzy Inference System (ANFIS) for the prediction of stock market return: The case of the Istanbul Stock Exchange. *Expert Syst. Appl.* **2010**, *37*, 7908–7912. [CrossRef]

Article

Comparison of Power Output Forecasting on the Photovoltaic System Using Adaptive Neuro-Fuzzy Inference Systems and Particle Swarm Optimization-Artificial Neural Network Model

Promphak Dawan [1], Kobsak Sriprapha [2], Songkiate Kittisontirak [2], Terapong Boonraksa [3], Nitikorn Junhuathon [4], Wisut Titiroongruang [1] and Surasak Niemcharoen [1,*]

[1] Department of Electrical Engineering, Faculty of Engineering, King Mongkut's Institute of Technology Ladkrabang, Bangkok 10520, Thailand; promphak.dawan@gmail.com (P.D.); ktwisut@kmitl.ac.th (W.T.)

[2] Solar energy technology laboratory, National Electronics, and Computer Technology Center, National Science and Technology Development Agency (NSTDA), Pathum Thani 12120, Thailand; kobsak.sriprapha@nectec.or.th (K.S.); songkiate.kittisontirak@nectec.or.th (S.K.)

[3] School of Electrical Engineering, Faculty of Engineering, Rajamangala University of Technology Rattanakosin, Nakhon Pathom 73170, Thailand; terapong.boo@rmutr.ac.th

[4] School of Electrical Engineering, Faculty of Engineering, Bangkok Thonburi University, Bankok 10170, Thailand; Nitikorn.ju@gmail.com

* Correspondence: surasak.ni@kmitl.ac.th

Received: 15 November 2019; Accepted: 27 December 2019; Published: 10 January 2020

Abstract: The power output forecasting of the photovoltaic (PV) system is essential before deciding to install a photovoltaic system in Nakhon Ratchasima, Thailand, due to the uneven power production and unstable data. This research simulates the power output forecasting of PV systems by using adaptive neuro-fuzzy inference systems (ANFIS), comparing accuracy with particle swarm optimization combined with artificial neural network methods (PSO-ANN). The simulation results show that the forecasting with the ANFIS method is more accurate than the PSO-ANN method. The performance of the ANFIS and PSO-ANN models were verified with mean square error (MSE), root mean square error (RMSE), mean absolute error (MAP) and mean absolute percent error (MAPE). The accuracy of the ANFIS model is 99.8532%, and the PSO-ANN method is 98.9157%. The power output forecast results of the model were evaluated and show that the proposed ANFIS forecasting method is more beneficial compared to the existing method for the computation of power output and investment decision making. Therefore, the analysis of the production of power output from PV systems is essential to be used for the most benefit and analysis of the investment cost.

Keywords: PVs power output forecasting; adaptive neuro-fuzzy inference systems; particle swarm optimization-artificial neural networks; solar irradiation

---

1. Introduction

At present, the world population has become more alert to the use of renewable energy due to the impact of the use of energy from coal, petroleum, and natural gas, emitting large amounts of carbon dioxide gas, which causes global warming [1]. There are many interesting renewable energy sources such as solar energy, wind energy, water energy, ocean tidal energy, geothermal energy, and biofuels. These renewable energy sources are non-polluting and do not negatively affect the environment [2].

Nowadays, many scientists and researchers have studied and developed the ability to use renewable energy, such as solar energy, wind energy, and water energy, and it is expected that by 2030, 100% of energy will come from such renewable resources. Therefore, there is a clear trend that these renewable energies will play an important role in Thailand as well [3,4]. The most popular renewable energy in Thailand is solar energy. Since Thailand is located near the equator, it receives high amounts of solar energy. The average energy that can be obtained nationwide is about 4 to 4.5 kilowatt-hours per square meter per day. It consists of approximately 50% of direct radiation and the rest is diffused radiation, which is caused by water droplets in the atmosphere (clouds), which are higher than the area away from the equator to the north-south [5]. According to energy consumption estimates by the International Energy Agency in the year 2011, the use of solar thermal energy may be the main energy for electricity generation in the world in the next 50 years, which can reduce greenhouse gas emissions that affect the environment [6]. Solar radiation is measured to evaluate the energy potential, with hourly measurements that are 295–2800 nm and 695–2800 nm and ultraviolet (UV) 295–385 nm at the National Observatory of Athens (NOA) [7]. The intensity of the solar radiation will vary according to the area, day, time, and season, it will be of very high intensity in the afternoon and the sky conditions and wind speed will also affect the intensity and distribution of solar radiation [8]. However, the PV power output has a significant limitation in terms of instability of the power system as its output varies over a wide range throughout the day according to the available solar radiation. The energy storage can improve the stability of the PV system until low solar irradiance. Now, it is well known that the storage is still expensive, so we should get the most out of it [9]. Today, the industrial internet of things (IIoT) plays an essential role in creating tools to help entrepreneurs or investment decision-makers know the benefits of investing in IIoT applications of machine learning and deep learning to provide a new way to develop complex system models, instead of using system physics models to describe the behavior of that system. Some algorithms can infer the operation of the model from the sample input data, in which these models are used to forecast the state of the system, and it is often called forecasting analysis [10].

The power forecasting of PV systems has a variety of methods depending on the data that is used, such as the PV power forecasting with the 1D5P method. This uses the parameters of the solar panel and the solar radiation intensity to calculate the PV power of the solar system and can also improve the accuracy of power output with a weight function that is appropriate for each area [11]. Later, the predictive method with the HIstorical SImilar MIning (HISIMI) method was used to predict the short-term power of the solar system, and it was adjusted by using genetic algorithms and using historical forecasting data [12]. There are discrepancies in the power output forecasting for renewable energy due to fluctuations in renewable energy. The learning methods of machine learning models are more accurate than traditional predictive models. Precise forecasting helps stakeholders decide to plan investment and install solar farms connected to grid systems [13,14]. The auto-regressive moving average (ARMA) model is used, including the exponentially weighted moving average (EWMA) improvements by Piorno et al. [15]. Biological systems and natural phenomena exhibit chaotic behaviors that have been applied for PV power output forecasting. Such as least-squares support vector machines (SVM) methods, it can predict experimental chaotic time series [16] The power forecasting methods with artificial neural networks (ANN) is one method that is providing precise values, using actual measurement data learning to predict future power [17]. The hybrid method forecasting using artificial neural networks has been the basis of fuzzy inference systems (ANFIS) [18] and predictions using particle swarm optimization methods combined with artificial neural networks (PSO-ANN) [19]. The PV power forecasting with these hybrid methods is more accurate than other methods, which are interesting for many researchers to improve the power efficiency of the PV system as well as to develop different forecast models. In this research, we wanted to focus on the ANFIS model because it is an interesting method of forecasting at present, and it is the forecasting method with deep learning techniques that have not been widely used forecasting on solar power generation systems. It is a forecasting technique that can be applied to other complex systems that provide accurate and quick predictions. Then the ANFIS model is applied and compared with the ANN-PSO method, which has similar forecasting techniques. PSO-ANN is one of the forecasting techniques that is more accurate.

As the number of iterations and the number of particle swarms increases, the accuracy becomes more and more accurate. Therefore, the researcher focused on these two forecasting techniques.

So, this work was to study the power output forecasting of the PV systems by the hybrid model and the fast-operating process model. Therefore, the forecasting model using the ANFIS was applied in the study. The simulation results were compared with the PSO-ANN model forecasting. This article is divided into the following six sections. The first section is the introduction. The second section is an explanation of energy efficiency analysis in Thailand. The third section explains the PV power output forecasting model. The fourth section will discuss the case studies of this research and PV power data analysis, and in this study, we will use case study data as a solar system in Thailand. The fifth section will mention simulation results and analysis, and the last section is conclusions.

2. Energy Efficiency Analysis in Thailand

2.1. Energy Sources in Thailand

Natural gas is the primary fuel for electricity generation in the country. If considering the installed production capacity, the power plants that use natural gas as a fuel are approximately 67 percent. Natural gas is a hydrocarbon compound consisting of hydrogen and carbon, which is caused by the accumulation of fossilized micro-organisms that are hundreds of millions of years old. It is also able to separate components into methane, ethane, propane, butane, pentane, petroleum with gasification, etc.

Natural gas is clean energy, and the cost of electricity with natural gas is lower than fuel oil. However, it is slightly higher than coal because the electricity generating system in Thailand has a high proportion of natural gas, so we want to diversify to use other forms of fuel. The advantage of natural gas fuel is that it is a petroleum fuel that can be used with high efficiency. It has complete combustion and highly secure usability because it is lighter than the air, therefore, it floats up when a natural gas leak occurs. Most of it used in Thailand is produced by itself from domestic sources, thus helping to reduce the import of other fuels and saving a lot of foreign currency.

Lignite coal is a natural fuel used in electricity generation, which is a combustible fuel mineral. It consists of four essential elements, which are carbon, hydrogen, nitrogen, and oxygen. It is the second most used fuel from natural gas. The advantage of lignite is that electricity production costs are lower than other fuels, whether natural gas and renewable energy. Coal has a large number of reserves, especially lignite and sub-bituminous coal, which are found mainly in Mae Moh District and Li District, Lampang Province, Mueang District, and Krabi Province. Currently, clean coal technology can be used, which can eliminate more than 99% of coal pollution.

Normally, two types of fuel, namely furnace oil and diesel oil, are used for electricity generation in Thailand. Due to price hike in the international market, such fuels are becoming more and more expensive, resulting in an increase in electricity prices. Additionally, fuel oil causes more pollution than diesel and natural gas. Therefore, fuel oil was used as a secondary fuel rather than a primary fuel.

Diesel is used as fuel for electricity generation of diesel power plants. In Thailand, there are only three locations. Nowadays, the price of diesel has dramatically increased, resulting in a high cost. It also causes more pollution than natural gas. The use of diesel fuel, therefore, is used as a secondary fuel rather than a primary fuel.

Solar energy is a natural energy that is clean and free from pollution. At this time, it is used widely around the world. It is renewable energy with high potential and can be used endlessly. Especially, the use of solar energy to produce electricity, which will help strengthen the electrical system of Thailand and also help reduce global warming. The advantage of solar energy is that it is the largest natural energy source. It is an energy that will never run out, and there is no fuel cost. Solar energy can be used in energy sources that do not have electricity and are far away from power transmission and distribution systems. It is clean energy that does not cause pollution from the electricity production process. There is also renewable energy from water power, wind power, biomass energy, biogas, and waste energy. Figure 1 shows the proportion of electricity production from fuel and renewable energy sources in Thailand. In Thailand, the production of electricity from biomass is the highest, followed by water and solar energy [20].

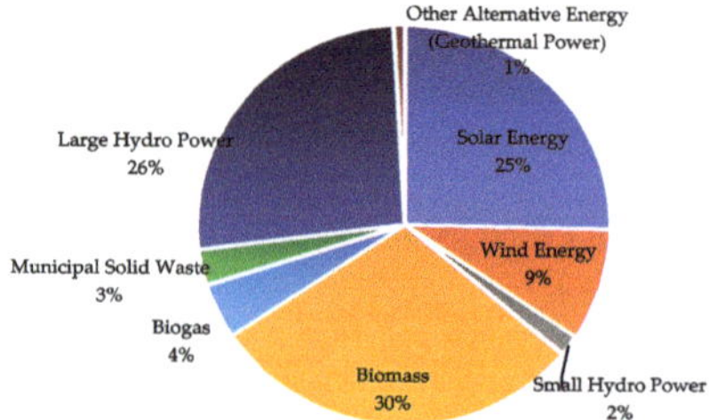

Figure 1. The proportion of electricity production from fuel and renewable energy sources in Thailand.

2.2. Solar Radiation

Solar radiation is the energy released from the sun, which hits the edge of the atmosphere called extraterrestrial solar radiation. In Thailand, ten provinces have the highest potential for the production of electricity from solar energy. Nakhon Ratchasima Province has the highest potential for solar electricity production in Thailand. However, the availability of high-intensity solar radiation will not guarantee the most efficient site for the installation of generation facilities, there are many other factors to be considered [21,22].

The solar potential in Thailand found that most areas receive the highest solar radiation between April and May in the range of 5.56–6.67 kWh/m^2/day. Moreover, the area that has the highest average annual solar radiation is in the Northeast. It covers parts of Nakhon Ratchasima, Buriram, Surin, Sisaket, Roi Et, Yasothon, Ubon Ratchathani, and Udon Thani and parts of the central region with an average annual solar radiation of 5.28–5.56 kWh/m^2/day. The area accounts for 14.3% of the country's total area. Additionally, 50.2% of the total area receives the average annual solar radiation during 5–5.28 kWh/m^2/day, and only 0.5% of the total area exposed to solar radiation is less than 4.45 kWh/m^2/day [23]. Figure 2 shows the average daily solar radiation intensity in Thailand.

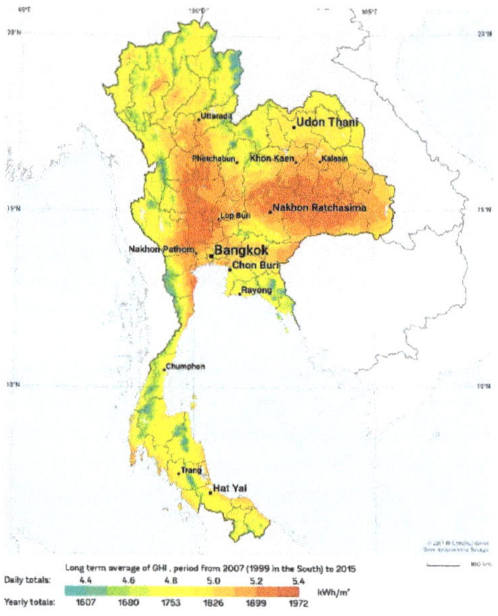

Figure 2. The average daily solar radiation intensity in Thailand [23].

3. PV Power Output Forecasting Model

The accurate forecasting of the PV power generation is essential for the estimation of cost and breakeven. Currently, there are many researchers, including this research, interested in studying the forecasting of power output produced by solar cell systems. In this section, we will explain the principles of the ANFIS and PSO-ANN forecasting models. The PSO-ANN model was used to compare the forecasting result, which has higher accuracy than other methods and can be applied to other areas of research [18]. The models of both methods are shown in the following topics.

3.1. ANFIS Model

Adaptive network-fuzzy inference systems (ANFIS) is a type of network adaptation based on the fuzzy inference systems (FIS), which is a theory adapted from the fuzzy logic theory. The fuzzy rules from the input and output data groups are created using the basis constructing of the neural network system. Artificial neural network (ANN) is a calculation model where its functions and methods are based on the structure of the human brain cells. The neural network follows graph topology in which neurons are nodes of the graph and weights are edges of the graph. It consists of such multilayers that should be a limit in order to the time of problem-solving. The weights changing in the connections between network layers are the training process of the network to achieve the expected output. Another model is neuro-fuzzy, which is a combination of fuzzy logic and neural networks to solve a variety of problems efficiently.

This theory is used in the analysis of problems consisting of information that is widely characterized by uncertainty. The structure of ANFIS will be a fuzzy inference system, which under consideration has two inputs (x and y), and one output is f for the first Sugeno fuzzy model [24,25]. This study uses the hybrid learning algorithm with the following principles. Figure 3 shows the fuzzy inference system [26].

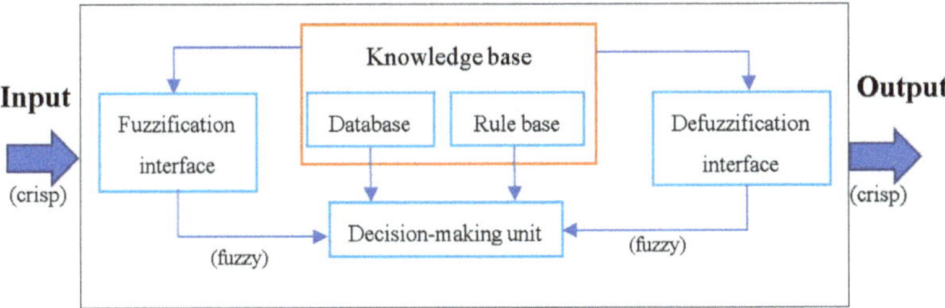

Figure 3. The fuzzy inference system [26].

Layer 1 consists of a member function of each input variable, which can be adjusted. The variables in this layer are also known as premise parameters and can be shown in Equation (1).

$$o_{1,i} = \mu_{Ai}(x) \tag{1}$$

Layer 2 is the calculation of all possible equations of the input vector relationships, which can be expressed in Equation (2). It has 4 ($2^2 = 4$) fuzzy rules.

$$o_{2,i} = wi(x, y) = \mu_{Ai}(x)\mu_{Bi}(y) \tag{2}$$

Layer 3 is normalized to find the input vector obtained from Layer 2, as shown in Equation (3).

$$o_{3,i} = \overline{w} = \frac{w_i}{w_1 + w_2} \tag{3}$$

Layer 4 is called standard perceptron, which can be written according to Equation (4), where (p, q, r) is called the consequent parameter.

$$o_{4,i} = \overline{w}_i f_i = \overline{w}_i (p_i x + q_i y + r_i) \tag{4}$$

Layer 5 is the calculation to find the output value in real numbers, and the result can be written in Equation (5).

$$o_{5,i} = \sum_i \overline{w}_i f_i \tag{5}$$

In this study, the determination of consequent parameters using the least-square estimator method and the reverse pattern study to modify the premise parameters using the gradient descent method. The structure of the ANFIS is shown in Figure 4. Using the ANFIS model to predict the PV power output in this article, it will use 2 inputs, panel temperature, and solar radiation. The model training with one output using the measured power output from the PV system. The structure of a 5-layer model identifies a fixed node, while a square refers to a modified node in which parameters are changed during adjustment or training.

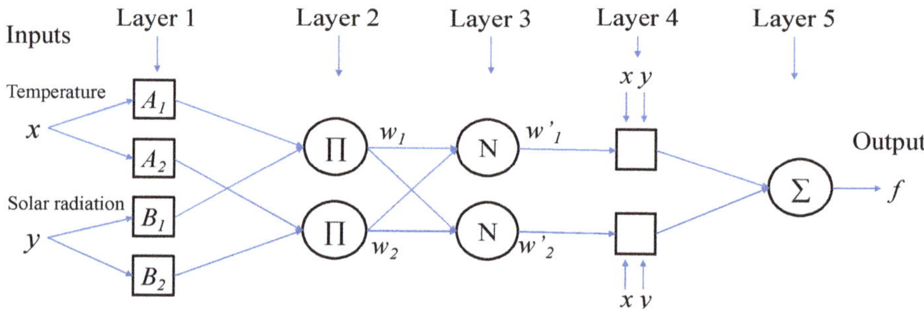

Figure 4. The structure of the adaptive neuro-fuzzy inference systems (ANFIS) model.

3.2. PSO-ANN Model

Particle swarm optimization (PSO) is a natural-inspired algorithm, especially the movement of fish and bird swarms. The change of both types is the movement of the small elements that move together in synchronous time. Fish or birds can move in a swarm, separate from the swarm, and then reunite into the swarm again. The movement of this particle swarm can be considered social behavior. Details of the PSO process were presented in 1995 by James Kennedy and Russell Eberhart [27].

A bird, which is comparable to one particle, and each particle remembers its current position, along with the direction and speed of its movement. Figure 5 shows the position and direction of particle movement. When each particle moves, each particle collects its best data (P_{best}) and compares it to find the best position of every particle (G_{best}). Every cycle at t time, the speed of movement is changed by using data of the best position of each particle and the best position of all particles [28,29].

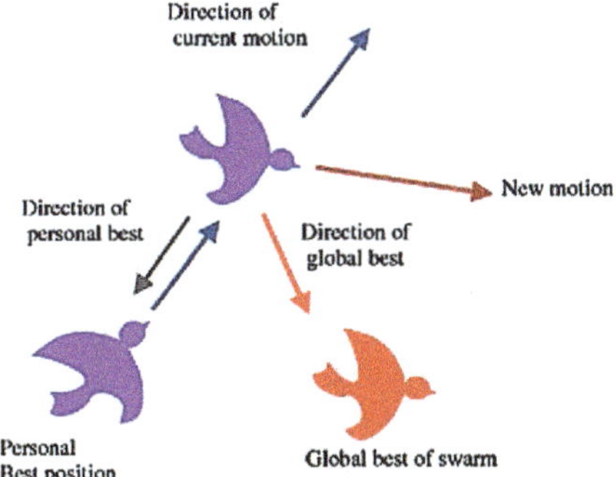

Figure 5. The position and direction of particle movement [28].

PSO has many features that are similar to evolutionary calculations, such as the genetic algorithm (GA). The initial population is randomly generated and used to find the best answer by adjusting that population in every calculation cycle. The solution of the system is represented by particles moving in the search space in the direction of the particles that are closest to the answer that is most appropriate at that time. PSO has been successful in many applications such as optimization of functions, training of artificial neural networks, and the fuzzy control system, including finding the power output or forecasting the suitable power of the photovoltaic power generation systems [30].

Artificial neural networks (ANN) is a simulation of the human nervous system with repeated learning. When learning from something that is repeated many times, it will be able to find a relationship from past learning [31,32]. Figure 6 shows the structure of an artificial neural network that is simulated from the human nervous system. This article will be a supervised learning network to help define the output target for the artificial neural network that uses the multilayer feedforward neural network. It is a back-propagation type which is rather complicated and non-linear. Each neural consists of weight and bias, which begin at random. There is also an activation function or transfer function, which helps to calculate the suitable values such as tan sigmoid, log-sigmoid, and linear. ANN consists of neurons, also called nodes in a circle (Node); the line connecting the nodes is called the weight (Weight) indicates the connection between nerve cells. The artificial neural network has three layers, the input layer, hidden layer, and the output layer.

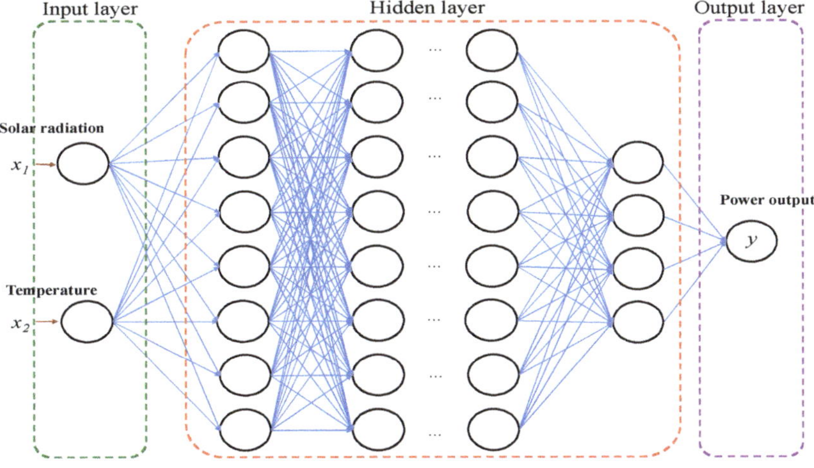

Figure 6. Forecast model based on artificial neural networks (ANN).

The procedure of the hybrid PSO-ANN forecast model is a combination of particle swarm optimization algorithm and BP_ANN algorithm using MATLAB® programs [33,34]. The first step is to determine the number of particles in the ANN structure, beginning with the sampling of particles showing weights, determining the position and speed of the particles. Next, simulate an artificial neural network and evaluate the suitability of initial particles. Find the best of G_{best} and P_{best}, calculate the fitness of each particle in the ANN structure. Find the best fitness in the group or calculation cycle, improve particle velocity and position. Then collect the best particle of the current particle and repeat it until you reach the maximum number of iterations you set [30,35]. Figure 7 shows a diagram of the operation of the hybrid PSO-ANN forecast model.

The parameters of the PSO algorithm are set before the optimization of the ANN model, which has two inputs: panel temperature and solar radiation. The training output is PV power output form the PV systems. Therefore, the structure of the ANN model is 2 8 8 8 4 1. For this article, use the number of particle swarm 100, the maximum number of iterations 100, lower and upper bound of variables −5 to 5, inertia weight 1, inertia weight damping ratio 0.99, personal learning coefficient 1.5, global learning coefficient 2.0, and 4 the number of neurons as shown in Table 1.

Table 1. PSO-ANN parameters.

Particle Swarm	Iterations	Lower-Upper Bound	Inertia Weight	Damping Ratio	Personal Learning Coefficient	Global Learning Coefficient	Number of Neurons
100	100	−5 to 5	1	0.99	1.5	2.0	4

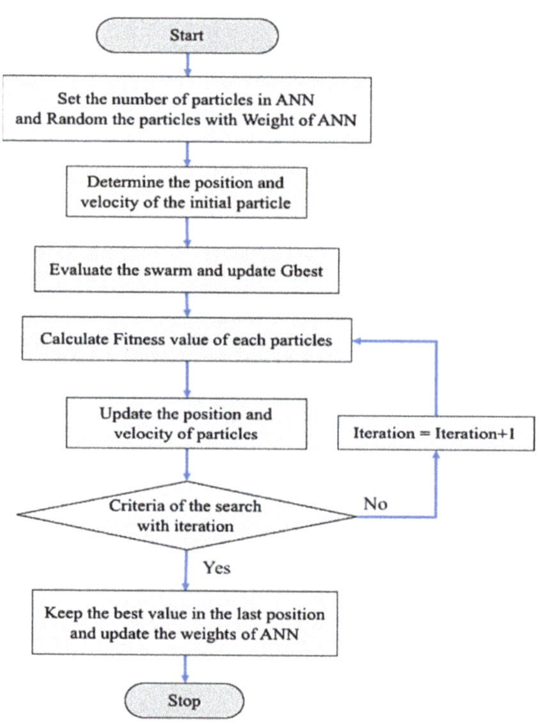

Figure 7. Diagram of the operation of the hybrid PSO-ANN forecast model.

3.3. Accuracy of the Simulation Results

After the simulation, the simulation results will be checked for accuracy, so all measurement forms will always have inaccuracies or uncertainties. An effective experiment must begin with the smallest data error; the percentage of the error can determine the accuracy and reliability of the experiment. Therefore, there must be a realistic and accurate quantity for comparison. If S is defined as the standard quantitative physics and E is the same physical quantitative value as S but obtained from the experiment, then the percentage error can be determined by the Equation (6).

$$\text{The percentage error} = \frac{|E - S|}{S} \times 100\% \tag{6}$$

Statistical analysis can find the best numerical solution of all data sets and determine the statistical error of the answer. The best replacement number is the average value or mean value [36]. In order to assess the accuracy of the prediction methods, the mean absolute percent error (MAPE) and root mean squared error (RMSE) are used as criteria for consideration, [31] which can be found in the following equation.

Mean square error (MSE) is an indication of the variance of the forecasting error which can be obtained by Equation (7)

$$MSE = \frac{1}{n} \sum_{t=1}^{n} \left(\frac{X_i - Y_i}{Y_i}\right)^2 \tag{7}$$

where Y_i is the measured power value, X_i is the predicted value, and n is the amount of data to be tested; it is the power in every period of measurement.

Root mean square error (RMSE) or standard error (SE) is shown by Equation (8).

$$RMSE = \sqrt{\frac{1}{n} \sum_{t=1}^{n} \left(\frac{X_i - Y_i}{Y_i}\right)^2} \tag{8}$$

Mean absolute error (MAP) is shown by Equation (9).

$$MAE = \frac{1}{n} \sum_{t=1}^{n} \left|\frac{X_i - Y_i}{Y_i}\right| \tag{9}$$

The mean absolute percent error (MAPE) is shown by Equation (10). It is also possible to calculate the forecast accuracy (Acc), which indicates how close the forecast the power value to the actual value, which can be obtained by Equation (11).

$$MAPE(\%) = \frac{1}{n} \sum_{t=1}^{n} \left|\frac{X_i - Y_i}{Y_i}\right| \times 100 \tag{10}$$

$$Acc = 100 - MAPE\ (\%) \tag{11}$$

4. PV Power Output Data Analysis

This research uses a case study area in the northeastern region of Thailand with the installation of a 14 MW solar cell system using a 330 W polycrystalline solar panel, the maximum solar irradiation of 1.142 kW/m^2, the highest ambient temperature of 39.4 °C, and the highest panel temperature of 57.44 °C. This data is obtained from a PV plant in the Nakhon Ratchasima province. The results of the measurement are used for one year as forecasting data. Figures 8 and 9 shows the annual solar irradiation and solar panels temperature.

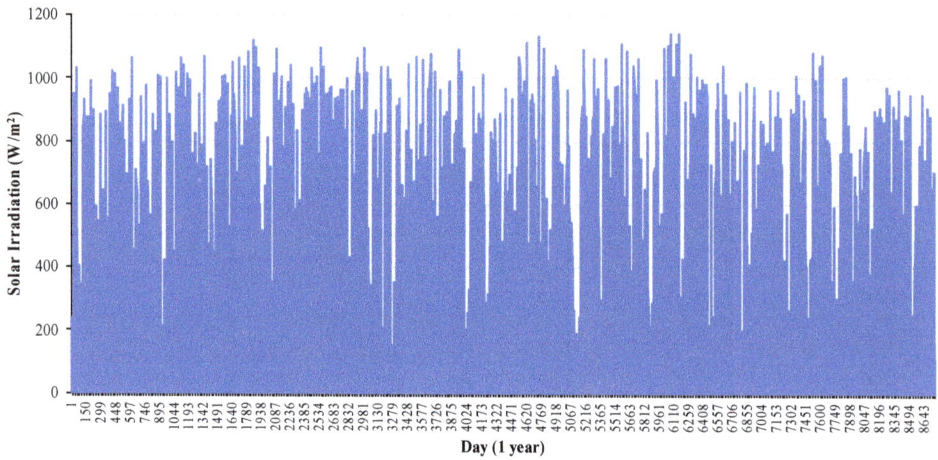

Figure 8. Annual solar irradiation of the case study area.

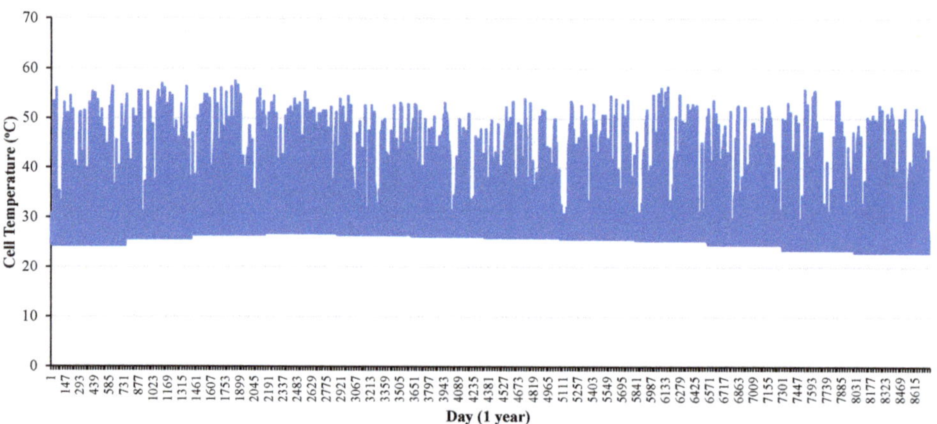

Figure 9. Annual solar panels temperature.

This PV power generation system will be connected to the grid system (Grid connected) and it will produce the electricity for the power distribution systems to be the source of the residential electricity load. In order to study the energy efficiency, we must also analyze the output power of this solar power generation system. Figure 10 shows the PV power output produced by the solar system case study and solar irradiance, for example 5-day data during April in 2018. The input data used for learning of the model is the temperature of the solar panel, ambient temperature, solar irradiance, and PV output power.

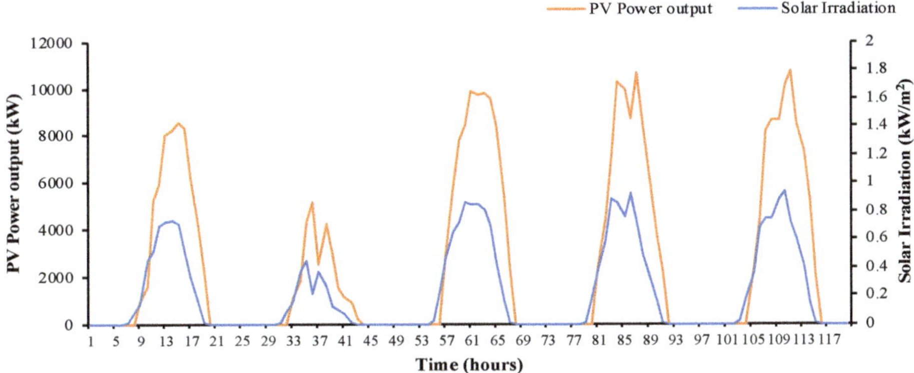

Figure 10. PV power output and solar irradiance.

5. Simulation Results and Discussion

A MATLAB program is used to simulate the PV power output forecasting by inputting one year of input data, which is the actual measured power output, solar irradiation, panel temperature, and ambient temperature. The PSO-ANN technique is used to simulate the calculation cycle of 100 cycles, the population of 100 population, and the number of neurons of 10 cells. The annual power output forecasts are shown in Figure 11, in which some of the predicted electrical power is less than zero because it is a random method. It has a maximum error of 1721 kW, as shown in Figure 12. In April it is the month in which the most energy was produced from solar energy and it is summer in Thailand, which has the most solar radiation intensity. Therefore, the forecasting results of April 1–7 were selected. Figure 13 shows a weekly PV power output forecasting with PSO-ANN. Figure 14 shows a weekly percentage error of PV forecasting using PSO-ANN. In a positive state, the predicted value is higher than the actual value, and in a negative state, the predicted value is less than the real value. It can be seen that the percentage error from the forecast is as high as 190.7% because the PV power forecasting value is much higher than the actual PV power of the system (actual PV power value of 238.5 kW and PV power forecasting value of 693.5 kW). The PSO-ANN is a technique for determining optimal values with sampling, so high errors may occur due to the input instability. The simulation results, when compared to the real power output, can be seen that at night without irradiation intensity, the PSO-ANN model has a rather high discrepancy. The calculation efficiency of the ANFIS method was 429.522 s.

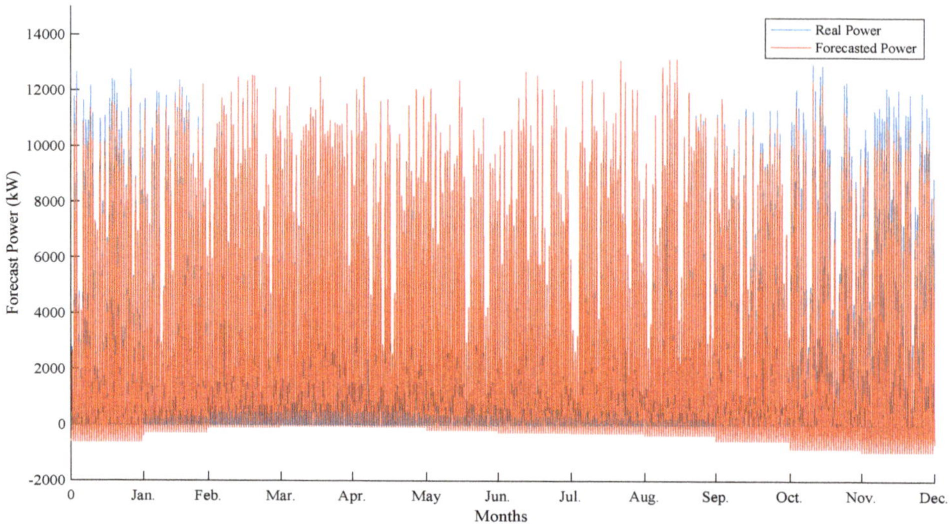

Figure 11. Annual PV power output with PSO-ANN forecasting.

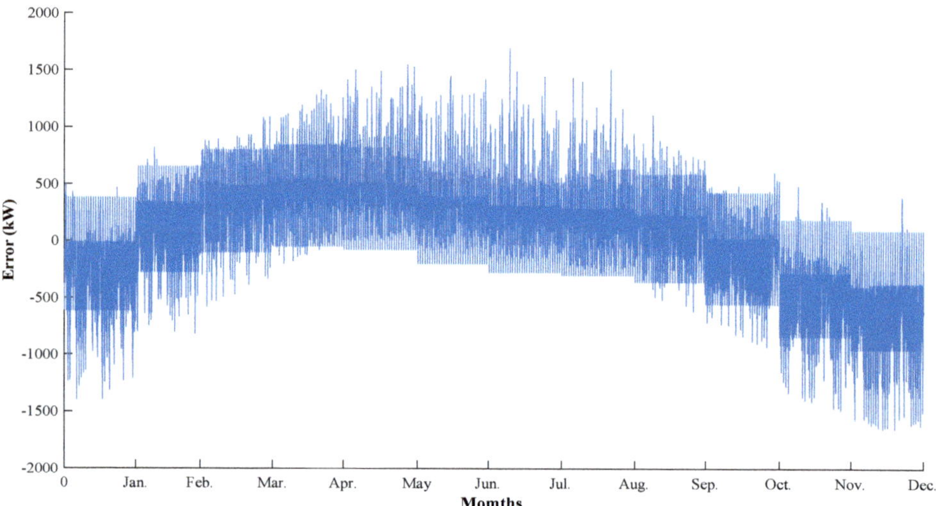

Figure 12. Annual PV power output error with PSO-ANN forecasting.

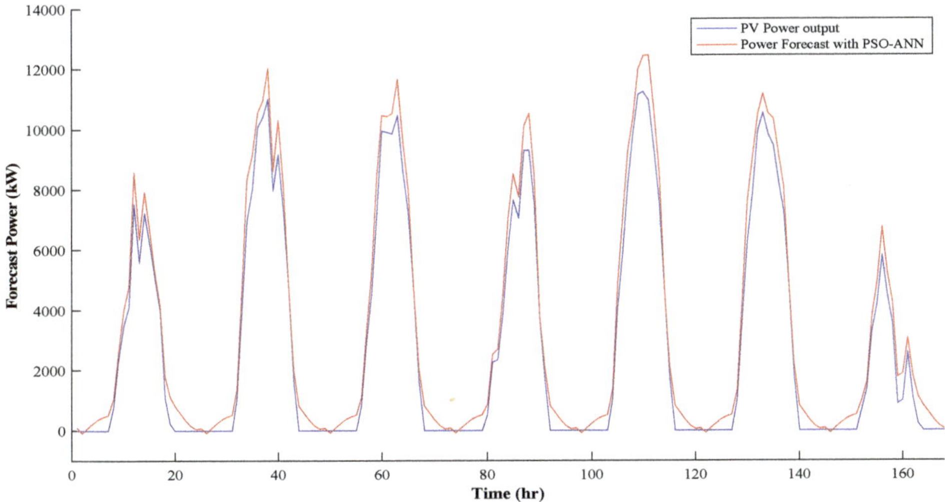

Figure 13. Weekly PV power output with PSO-ANN forecasting.

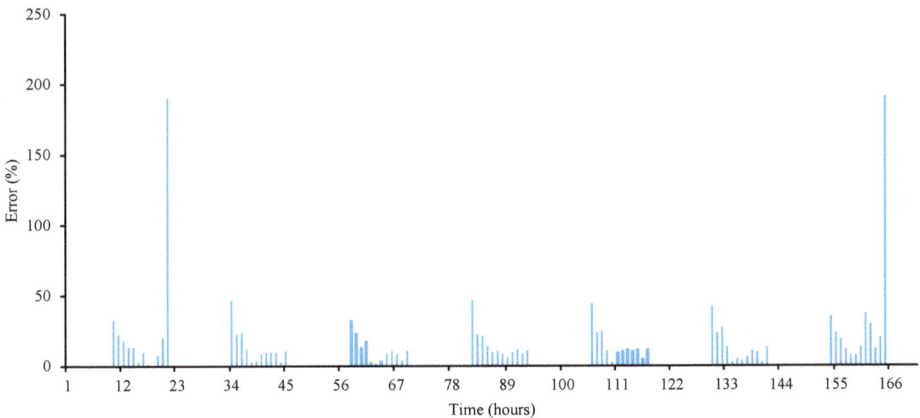

Figure 14. Weekly PV power output percentage error with PSO-ANN forecasting.

The simulation of PV power output forecasting with ANFIS is shown in Figure 15, which is a comparison between the actual PV power output and the PV power forecasting. Figure 16 shows the annual error of PV power output forecasting, which a maximum of 1742.2 kW. The maximum error occurs in the period from October to December. Figure 17 shows a weekly PV power output forecasting with ANFIS, and Figure 18 shows a weekly percentage error of PV power forecasting using ANFIS. It can be seen that the forecasted results at night have less error than the PSO-ANN method. Figure 19 shows the comparison of power output forecasting, which shows that the PV power forecasting using the ANFIS method provides more accurate forecasting results than the PSO-ANN method. The calculation efficiency of the ANFIS method was 3.5675 s.

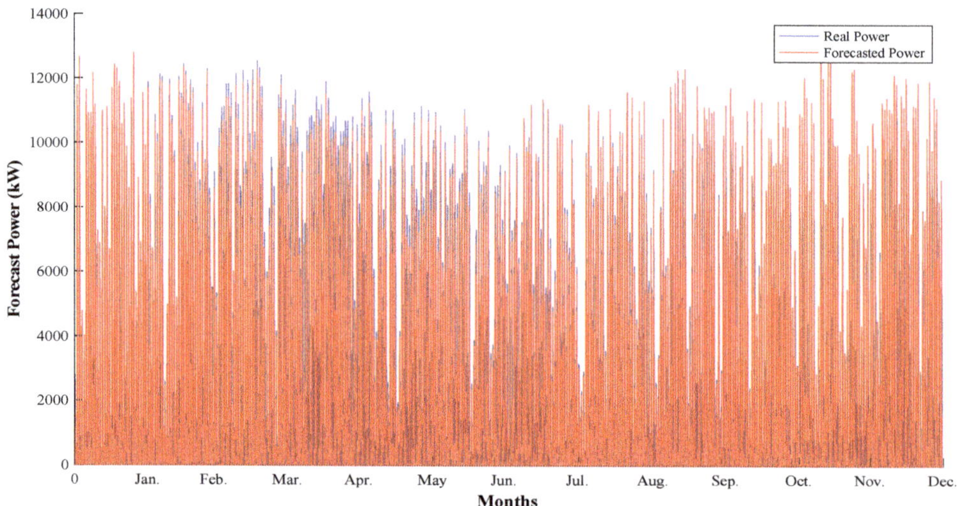

Figure 15. Annual PV power output with ANFIS forecasting.

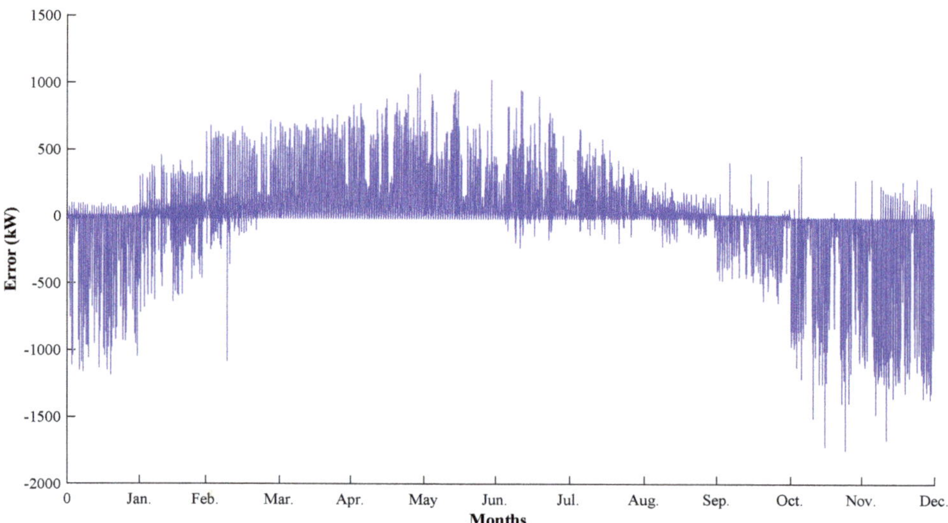

Figure 16. Annual power output error with ANFIS forecasting.

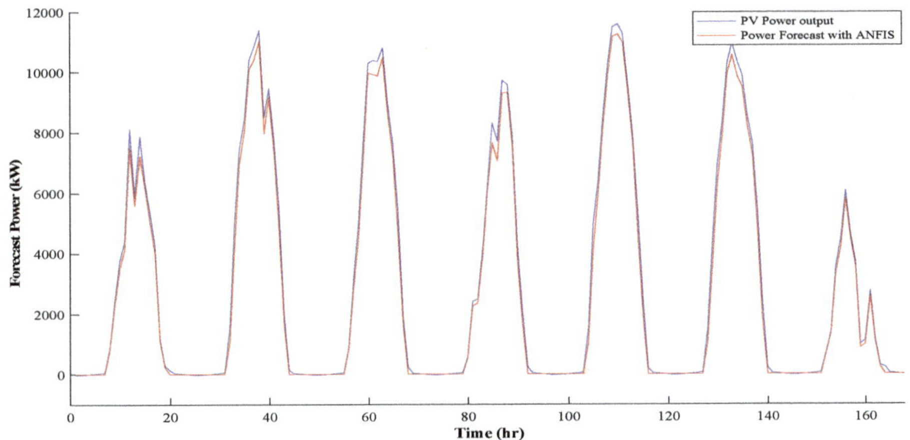

Figure 17. Weekly PV power output with ANFIS forecasting.

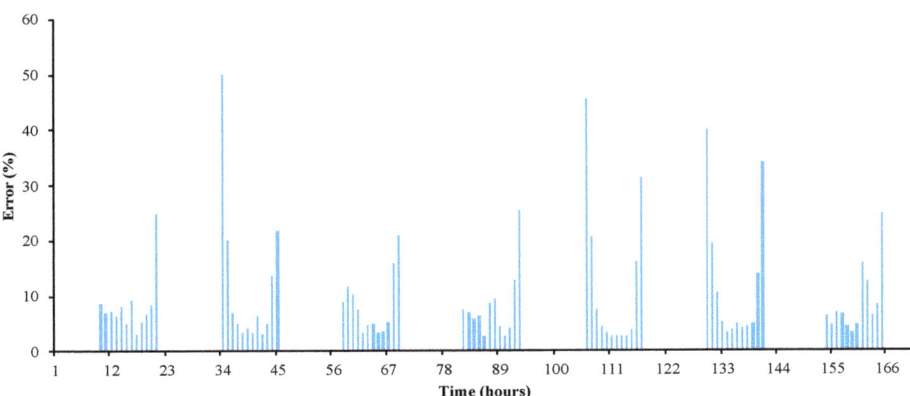

Figure 18. Weekly PV power output percentage error with ANFIS forecasting.

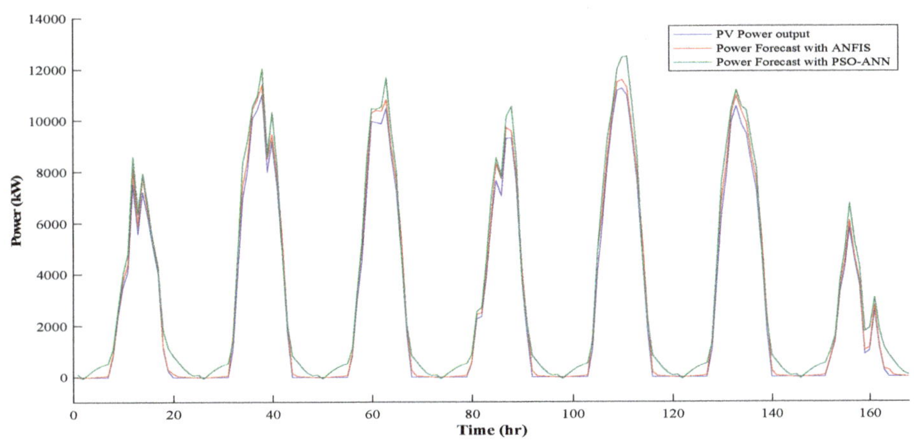

Figure 19. Comparison of PV power output forecasting.

Table 2 shows the calculation results to assess the accuracy of the simulation results of the PV power forecasting model using the PSO-ANN model and ANFIS model. The indicators such as MSE,

RMSE, MAD, MAPE, and Accuracy are used for evaluating forecasting performance. The simulation results show that the ANFIS model has less error than the PSO-ANN model and has 99.8532% greater accuracy. RMSE is another indicator used to check the accuracy of the simulation results. Forecasting results with the ANFIS model provided an RMSE value of only 0.1184, which is less than the PSO-ANN model. When compared with the results of the PV power output forecasting of D. Lee and K. Kim [37], they predict the power output with long- and short-term memory (LSTM) -based models with an RMSE of 0.563 with summer, so it shows that the simulation results with the ANFIS model have greater accuracy. The calculation efficiency of the ANFIS model takes less time than the PSO-ANN model. Therefore, this article shows that the ANFIS model is more efficient for the PV power forecasting than the PSO-ANN model, and it also takes quick calculations.

Table 2. Performance and forecast accuracy of the model.

Models	MSE	RMSE	MAE	MAPE (%)	Accuracy (%)	Calculation Efficiency (s)
PSO-ANN	0.3234	0.5687	8.8233	1.0842	98.9157	429.522
ANFIS	0.0140	0.1184	1.1952	0.1468	99.8532	3.5675

6. Conclusions

In this article, the PV power output is forecast using one-year of electricity production data from a solar power plant in the northeast Thailand area. A comparison of the PV power output forecasting using the ANFIS and PSO-ANN method was undertaken. The performance of the ANFIS and PSO-ANN models were verified accurate with MSE, RMSE, MAP, and MAPE. The accuracy of the ANFIS model is 99.8532%, and the PSO-ANN method is 98.9157%. The calculation efficiency of the ANFIS model takes less time than the PSO-ANN model. The simulation results show that the ANFIS method has more accurate simulation results than the PSO-ANN method. For the most efficient use of PV power generation systems, it is necessary to analyze the energy consumption of the user load (household loads, industrial loads, department store loads). The forecasting results of the PSO-ANN model has more discrepancies than the ANFIS method at night. Therefore, nighttime inputs may be omitted before the simulation. The ANFIS model is an interesting method for forecasting at present, and it uses deep learning techniques to solve the problem. It is a forecasting technique that can compare to other complex systems which provide accurate and quickly predictions. The power output forecasting is essential for planning the installation of a PV power system. The PV power output forecasting using the ANFIS model is another method that can predict and analyze the energy, cost, and cost-effectiveness that will occur in the future. Future work will study the PV power output forecasting in a new method and a more efficient way and improve the input data with small deviations, which can make the simulation results more accurate. Additionally, to analyze the simulation results using reliable and widely used methods.

Author Contributions: P.D. and T.B. conducted the initial design, conceptualization, writing, editing, and developing the proposed forecasting method; N.J., K.S., and S.K. contributed the resources, data curation, and analysis; W.T. and S.N. conceived the theoretical approaches, review, and investigation. All authors have read and agreed to the published version of the manuscript.

Funding: The APC was funded by King Mongkut's Institute of Technology Ladkrabang (KMITL).

Acknowledgments: The author's thanks to King Mongkut's Institute of Technology Ladkrabang for supporting this research. The authors would like to thank the solar energy technology laboratory, National Electronics and Computer Technology Center, National Science and Technology Development Agency (NSTDA) for supporting the data for this research.

Conflicts of Interest: The authors declare no conflict of interest.

References

1. Alternative Energy. Available online: https://en.wikipedia.org/wiki/Alternative_energy (accessed on 10 July 2019).
2. Zehner, O. *Green Illusions*; University of Nebraska Press: Lincoln, NE, USA; London, UK, 2012; pp. 1–169; 331–342.
3. Jacobson, M.Z.; Delucchi, M.A. A path to sustainable energy by 2030. *Sci. Am.* **2009**, *301*, 58–65. [CrossRef] [PubMed]
4. Jacobson, M.Z.; Delucchi, M.A. Providing all global energy with wind, water, and solar power, Part I: Technologies, energy resources, quantities and areas of infrastructure, and materials. *Energy Policy* **2011**, *39*, 1154–1169. [CrossRef]
5. Inthacha, S. The Climatology of Thailand and Future Climate Change Projections Using the Regional Climate Model Precis. Ph.D. Thesis, University of East Anglia, Norwich, UK, May 2011.
6. Prospect of Limiting the Global Increase in Temperature to 2·°C is Getting Bleaker. Available online: https://www.iea.org/newsroom/news/2011/may/2011-05-30-.html (accessed on 10 July 2019).
7. Papaioannou, G.; Nikolidakis, G.; Asimakopoulos, D.; Retalis, D. Photosynthetically active radiation in Athens. *Agric. For. Meteorol.* **1996**, *81*, 287–298. [CrossRef]
8. Codato, G.; Oliveira, A.P.; Soares, J.; Escobedo, J.F.; Gomes, E.N.; Pai, A.D. Global and diffuse solar irradiances in urban and rural areas in southeast Brazil. *Theor. Appl. Climatol.* **2017**, *93*, 57–73. [CrossRef]
9. Janjai, S. *Solar Radiation*; Department of Physics, Faculty of Science, Silpakorn University Campus: Nakhon Pathom, Thailand, 2014.
10. The Industrial Internet of Things Volume T3: Analytics Framework. Available online: https://www.iiconsortium.org/pdf/IIC_Industrial_Analytics_Framework_Oct_2017.pdf (accessed on 1 December 2019).
11. Kittisontirak, S.; Dawan, P.; Atiwongsangthong, N.; Titiroongruang, W.; Chinnavornrungsee, P.; Hongsingthong, A.; Manosukritkul, P. A novel power output model for photovoltaic system. In Proceedings of the International Electrical Engineering Congress (iEECON), Pattaya, Thailand, 8–10 March 2017; pp. 209–212.
12. Monteiro, C.; Santos, T.; Fernandez-Jimenez, L.; Ramirez-Rosado, I.; Terreros-Olarte, M. Short-term power forecasting model for photovoltaic plants based on historical similarity. *Energies* **2013**, *6*, 2624–2643. [CrossRef]
13. Mohammed, A.A.; Aung, Z. Ensemble learning approach for probabilistic forecasting of solar power generation. *Energies* **2016**, *9*, 1017. [CrossRef]
14. Agoua, X.G.; Girard, R.; Kariniotakis, G. Probabilistic models for spatio-temporal photovoltaic power forecasting. *IEEE Trans. Sustain. Energy* **2019**, *10*, 780–789. [CrossRef]
15. Piorno, J.R.; Bergonzini, C.; Atienza, D.; Rosing, T.S. Prediction and management in energy harvested wireless sensor nodes. In Proceedings of the 1st International Conference on Wireless Communication, Vehicular Technology, Information Theory and Aerospace & Electronic Systems Technology, Aalborg, Denmark, 17–20 May 2009.
16. Pano-Azucena, A.; Tlelo-Cuautle, E.; Tan, S.; Ovilla-Martinez, B.; Fraga, L.D.L. FPGA-based implementation of a multilayer perceptron suitable for chaotic time series prediction. *Technologies* **2018**, *6*, 90. [CrossRef]
17. Abuella, M.; Chowdhury, B. Solar power forecasting using artificial neural networks. In Proceedings of the North American Power Symposium (NAPS), Charlotte, NC, USA, 4–6 October 2015.
18. Omid, M.; Ramedani, Z.; Keyhani, A.R. Forecasting of daily solar radiation using neuro-fuzzy approach. In Proceedings of the 5th International Mechanical Engineering Forum, Prague, Czech Republic, 20–22 June 2012; pp. 728–742.
19. Hoballah, A.; Erlich, I. PSO-ANN approach for transient stability constrained economic power generation. In Proceedings of the IEEE Bucharest Power Tech Conference, Bucharest, Romania, 28 June–2 July 2009; pp. 1–6.
20. Department of Alternative Energy Development and Efficiency. Ministry of Energy in Thailand. Energy Situation. Available online: https://www.dede.go.th/download/stat62/sit_2_61_dec.pdf (accessed on 10 July 2019).
21. Solar Energy Distribution at the Top of the Atmosphere and at the Surface of the Earth. Available online: http://www.physics.usyd.edu.au/teach_res/hsp/sp/mod7/m7emrSpectra.pdf (accessed on 20 October 2019).
22. Ministry of Energy. Promotion of Using Hot Water from Solar Energy, Power Point Presentation. Available online: https://www.slideserve.com/pete/outline (accessed on 10 July 2019).

23. Solar Resource Maps of Thailand. Available online: https://solargis.com/maps-and-gis-data/download/Thailand (accessed on 10 July 2019).

24. Jang, J.-S.R. ANFIS: Adaptive network-based fuzzy inference system. *IEEE Trans. Syst. Cybern.* **1993**, *23*, 665–685. [CrossRef]

25. Aghbashlo, M.; Hosseinpour, S.; Mujumdar, A.S. Artificial neural network-based modeling and controlling of drying systems. *Intell. Control Dry.* **2018**, *1*, 155–172.

26. Brahim, K.; Zell, A. ANFIS-SNNS: Adaptive network fuzzy inference system in the stuttgart neural network simulator. *Fuzzy Syst. Comput. Sci.* **1994**, *1*, 117–127.

27. Kennedy, J.; Eberhart, R. Particle swarm optimization. In Proceedings of the IEEE International Conference on Neural Networks, Perth, Australia, 27 November–1 December 1995; pp. 1942–1948.

28. Jumpasri, N.; Pinsuntia, K.; Woranetsuttikul, K.; Nilsakorn, T.; Khan-Ngern, W. Improved particle swarm optimization algorithm using average model on MPPT for partial shading in PV array. In Proceedings of the International Electrical Engineering Congress (IEECON), Chonburi, Thailand, 19–21 March 2014.

29. Esmin, A.A.A.; Coelho, R.A.; Matwin, S. A review on particle swarm optimization algorithm and its variants to clustering high-dimensional data. *Artif. Intell. Rev.* **2013**, *44*, 23–45. [CrossRef]

30. Abdullah, A.G.; Suranegara, G.M.; Hakim, D.L. Hybrid PSO-ANN application for improved accuracy of short-term load forecasting. *IEEE Trans. Power Syst.* **2014**, *9*, 446–451.

31. Nespoli, A.; Ogliari, E.; Leva, S.; Pavan, A.M.; Mellit, A.; Lughi, V.; Dolara, A. Day-ahead photovoltaic forecasting: A comparison of the most effective techniques. *Energies* **2019**, *12*, 1621. [CrossRef]

32. Martins, R.P.; Ferreira, V.H.; Lopes, T.T. Artificial neural network for probabilistic forecasting of the output power of photovoltaic systems. In Proceedings of the Simposio Brasileiro De Sistemas Eletricos (SBSE), Niteroi, Brazil, 12–16 May 2018.

33. Le, L.T.; Nguyen, H.; Dou, J.; Zhou, J. A comparative study of PSO-ANN, GA-ANN, ICA-ANN, and ABC-ANN in estimating the heating load of buildings' energy efficiency for smart city planning. *Appl. Sci.* **2019**, *9*, 2630. [CrossRef]

34. Houria, B.; Mahdi, K.; Zohra, T.F. PSO-ANN's based suspended sediment concentration in Ksob basin, Algeria. *J. Eng. Technol. Res.* **2014**, *6*, 129–136.

35. Said, S.Z.; Thiaw, L. Performance of artificial neural network and particle swarm optimization technique based maximum power point tracking for photovoltaic systems under different environmental conditions. *J. Phys. Conf. Ser.* **2018**, *1049*, 012047. [CrossRef]

36. Yousif, J.H.; Kazem, H.A.; Alattar, N.N.; Elhassan, I.I. A comparison study based on artificial neural network for assessing PV/T solar energy production. *Case Stud. Therm. Eng.* **2019**, *13*, 100407. [CrossRef]

37. Lee, D.; Kim, K. Recurrent neural network-based hourly prediction of photovoltaic power output using meteorological information. *Energies* **2019**, *12*, 215. [CrossRef]

Article

Multi-Step Solar Irradiance Forecasting and Domain Adaptation of Deep Neural Networks

Giorgio Guariso [1], Giuseppe Nunnari [2] and Matteo Sangiorgio [1,*

[1] Dipartimento di Elettronica, Informazione e Bioingegneria, Politecnico di Milano, 20133 Milan, Italy; giorgio.guariso@polimi.it
[2] Dipartimento di Ingegneria Elettrica, Elettronica e Informatica, Università degli Studi di Catania, 95125 Catania, Italy; giuseppe.nunnari@dieei.unict.it
* Correspondence: matteo.sangiorgio@polimi.it

Received: 2 June 2020; Accepted: 27 July 2020; Published: 2 August 2020

Abstract: The problem of forecasting hourly solar irradiance over a multi-step horizon is dealt with by using three kinds of predictor structures. Two approaches are introduced: Multi-Model (*MM*) and Multi-Output (*MO*). Model parameters are identified for two kinds of neural networks, namely the traditional feed-forward (FF) and a class of recurrent networks, those with long short-term memory (LSTM) hidden neurons, which is relatively new for solar radiation forecasting. The performances of the considered approaches are rigorously assessed by appropriate indices and compared with standard benchmarks: the clear sky irradiance and two persistent predictors. Experimental results on a relatively long time series of global solar irradiance show that all the networks architectures perform in a similar way, guaranteeing a slower decrease of forecasting ability on horizons up to several hours, in comparison to the benchmark predictors. The domain adaptation of the neural predictors is investigated evaluating their accuracy on other irradiance time series, with different geographical conditions. The performances of FF and LSTM models are still good and similar between them, suggesting the possibility of adopting a unique predictor at the regional level. Some conceptual and computational differences between the network architectures are also discussed.

Keywords: feed-forward neural networks; recurrent neural networks; LSTM cell; performances evaluation; clear sky irradiance; persistent predictor

1. Introduction

As is well known, the key challenge with integrating renewable energies, such as solar power, into the electric grid is that their generation fluctuates. A reliable prediction of the power that can be produced using a few hours of horizon is instrumental in helping grid managers in balancing electricity production and consumption [1–3]. Other useful applications could benefit from solar radiation forecasting, such as the management of charging stations [4,5]. Indeed, such a kind of micro-grid is developing in several urban areas to provide energy for electric vehicles by using Photo Voltaic (PV) systems [6]. To optimize all these applications, a classical single-step-ahead forecast is normally insufficient and a prediction over multiple steps is necessary, even if with decreasing precision.

Reviews concerning machine learning approaches to forecasting solar radiation were provided, for instance, in [7–10], but the recent scientific literature has seen a continuous growth of papers presenting different approaches to the problem of forecasting solar energy. According to Google Scholar, the papers dealing with solar irradiance forecast were about 5000 in 2009 and became more than 15,000 in 2019, with a yearly increase of more than 15% in the last period. The growth of papers utilizing neural networks as a forecasting tool has been quite more rapid since they grew up at a pace of almost 30% a year in the last decade and they constituted about half of those cataloged in 2019.

The short-term prediction of solar radiation and/or photovoltaic power output has expanded from traditional ARMA (Auto-Regressive Moving Average) models and their evolution such as ARMA-GARCH (Generalized Autoregressive with Conditional Heteroscedasticity) models [11], to Bayesian statistical models [12], and Naïve Bayes Classifiers [13], and to Neuro-Fuzzy Systems [14–16], and Fuzzy with Genetic Algorithms models [17]. More recently, other approaches such as Markov Switching Models [18], Support Vector Machines [19–21], wavelets transformation [22], kernel density estimation [23], and tree-based algorithms [24] have been suggested. Traditional Feed Forward (FF) neural network for ultra-short-time horizon (one-minute resolution) was proposed in [25]. Specific classes of neural networks have also been used, such as TDNN (Time Delay Neural networks) [26], Recurrent [27–29], Convolutional Neural Networks [30] and other configurations [31–35]. Other works specifically concentrate on the use of Long Short-Term Memory (LSTM) networks [36–38]. Specifically, Abdel-Nasser and Mahmoud's (2017) [36] use of the Long Short-Term Memory Recurrent Neural Network (LSTM-RNN) to accurately forecast the output power of PV systems was proposed, based on the idea that such a network can model the temporal changes in PV output power due of their recurrent architecture and memory units. The proposed method was evaluated using hourly datasets of different sites for a year. The authors have implemented five different model structures in order to predict the average hourly time series one-step ahead. The accuracy of the models was assessed comparing the RMSE error in comparison with other techniques, namely Multiple Linear Regression (MLR) models, Bagged Regression Trees (BRT) and classical Feed-Forward (FF) Neural Networks (NN), showing that one of the five studied models outperforms the others. One of the limitations of this attempt seems to be the very short time horizon considered (1 h). On the contrary, several regression-based trees, were considered by Fouilloy et al. (2018) [39] for prediction horizons from 1 to 6 hours ahead. They compared the performance of these ML approaches with the more traditional ARMA and Multi-layer Perceptrons (MLP), over three different data sets recorded in France and Greece. The performances, in terms of nRMSE and MAE, seem to point out that among the methods considered there is no prevalent one and their ranking depends on the variability of the data set. The recent paper [40] provides a large bibliography on neural network applications to solar power forecast. The paper by Husein et al. [41] also analyses several preceding studies and compares the results of LSTM-RNN and FF networks for hourly ahead solar irradiance forecasting using only indirect weather measures (namely, dry-bulb temperature, dew-point temperature, and relative humidity), even if irradiance data are still needed for network training. Various deep learning approaches, namely Gated Recurrent Units (GRUs), LSTM, RNN and FF neural networks were considered by Aslam et al. [42] to represent the statistical relations of solar radiation data over a very long horizon (one year ahead).

What clearly emerges is that the very large majority of these studies concentrate on a single (very short to very long) forecast step. However, the implementation of advanced control methods, such as model predictive control requires the availability of forecasted values for some steps ahead. This paper thus contributes to the field by clearly defining the possible model structures for a multi-step prediction and comparing the performances of feed-forward and LSTM recurrent networks in the problem of multi-step ahead forecasting of the solar irradiance at a station in Northern Italy, which is characterized by a relatively high annual and daily variability. The forecasting approaches are rigorously evaluated using appropriate indices and compared with standard benchmarks: the clear sky irradiance and two persistent predictors.

Furthermore, the adaptation capability of the neural predictors is investigated evaluating their accuracy on three locations with different geographical conditions with respect to that used for the training.

The paper is organized as follows: In Section 2, the basic characteristics of the multi-step feed-forward and LSTM forecasting model are presented together with the different ways of formulating and approaching the prediction task. Some features of solar irradiance time series are pointed out through statistical and spectral analysis. In addition, the benchmark predictors and the evaluation metrics are presented. Section 3 shows the results of solar irradiance forecasting performed by autoregressive neural models, along with the domain adaptation capability of such predictors. Section 4 reports a discussion on the intrinsic differences in the identification phase of feed-forward and LSTM neural predictors. In Section 5, some concluding remarks are drawn.

2. Materials and Methods

2.1. Structure of the Multi-Step Neural Predictors

The time-series forecasting task is inherently dynamic because the variable to be predicted can be seen as the output of a dynamical system. Recurrent neural networks, which are themselves dynamical systems due to the presence of one or more internal states, are a natural choice. On the other hand, the more traditional feed-forward (FF) nets are static and can reproduce any arbitrarily complex mapping from the input to the output space. Static models can be adapted to deal with dynamic tasks in different ways.

Thinking for instance to an hourly autoregressive forecast, we can consider three different ways to predict the h step-ahead hourly sequence $\left[\hat{I}(t+h-1),\ \ldots,\hat{I}(t+1),\hat{I}(t)\right]$ basing on the observations of the last d steps (i.e., $[I(t-1),\ I(t-2),\ \ldots,\ I(t-d)]$) with a FF network. Here, the symbols $\hat{I}(t)$ and $I(t)$ indicate the forecasted and observed solar irradiance, respectively.

2.1.1. The Recursive (*Rec*) Approach

Rec approach repeatedly uses the same one-step-ahead model, assuming as input for the second step the forecast obtained one step before [43]. Only one model, f_{Rec}, is needed in this approach, whatever the length of the forecasting horizon and this explains the large diffusion of this approach.

$$\hat{I}(t) = f_{Rec}(I(t-1),\ I(t-2),\ \ldots,\ I(t-d)) \tag{1}$$

$$\hat{I}(t+1) = f_{Rec}\big(\hat{I}(t),\ I(t-1),\ \ldots,\ I(t-d+1)\big) \tag{2}$$

Equations (1) and (2) show the first two steps of the procedure, which has to be repeated h times.

2.1.2. The Multi-Model (*MM*) Approach

Although the *Rec* model structure is the most natural one also for h-steps ahead prediction, other alternatives are possible and in this work, we have explored two new structures, here respectively referred as multi-model (*MM*) and multi-output (*MO*), which are illustrated below. The peculiarity of the *MM* model is that, unlike the f_{Rec}, its parameters are optimized for each prediction time horizon, following the framework expressed by Equations (3)–(5). *MM* thus requires h different models to cover the prediction horizon.

It is trivial to note that for $h = 1$ the model works exactly as a *Rec* model, but it represents a generalization when $h > 1$.

$$\hat{I}(t) = f_{MM,1}(I(t-1),\ I(t-2),\ \ldots,\ I(t-d)) \tag{3}$$

$$\hat{I}(t+1) = f_{MM,2}(I(t-1),\ I(t-2),\ \ldots,\ I(t-d)) \tag{4}$$

$$\hat{I}(t+h-1) = f_{MM,h}(I(t-1),\ I(t-2),\ \ldots,\ I(t-d)) \tag{5}$$

2.1.3. The Multi-Output (*MO*) Approach

The *MO* model [44] expressed by Equation (6) can be considered a trade-off between the *Rec* and the *MM* since it proposes to develop a single model with a vector output composed by the h values predicted for each time step.

$$\left[\hat{I}(t+h-1),\ \ldots,\hat{I}(t+1),\hat{I}(t)\right] = f_{MO}(I(t-1),\ I(t-2),\ \ldots,\ I(t-d)) \tag{6}$$

The number of parameters of the *MO*, is a bit higher than the *Rec* (a richer output layer has to be trained), but much lower than the *MO*. However, with respect to the *Rec* it could appear, at least in principle, more accurate, since it allows its performance to be optimized over a higher dimensional space. A *MO* in comparison to the *Rec* offers the user a synoptic representation of the various prediction scenarios and therefore can be more attractive from the application point of view. Furthermore, in case the *Rec* model has some exogenous inputs, one must limit the forecasting horizon h to the maximum delay between the external input and the output. Nothing changes, on the contrary, for the other two approaches.

2.2. Model Identification Strategies

Neural networks are among the most popular approaches for identifying nonlinear models starting from time series. Probably this popularity is due to the reliability and efficiency of the optimization algorithms, capable of operating in the presence of many parameters (many hundreds or even thousands, as in our case). Two different kinds of neural networks have been used in this work. One of the purposes behind this work was in fact to explore, on rigorous experimental bases, if more complex but more promising neural network architectures, such as the LSTM neural networks, could offer greater accuracy for predicting solar radiation, compared to simpler and more consolidated architectures such as feed-forward (FF) neural networks.

LSTM cells were originally proposed by Hochreiter and Schmidhuber in 1997 [45] as a tool to retain a memory of past errors without increasing the dimension of the network explosively. In practice, they introduced a dynamic within the neuron that can combine both long- and short-term memory. In classical FF neural networks, the long-term memory is stored in the values of parameters that are calibrated on past data. The short-term memory, on the other side, is stored in the autoregressive inputs, i.e., the most recent values. This means that the memory structure of the model is fixed. LSTM networks are different because they allow balancing the role of long and short-term in a continuous way to best adapt to the specific process.

Each LSTM cell has three gates (input, output, and forget gate) and a two-dimensional state vector $s(t)$ whose elements are the so-called cell state and hidden state [46,47]. The cell state is responsible for keeping track of the long-term effects of the input. The hidden state synthesizes the information provided by the current input, the cell state, and the previous hidden state. The input and forget gates define how much a new input and the current state, respectively, affect the new state of the cell, balancing the long- and short-term effects. The output gate defines how much the output depends on the current state.

Recurrent nets with LSTM cells appear particularly suitable for solar radiation forecast since the underlying physical process is characterized by both slow (the annual cycle) and fast (the daily evolution) dynamics.

An LSTM network can be defined as in Equations (7)–(8), namely as a function computing an output and an updated state at each time step.

$$\left[\hat{I}(t), s(t)\right] = f_{LSTM}(I(t-1),\ s(t-1)) \tag{7}$$

$$\left[\hat{I}(t+1), s(t+1)\right] = f_{LSTM}\left(\hat{I}(t),\ s(t)\right) \tag{8}$$

The predictor iteratively makes use of f_{LSTM} starting from an initial state $s(t-d-1)$ of LSTM cell and processes the input sequence $[I(t-1),\ I(t-2),\ \ldots,\ I(t-d)]$, updating the neurons internal states in order to store all the relevant information contained in the input. The LSTM net is thus able to consider iteratively all the inputs and can directly be used to forecast several steps ahead (h) at each time t. In a sense, these recurrent networks unify the advantages of the FF recursive and multi-output approaches mentioned above. They explicitly take into account the sequential nature of the time series as the FF recursive, and are optimized on the whole predicted sequence as the FF multi-output.

The four neural predictors presented above have been developed through an extensive trial-and-error procedure implemented on an Intel i5-7500 3.40 GHz processor with a GeForce GTX 1050 Ti GPU 768 CUDA Cores. FF nets have been coded using the Python library Keras with Tensorflow as backend [48]. For LSTM networks, we used Python library PyTorch [49]. The hyperparameters of the neural nets were tuned by systematic grid search together with the number of neurons per layer, the number of hidden layers, and all the other features defining the network architecture.

2.3. Preliminary Analysis of Solar Data

The primary dataset considered for this study was recorded from 2014 to 2019 by a Davis Vantage 2 weather station installed and managed by the Politecnico di Milano, Italy, at Como Campus. The station is continuously monitored and checked for consistence as part of the dense measurement network of the Centro Meteorologico Lombardo (www.centrometeolombardo.com). Its geographic coordinates are: *Lat* = 45.80079, *Lon* = 9.08065 and *Elevation* = 215 m a.s.l. Together with the solar irradiance $I(t)$, the following physical variables are recorded every 5 min: air temperature, relative humidity, wind speed and direction, atmospheric pressure, rain, and the UV index. However, as explained in Section 2.2, the current study only adopts purely autoregressive models; namely, the forecasted values of solar irradiance $\hat{I}(t)$ are computed only based on preceding values.

A detail of the time series recorded at 00:00 each hour is shown in Figure 1a. We can interpret this time series as the sum of three different components: the astronomical condition (namely the position of the sun), that produces the evident annual cycle; the current meteorological situation (the attenuation due to atmosphere, including clouds); and the specific position of the receptor that may be shadowed by the passage of clouds in the direction of the sun. The first component is deterministically known, the second can be forecasted with a certain accuracy, while the third is much trickier and may easily vary within minutes without a clear dynamic.

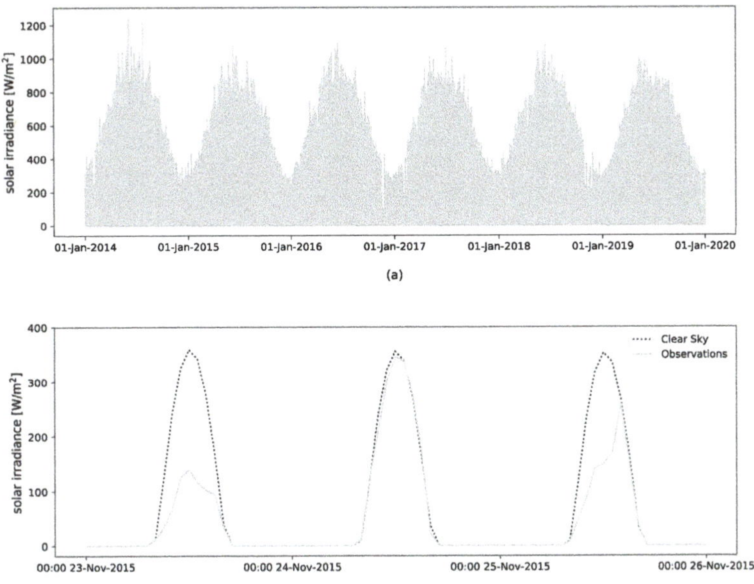

Figure 1. Hourly solar irradiance time series (**a**), compared with clear sky values for a few specific days (**b**).

The expected global solar radiation in average clear sky conditions $I_{Clsky}(t)$ (see Figure 1b) was computed by using the Ineichen and Perez model, as presented in [50] and [51]. The Python code that implements this model is part of the SNL PVLib Toolbox, provided by the Sandia National Labs PV Modeling Collaborative (PVMC) platform [52].

2.3.1. Fluctuation of Solar Radiation

Solar radiation time series, as well as other geophysical signals, belong to the class of the so-called $1/f$ noises (also known as pink noise), i.e., long-memory processes whose power density spectrum exhibit a slope, α, ranging in [0.5,1.5]. In other words, they are random processes lying between white noise processes, characterized by $\alpha = 0$, and random walks, characterized by $\alpha = 2$ (see, for instance [53]). Indeed, the slope of hourly solar irradiance recorded at Como is about $\alpha = 1.1$ (Figure 2), while the daily average time series exhibits a slope of about $\alpha = 0.6$, meaning that solar radiation at daily scale is more similar to a white process.

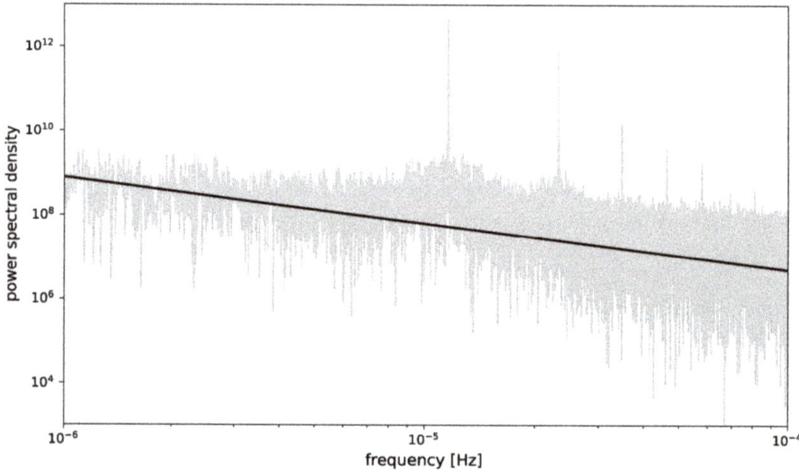

Figure 2. Power spectral density of the hourly solar irradiance (grey) and trend $1/f^\alpha$ characterized by $\alpha = 1.1$ (black).

Figure 2 also shows that the power spectral density has some peaks corresponding to periodicity of 24 hours ($1.16 \cdot 10^{-5}$ Hz) and its multiples.

2.3.2. Mutual Information

To capture the nonlinear dependence of solar irradiance time series from its preceding values, we computed the so-called mutual information $M(k)$, defined as in Equation (9) [54].

$$M(k) = -\sum_{i,j} p_{ij}(k) \cdot ln \frac{p_{ij}(k)}{p_i p_j} \tag{9}$$

In this expression, for some partition of the time series values, p_i is the probability to find a time series value in the i-th interval, and $p_{ij}(k)$ is the joint probability that an observation falls in the i-th interval, and the observation k time steps later falls into the j-th interval. The partition of the time series can be made with different criteria, for instance by dividing the range of values between the minimum and maximum in a predetermined number of intervals or by taking intervals with equal probability distribution [55]. In our case, we chose to divide the whole range of values into 16 intervals. The normalized mutual information of solar irradiance time series at Como is shown in Figure 3. In the case of Como hourly time series, it gradually decays reaching zero after about six lags. Moreover, it can be observed that the mutual information for the daily values decays more rapidly, thus confirming the greatest difficulty in forecasting solar radiation at a daily scale.

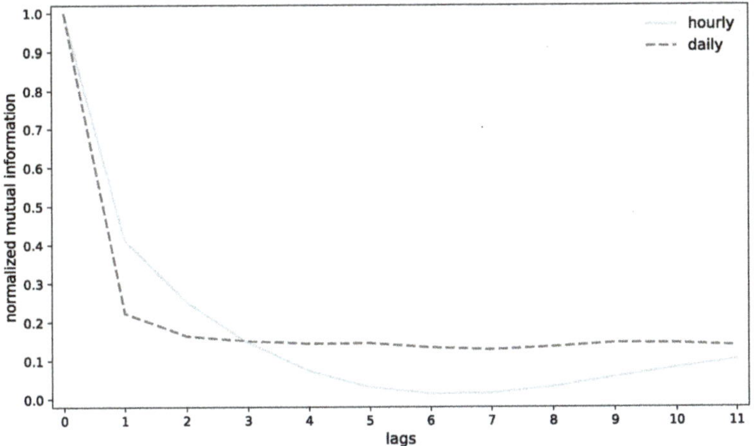

Figure 3. Normalized mutual information of hourly and daily solar irradiance at Como.

2.4. Benchmark Predictors of Hourly Solar Irradiance

Multi-step ahead forecasting of the global solar irradiance $I(t)$ at hourly scale was performed by using models of the form (1–8) defined above.

The performance of such a predictor has been compared with that of some standard baseline models. More specifically, we computed:

- The "clear sky" model, *Clsky* in the following, computed as explained in Section 2.2, which represents the average long-term cycle;
- The so-called *Pers24* model expressed as $\hat{I}(t) = I(t - 24)$, which represents the memory linked to the daily cycle;
- A classical persistent model, *Pers* in what follows, where $\hat{I}(t + k) = I(t)$, $k = 1, 2, \ldots h$, representing the component due to a very short-term memory.

2.5. Performance Assessment Metrics

The performances of the various predictors were assessed by computing the following error indices [56]: *Bias* (10), Mean Absolute Error—*MAE* (11), Root Mean Square Error—*RMSE* (12), and R^2, also known as Nash-Sutcliffe Efficiency—*NSE* (13).

$$Bias = \frac{1}{T} \sum_{t=1}^{T} \left(I(t) - \hat{I}(t) \right) \tag{10}$$

$$MAE = \frac{1}{T} \sum_{t=1}^{T} \left| I(t) - \hat{I}(t) \right| \tag{11}$$

$$RMSE = \sqrt{\frac{1}{T} \sum_{t=1}^{T} \left(I(t) - \hat{I}(t) \right)^2} \tag{12}$$

$$NSE = 1 - \frac{\sum_{t=1}^{T} \left(I(t) - \hat{I}(t) \right)^2}{\sum_{t=1}^{T} \left(I(t) - \bar{I} \right)^2} \tag{13}$$

where T is the length of the time series, while $\bar{I}$ is the average of the observed data.

As concerning to the NSE, an index originally developed for evaluating hydrological models [57], it is worth bearing in mind that it can range from $-\infty$ to 1. An efficiency of 1 ($NSE = 1$) means that the model perfectly interprets the observed data, while an efficiency of 0 ($NSE = 0$) indicates that the model predictions are as accurate as the mean of the observed data. It is worth stressing here that, in general, a model is considered sufficiently accurate if $NSE > 0.6$.

Regardless of what performance index is considered, it is worth noticing that the above classical indicators may overestimate the actual performances of models when applied to the complete time series. When dealing with solar radiation, there is always a strong bias due to the presence of many zero values. In the case at hand, they are about 57% of the sample due to some additional shadowing of the nearby mountains and to the sensitivity of the sensors. When the recorded value is zero, also the forecast is zero (or very close) and all these small errors substantially reduce the average errors and increase the *NSE*. Additionally, forecasting the solar radiation during the night is useless, and the power network dispatcher may well turn the forecasting model off. In order to overcome this deficiency, which unfortunately is present in many works in the current literature, and allow the models' performances to be compared when they may indeed be useful, we have also computed the same indicators considering only values above 25 Wm^{-2} (*daytime* in what follows), a small value normally reached before dawn and after the sunset. These are indeed the conditions when an accurate energy forecast may turn out to be useful.

Since the *Clsky* represents what can be forecasted even without using any information about the current situation, it can be assumed as a reference, and the skill index S_f (14) can be computed to measure the improvement gained using the f_{LSTM} and the f_{FF} models:

$$S_f = 1 - \frac{RMSE_f}{RMSE_{Clsky}} \tag{14}$$

3. Results

3.1. Forecasting Perfomances

The comparison of the multi-step forecast of solar irradiance with LSTM and FF networks was performed setting the delay parameter d to 24, despite the mutual information indicated that it would be enough to assume $d = 6$. This choice is motivated by the intrinsic periodicity of solar radiation at an hourly scale [54]. We have experimentally observed that this choice gives more accurate estimates for all the models considered. This probably helps the model in taking into account the natural persistence of solar radiation (see the comments about the performance of the *Pers24* model, below). The length of the forecasting horizon h, representing the number of steps ahead we predict in the future, was varied from 1 to 12. Data from 2014 to 2017 were used for network training, 2018 for validating the architecture, and 2019 for testing.

The average performances computed on the first 3 hours of the forecasting horizon of the *Pers*, *Pers24*, and *Clsky* models for the test year 2019 are shown in Tables 1 and 2. The *Pers* predictor performance rapidly deteriorate when increasing the horizon. They are acceptable only in the short term: after 1 h, the *NSE* is equal to 0.79 (considering whole day samples) and 0.59 (daytime samples only), but after two hours *NSE* decreases to 0.54 (whole day) and 0.14 (daytime only), and six steps ahead the *NSE* becomes -0.93 (whole day) and -1.64 (daytime). The *Pers24* and *Clsky* preserve the same performances for each step ahead since they are independent on the horizon. Such models inherently take into account the presence of the daily pseudo-periodic component, which affects hourly global solar radiation.

Table 1. Average performances of the *Pers*, *Pers24*, and *Clsky* predictors on the first 3 hours (whole day).

Index	Pers	Pers24	Clsky
Bias	0.07	0.01	−46.20
MAE	87.78	59.24	74.07
RMSE	158.40	136.63	146.39
NSE	0.51	0.63	0.58

Table 2. Average performances of the *Pers*, *Pers24*, and *Clsky* predictors on the first 3 hours (daytime samples only).

Index	Pers	Pers24	Clsky
Bias	28.82	7.08	−89.91
MAE	175.89	131.71	154.35
RMSE	228.24	204.91	212.97
NSE	0.11	0.28	0.22

The *Pers24* predictor appears to be superior to the *Clsky* (lower error indicators, higher *NSE*) confirming that the information of the last 24 h is much more relevant for a correct prediction than the long-term annual cycle. Indeed, the sun position does not change much between one day and the following, and the meteorological conditions have a certain, statistically relevant tendency to persist. Additionally, the *Pers24* predictor is the only one with practically zero bias, since the small difference that appears in the first column of Table 1 is simply due to the differences between 31/12/2018 (which is used to compute the predicted values of 1/1/2019) and 31/12/2019. The clear sky model that operates by definition in the absence of cloud cover, overestimates the values above the threshold by 89.86 Wm^{-2}, on average.

From Table 2, it appears that *Pers*, *Pers24* and *Clsky*, are not reliable models, on average, especially if the *NSE* is evaluated by using daytime samples only.

Figure 4 reports the results obtained with the three different FF approaches (Figure 4a–c) and the LSTM forecasting model (Figure 4d) in terms of *NSE* (both in the whole day and in daytime samples only).

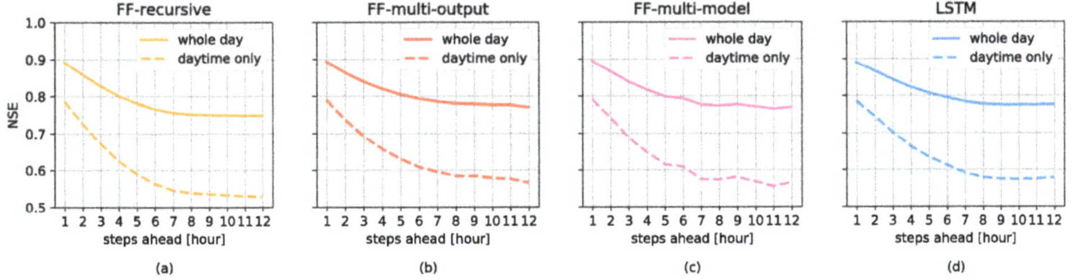

Figure 4. NSE of hourly solar irradiance at Como obtained with recursive FF (**a**), multi-output FF (**b**), multi-model FF (**c**), and LSTM (**d**). Solid line represents the performance in the whole day, dashed line in daytime only. Values refer to the test year 2019.

Generally speaking, Figure 4 shows that all the considered neural predictors exhibit an NSE which reaches an asymptotic value around six steps ahead. This is coherent with the previous analysis about the mutual information (see Figure 3), which, at an hourly scale, is almost zero after six lags.

If the evaluation is carried out considering whole day samples, all the models would have to be considered reliable enough since *NSE* is only slightly below 0.8, even for prediction horizons of 12 h. On the contrary, if the evaluation is made considering daytime samples only, it clearly appears that models are reliable for a maximum of 5 h ahead, as for higher horizons the *NSE* value typically falls below 0.6. Therefore, removing the nighttime values of the time series is decisive for a realistic assessment of a solar radiation forecasting model, that would otherwise be strongly biased.

Going in deeper details, the following considerations can be made.

The FF recursive approach performs slightly worse, particularly as measured by *NSE* and specifically after a forecasting horizon of 5 h. The FF multi-output and multi-model approaches show performances similar to the LSTM. Additionally, one can note that the performance regularly decreases with the length of the horizon for the FF recursive approach and LSTM net, since they explicitly take into account the sequential nature of the task. Conversely, the FF multi-output and the multi-model ones have irregularities, particularly the latter being each predictor for each specific time horizon completely independent on those on a shorter horizon. If perfect training was possible, such irregularities might perhaps be reduced, but they cannot be completely avoided, particularly on the test dataset, because they are inherent to the approach that considers each predicted value as a separate task.

For all the considered benchmarks and neural predictors, the difference between the whole time series (average value 140.37 Wm^{-2}) and the case with a threshold (daytime only), that excludes nighttime values (average 328.62 Wm^{-2}), emerges clearly, given that during all nights the values are zero or close to it, and thus the corresponding errors are also low.

FF nets and LSTM perform similarly also considering the indices computed along the first 3 h of the forecasting horizon as shown in Tables 3 and 4, for the whole day and daytime, respectively.

Table 3. Average performances of FF and LSTM predictors on the first 3 hours (whole day).

Index	FF-Recursive	FF-Multi-Output	FF-Multi-Model	LSTM
Bias	−1.49	−0.28	0.17	−4.19
MAE	40.26	40.39	39.26	45.91
RMSE	84.33	82.60	82.29	82.09
NSE	0.86	0.87	0.87	0.87
S	0.42	0.44	0.44	0.44

Table 4. Average performances of FF and LSTM predictors on the first 3 hours (daytime samples only).

Index	FF-Recursive	FF-Multi-Output	FF-Multi-Model	LSTM
Bias	3.79	6.59	6.46	12.11
MAE	86.31	84.63	84.75	86.01
RMSE	125.35	122.87	122.74	121.78
NSE	0.73	0.74	0.74	0.75
S	0.41	0.42	0.42	0.43

All the neural predictors provide a definite improvement in comparison to the *Pers*, *Pers24*, and *Clsky* models. Looking, for instance, at the *NSE* the best baseline predictor is the *Pers24*, scoring 0.63 (whole day) and 0.28 (daytime only). The corresponding values for the neural networks exceed 0.86 and 0.73, respectively.

An in-depth analysis should compare the neural predictors performance at each step with the best benchmark for that specific step. The latter can be considered as an ensemble of benchmarks composed by the the *Pers* model, the most performing one step ahead (*NSE* equal to 0.79), and the *Pers24* on the following steps (*NSE* equal to 0.63 for *h* from 2 to 12). Under this perspective, the neural nets clearly outperform the considered baseline since their *NSE* score varies from 0.90 to 0.75 (see the solid lines in Figure 4 referring to the whole day). The same analysis performed excluding nighttime values leads to quite similar results, confirming that the neural networks always provide a performance much higher than the benchmarks considered here.

An additional way to examine the model performances is presented in Table 5. We report here the NSE of the predictions of the LSTM network on three horizons, namely 1, 3, and 6 hours ahead.

The sunlight (i.e., above 25 W/m^2) test series is partitioned into three classes: cloudy, partly cloudy and sunny days, which constitute about 30, 30, and 40% of the sample, respectively. More precisely, cloudy days are defined as those when the daily average irradiance is below 60% of the clear sky index and sunny days those that are above 90% (remember that the clear sky index already accounts for the average sky cloudiness).

Table 5. LSTM performances in terms of NSE for cloudy, partly cloudy and sunny days (daytime samples only).

Index	1 Hour Ahead	3 Hours Ahead	6 Hours Ahead
Cloudy	0.44	0.06	−0.45
Partly cloudy	0.65	0.59	0.59
Sunny	0.89	0.83	0.73

It is quite apparent that the performance of the model decreases consistently from sunny, to partly cloudy, to cloudy days. This result is better illustrated in Figure 5 where the 3-hour-ahead predictions are shown for three typical days. In the sunny day, on the right, the process is almost deterministic (governed mainly by astronomical conditions), while the situation is completely different in a cloudy day. In the last case, the forecasting error is of the same order of the process (NSE close to zero) and it can be even larger at 6 hours ahead. This determines the negative NSE value shown in Table 5.

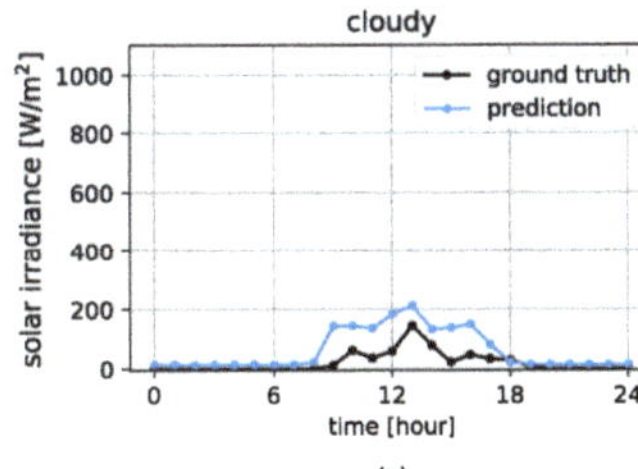

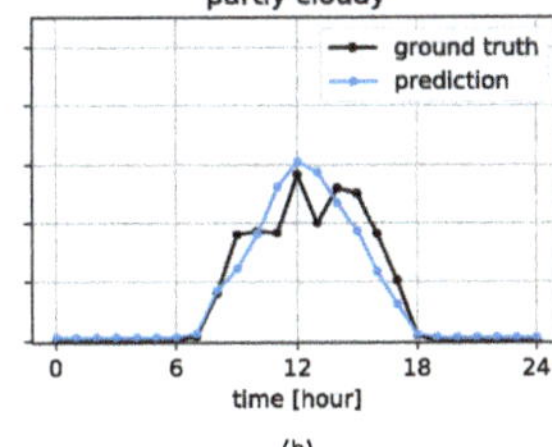

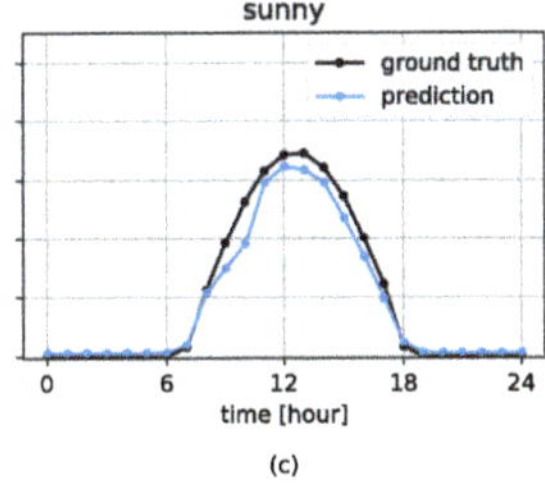

Figure 5. Three-hour-ahead LSTM predictions versus observations in three typical days with different cloudiness: a cloudy day (**a**), a partly cloudy day (**b**), and a sunny day (**c**). Values refer to the test year (2019).

3.2. Domain Adaptation

Besides the accuracy of the forecasted values, another important characteristic of the forecasting models is their generalization capability, often mentioned as domain adaptation in the neural networks literature [58]. This means the possibility of storing knowledge gained while solving one problem and applying it to different, though similar, datasets [59].

To test this feature, the FF and LSTM networks developed for the Como station (source domain) have been used, without retraining, on other sites (target domains) spanning more than one degree of latitude and representing quite different geographical settings: from the low and open plain at 35 m a.s.l. to up the mountains at 800 m a.s.l. In addition, the test year has been changed because solar radiation is far from being truly periodical and some years (e.g., 2017) show significantly higher values than others (e.g., 2011). This means quite different solar radiation encompassing a difference of about 25% between yearly average values.

Figure 6 shows the *NSE* for the multi-output FF and LSTM networks for three additional stations. All the graphs reach somehow a plateau after six steps ahead, as suggested by the mutual information computed on Como station, and the differences between FF and LSTM networks appear very small or even negligible in almost all the other stations. Six hours ahead, the difference in *NSE* between Como, for which the networks have been trained, in the test year (2019) and Bema in 2017, which appears to be the most different dataset, is only about 3% for both FF models and LSTM.

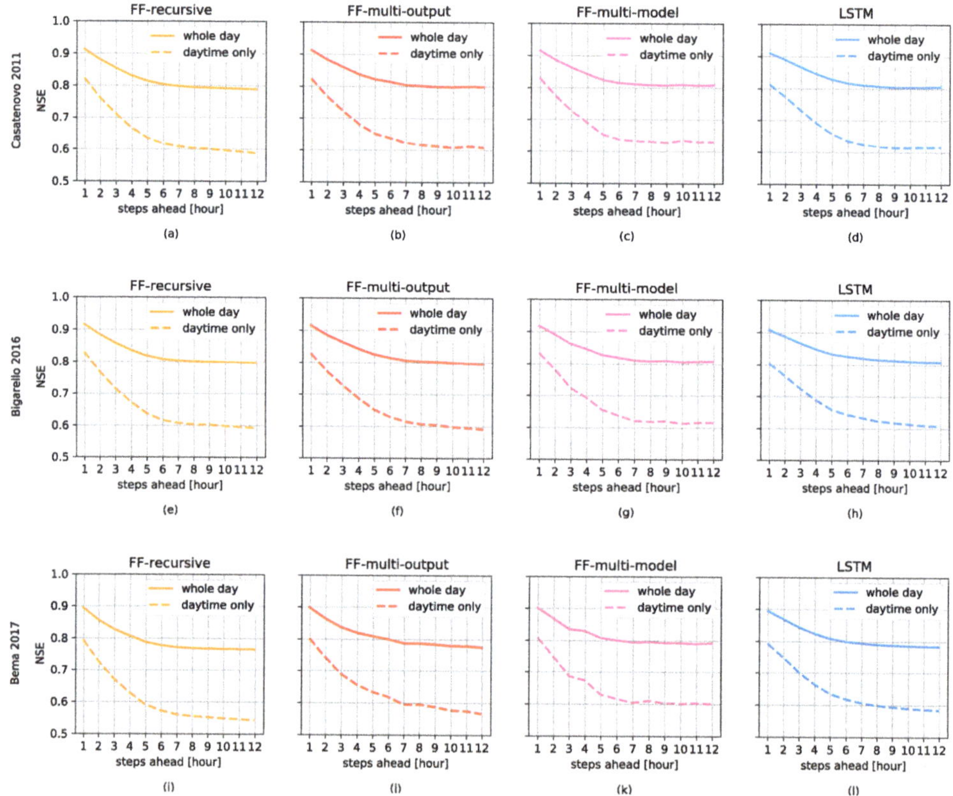

Figure 6. *NSE* of hourly solar irradiance forecast at Casatenovo (**a–d**) in 2011, Bigarello (**e–h**) in 2016, and Bema (**i–l**) in 2017. Each column is relative to a different neural predictor. Solid line represents the performance in the whole day, dashed line in daytime only.

As a further trial, both FF models and LSTM have been tested on a slightly different process, i.e., the hourly average solar radiation recorded at the Como station. While the process has the same average of the original dataset, the variability of the process is different since its standard deviation decreased of about 5%. The averaging process indeed filters the high frequencies. Forecasting results are shown in Figure 7. Additionally for this process, the neural models perform more or less as for the hourly values for which they have been trained. The accuracy of both LSTM and FF networks improves by about 0.02 (or 8%) in terms of standard *MAE* and slightly less in terms of *NSE*, in comparison to the original process. For a correct comparison with Figure 4, however, it is worth bearing in mind that the 1-hour-ahead prediction corresponds to $(t + 2)$ in the graph, since the average computed at hour $(t + 1)$ is that from (t) to $(t + 1)$ and, thus, includes values that are only 5 minutes ahead of the instant at which the prediction is formulated.

An ad-hoc training on each sequence would undoubtedly improve the performance, but the purpose of this section is exactly to show the potential of networks calibrated on different stations, to evaluate the possibility of adopting a predictor developed elsewhere when a sufficiently long series of values is missing. The forecasting models we developed for a specific site could be used with acceptable accuracy for sites were recording stations are not available or existing time series are not long enough. This suggest the possibility of developing a unique forecasting model for the entire region.

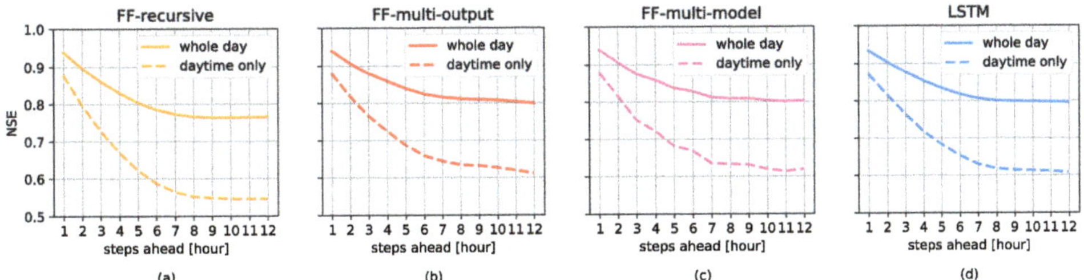

Figure 7. *NSE* of hourly average solar irradiance at Como obtained with recursive FF (**a**), multi-output FF (**b**), multi-model FF (**c**), and LSTM (**d**). Solid line represents the performance in the whole day, dashed line in daytime only. Values refer to the test year (2019).

4. Some Remarks on Network Implementations

The development of many successful deep learning models in various applications has been made possible by three joint factors. First, the availability of big data, which are necessary to identify complex models characterized by thousands of parameters. Second, the intuition of making use of fast parallel processing units (GPUs) able to deal with the high computational effort required. Third, the availability of an efficient gradient-based method to train these kinds of neural networks.

This latter are the well-known backpropagation (BP) techniques, which allow efficient computing of the gradients of the loss function with respect to each model weight and bias. To apply the backpropagation of the gradient, it is necessary to have a feed-forward architecture (i.e., without self-loops). In this case, the optimization is extremely efficient since the process can be entirely parallelized, exploiting the GPU.

When the neural architecture presents some loops, as it happens in recurrent cells, the BP technique has to be slightly modified in order fit the new situation. This can be done by unfolding the neural networks through time, to remove self-loops. This extension of BP technique is known as backpropagation through time (BPTT) in the machine learning literature. The issue with BPTT is that the unfolding process should in principle lasts for an infinite number of steps, making the technique useless for practical purposes. For this reason, it is necessary to limit the number of unfolding, considering only the time steps that contain useful information for the prediction (in this case, we say that the BPTT is truncated). As it is easy to understand, BPTT is not as efficient as the traditional BP, because it is not possible to fully parallelize the computation. As a consequence, we are not able to fully exploit the GPU's computing power, resulting in a slower training. The presence of recurrent units also produces much complex optimization problems due to the presence of a significant number of local optima [60].

Figure 8 shows the substantial difference in the evolution of the training process. As usual, the mean of the quadratic errors of the FF network slowly decreases toward a minimum while this function shows sudden jumps followed by several epochs of stationarity in the case of LSTM. The training algorithm of LSTM can avoid being trapped for too many epochs into local minima, but on the other side, these local minima are more frequent.

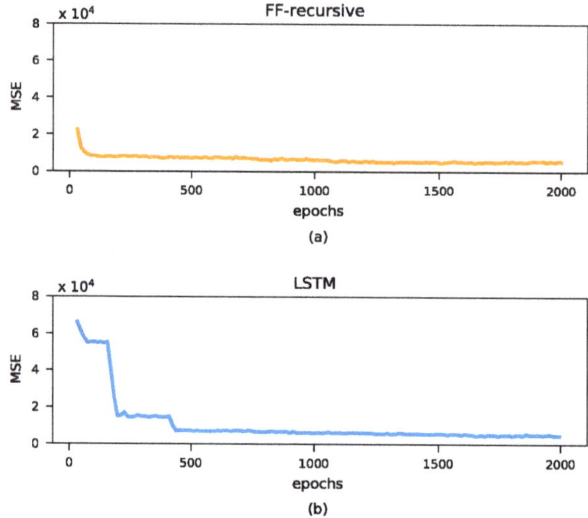

Figure 8. Evolution of the *MSE* across the training epochs for recursive FF (**a**) and LSTM (**b**) predictors.

For the sake of comparison, we trained the four neural architectures using the same hyperparameters grid, considering the ranges reported in Table 6. As training algorithm, we used the Adam optimizer. Each training procedure has been repeated three times to avoid the problems of an unlucky weights initialization.

Table 6. Hyperparameters values considered in the grid-search tuning process for FF and LSTM predictors.

Hyperparameter	Search Range	Optimal Values			
		FF-Recursive	FF-Multi-Output	FF-Multi-Model	LSTM
Hidden layers	3–5	3	5	5	3
Neurons per layer	5–10	5	10	10	5
Learning rate	10^{-2}–10^{-3}	10^{-3}	10^{-2}	10^{-2}	10^{-3}
Decay rate	0–10^{-4}	0	10^{-4}	10^{-4}	10^{-4}
Batch size	128–512	512	128	512	512

While performing similarly under many viewpoints, the FF and LSTM architectures show significant differences if we consider the sensitivity to the hyperparameters values. Figure 9 shows the sensitivity bands obtained for each architecture. The upper bound represents, for each step ahead, the best *NSE* score achieved across the hyperparameters combinations. The cases in which the optimization process fails due to a strongly inefficient initialization of the weights have been excluded.

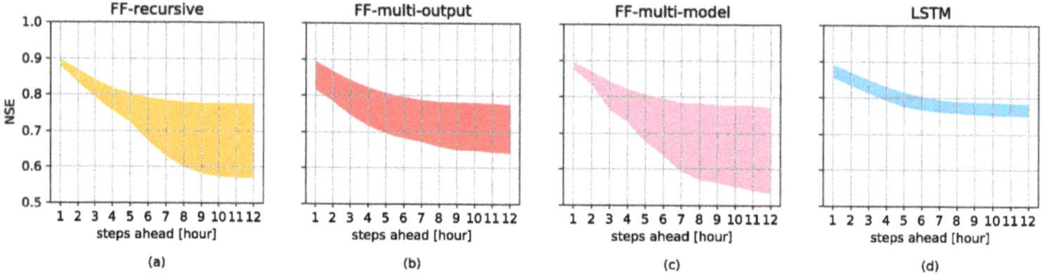

Figure 9. Sensitivity bands obtained with FF-recursive (**a**), FF-multi-output (**b**), FF-multi-model (**c**), and LSTM (**d**) on the 12-step-ahead prediction. Values refer to the test year (2019).

The variability of performances of LSTM across the hyperparameter space is quite limited if compared with that of FF architectures. This is probably because the LSTM presents a sequential structure and is optimized on the whole 12-step-forecasting horizon. The recursive FF model is identified as the optimal one-step ahead predictor, and thus there are some cases characterized by poor performances in the rest of the horizon. The multi-output and multi-model FF seem to suffer of the same problem because, as already pointed out, they predict the values at each time step as independent variables.

5. Conclusions

The availability of accurate multi-step ahead forecast of solar irradiance (and, hence, power) is of extreme importance for the efficient balancing and management of power networks since they allow the implementation of accurate and efficient control procedures, such as model predictive control.

The results reported in this study further confirm the well-known accuracy of FF and LSTM networks for the above purpose and, more in general, in predicting time series related to environmental variables. Another interesting conclusion is that among the *Rec*, *MM* and *MO* models, the *Rec* is the one that exhibits the lowest performances. A rough explanation probably lies in the fact that its parameters are optimized over a time horizon of 1 step, but then artificially used for a longer horizon, thus propagating the error. Therefore one of the merits of this study is to clarify that a common practice, namely to identify a model of the type $x(t+1) = f(x(t))$, to then predict $x(t+h)$ is not the best choice. To this end, the proposed *MM* and *MO* may represent alternatives that are more appropriate.

However, such good performances are obtained at a cost in terms of data and time required to train the network. In actual applications, one has to trade these costs versus the improvement in the precision of the forecasting. In this respect, the solution which appears to be a trade-off between accuracy and complexity seems to be the *MO*. Indeed, it can reach a very good performance with a minimum increment of the number of parameters compared to the recursive approach. The *MM* approach performs slightly better than the *MO* one but requires a different training (and possibly, a different architecture) for each forecasting horizon, which, in the present study, means training over a million parameters.

In more general terms, the selection of the forecasting model should be made by looking at the comparative advantages that a better precision provides versus the effort to obtain such a precision. Though the economic cost and the computation time required to synthesize even a very complex LSTM network are already rather low and still decreasing, one may also consider adopting a classical FF neural network model, which outperforms the traditional *Pers24* model and is much easier to train with respect to the corresponding LSTM.

Another of the peculiarities of this work has been showing how performance indices are strongly affected by the presence of null values and, in this respect, nighttime samples should be removed for a correct assessment of model performances.

Both FF and LSTM networks developed in this study have proved to be able to forecast solar radiation in other stations of a relatively wide domain with a minimal loss of precision and without the need to retrain them. This opens the way to the development of county or regional predictors, valid for ungauged locations with different geographical settings. Such precision may perhaps be further improved by using other meteorological data as input to the model, thus extending the purely autoregressive approach adopted in this study.

Author Contributions: Conceptualization, G.G., G.N. and M.S.; methodology, G.G., G.N. and M.S.; software, M.S.; validation, G.G., G.N. and M.S.; data curation, G.G., G.N. and M.S.; writing—original draft preparation, G.G., G.N. and M.S.; writing—review and editing, G.G., G.N. and M.S.; visualization, M.S.; supervision, G.G. and G.N. All authors have read and agreed to the published version of the manuscript.

Funding: This research received no external funding.

Conflicts of Interest: The authors declare no conflict of interest.

References

1. Shah, A.S.B.M.; Yokoyama, H.; Kakimoto, N. High-precision forecasting model of solar irradiance based on grid point value data analysis for an efficient photovoltaic system. *IEEE Trans. Sustain. Energy* **2015**, *6*, 474–481. [CrossRef]
2. Urrego-Ortiz, J.; Martínez, J.A.; Arias, P.A.; Jaramillo-Duque, Á. Assessment and Day-Ahead Forecasting of Hourly Solar Radiation in Medellín, Colombia. *Energies* **2019**, *12*, 4402. [CrossRef]
3. Mpfumali, P.; Sigauke, C.; Bere, A.; Mulaudzi, S. Day Ahead Hourly Global Horizontal Irradiance Forecasting—Application to South African Data. *Energies* **2019**, *12*, 3569. [CrossRef]
4. Mohamed, A.; Salehi, V.; Ma, T.; Mohammed, O. Real-time energy management algorithm for plug-in hybrid electric vehicle charging parks involving sustainable energy. *IEEE Trans. Sustain. Energy* **2014**, *5*, 577–586. [CrossRef]
5. Bhatti, A.R.; Salam, Z.; Aziz, M.J.B.A.; Yee, K.P.; Ashique, R.H. Electric vehicles charging using photovoltaic: Status and technological review. *Renew. Sustain. Energy Rev.* **2016**, *54*, 34–47. [CrossRef]
6. Sáez, D.; Ávila, F.; Olivares, D.; Cañizares, C.; Marín, L. Fuzzy prediction interval models for forecasting renewable resources and loads in microgrids. *IEEE Trans. Smart Grid* **2015**, *6*, 548–556.
7. Yadav, A.K.; Chandel, S.S. Solar radiation prediction using Artificial Neural Network techniques: A review. *Renew. Sustain. Energy Rev.* **2014**, 772–781. [CrossRef]
8. Wan, C.; Zhao, J.; Song, Y.; Xu, Z.; Lin, J.; Hu, Z. Photovoltaic and solar power forecasting for smart grid energy management. *CSEE J. Power Energy Syst.* **2015**, *1*, 38–46. [CrossRef]
9. Voyant, C.; Notton, G.; Kalogirou, S.; Nivet, M.; Paoli, C.; Motte, F.; Fouilloy, A. Machine learning methods for solar radiation forecasting: A review. *Renew. Energy* **2017**, *105*, 569–582. [CrossRef]
10. Sobri, S.; Koohi-Kamali, S.; Rahim, N.A. Solar photovoltaic generation forecasting methods: A review. *Energy Convers. Manag.* **2018**, *156*, 459–497. [CrossRef]
11. Sun, H.; Yan, D.; Zhao, N.; Zhou, J. Empirical investigation on modeling solar radiation series with ARMA–GARCH models. *Energy Convers. Manag.* **2015**, *92*, 385–395. [CrossRef]
12. Lauret, P.; Boland, J.; Ridley, B. Bayesian statistical analysis applied to solar radiation modelling. *Renew. Energy* **2013**, *49*, 124–127. [CrossRef]
13. Kwon, Y.; Kwasinski, A.; Kwasinski, A. Solar Irradiance Forecast Using Naïve Bayes Classifier Based on Publicly Available Weather Forecasting Variables. *Energies* **2019**, *12*, 1529. [CrossRef]
14. Piri, J.; Kisi, O. Modelling solar radiation reached to the earth using ANFIS, NN-ARX, and empirical models (case studies: Zahedan and Bojnurd stations). *J. Atmos. Solar Terr. Phys.* **2015**, *123*, 39–47. [CrossRef]
15. Gholipour, A.; Lucas, C.; Araabi, B.N.; Mirmomeni, M.; Shafiee, M. Extracting the main patterns of natural time series for long-term neurofuzzy prediction. *Neural Comput. Appl.* **2007**, *16*, 383–393. [CrossRef]
16. Dawan, P.; Sriprapha, K.; Kittisontirak, S.; Boonraksa, T.; Junhuathon, N.; Titiroongruang, W.; Niemcharoen, S. Comparison of Power Output Forecasting on the Photovoltaic System Using Adaptive Neuro-Fuzzy Inference Systems and Particle Swarm Optimization-Artificial Neural Network Model. *Energies* **2020**, *13*, 351. [CrossRef]
17. Kisi, O. Modeling solar radiation of Mediterranean region in Turkey by using fuzzy genetic approach. *Energy* **2014**, *64*, 429–436. [CrossRef]
18. Shakya, A.; Michael, S.; Saunders, C.; Armstrong, D.; Pandey, P.; Chalise, S.; Tonkoski, R. Solar irradiance forecasting in remote microgrids using Markov switching model. *IEEE Trans. Sustain. Energy* **2017**, *8*, 895–905. [CrossRef]
19. Sharma, N.; Sharma, P.; Irwin, D.; Shenoy, P. Predicting solar generation from weather forecasts using machine learning. In Proceedings of the IEEE International Conference on Smart Grid Communications (SmartGridComm), Brussels, Belgium, 17–20 October 2011; pp. 528–533.
20. Hassan, M.Z.; Ali, M.E.K.; Ali, A.S.; Kumar, J. Forecasting Day-Ahead Solar Radiation Using Machine Learning Approach. In Proceedings of the 4th Asia-Pacific World Congress on Computer Science and Engineering (APWC on CSE), Mana Island, Fiji, 11–13 December 2017; pp. 252–258.
21. Bae, K.Y.; Jang, H.S.; Sung, D.K. Hourly solar irradiance prediction based on support vector machine and its error analysis. *IEEE Trans. Power Syst.* **2017**, *32*, 935–945. [CrossRef]

22. Capizzi, G.; Napoli, C.; Bonanno, F. Innovative second-generation wavelets construction with recurrent neural networks for solar radiation forecasting. *IEEE Trans. Neural Netw. Learn. Syst.* **2012**, *23*, 1805–1815. [CrossRef]

23. Lotfi, M.; Javadi, M.; Osório, G.J.; Monteiro, C.; Catalão, J.P.S. A Novel Ensemble Algorithm for Solar Power Forecasting Based on Kernel Density Estimation. *Energies* **2020**, *13*, 216. [CrossRef]

24. McCandless, T.; Dettling, S.; Haupt, S.E. Comparison of Implicit vs. Explicit Regime Identification in Machine Learning Methods for Solar Irradiance Prediction. *Energies* **2020**, *13*, 689. [CrossRef]

25. Crisosto, C.; Hofmann, M.; Mubarak, R.; Seckmeyer, G. One-Hour Prediction of the Global Solar Irradiance from All-Sky Images Using Artificial Neural Networks. *Energies* **2018**, *11*, 2906. [CrossRef]

26. Ji, W.; Chee, K.C. Prediction of hourly solar radiation using a novel hybrid model of ARMA and TDNN. *Solar Energy* **2011**, *85*, 808–817. [CrossRef]

27. Zhang, N.; Behera, P.K. Solar radiation prediction based on recurrent neural networks trained by Levenberg-Marquardt backpropagation learning algorithm. In Proceedings of the 2012 IEEE PES Innovative Smart Grid Technologies (ISGT), Washington, DC, USA, 16–20 January 2012; pp. 1–7.

28. Zhang, N.; Behera, P.K.; Williams, C. Solar radiation prediction based on particle swarm optimization and evolutionary algorithm using recurrent neural networks. In Proceedings of the IEEE International Systems Conference (SysCon), Orlando, FL, USA, 15–18 April 2013; pp. 280–286.

29. Jaihuni, M.; Basak, J.K.; Khan, F.; Okyere, F.G.; Arulmozhi, E.; Bhujel, A.; Park, J.; Hyun, L.D.; Kim, H.T. A Partially Amended Hybrid Bi-GRU—ARIMA Model (PAHM) for Predicting Solar Irradiance in Short and Very-Short Terms. *Energies* **2020**, *13*, 435. [CrossRef]

30. Zang, H.; Cheng, L.; Ding, T.; Cheung, K.W.; Liang, Z.; Wei, Z.; Sun, G. Hybrid method for short-term photovoltaic power forecasting based on deep convolutional neural network. *IET Generation Transm. Distrib.* **2018**, *12*, 4557–4567. [CrossRef]

31. Yona, A.; Senjyu, T.; Funabashi, T.; Kim, C. Determination method of insolation prediction with fuzzy and applying neural network for long-term ahead PV power output correction. *IEEE Trans. Sustain. Energy* **2013**, *4*, 527–533. [CrossRef]

32. Yang, H.; Huang, C.; Huang, Y.; Pai, Y. A weather-based hybrid method for 1-day ahead hourly forecasting of PV power output. *IEEE Trans. Sustain. Energy* **2014**, *5*, 917–926. [CrossRef]

33. Liu, J.; Fang, W.; Zhang, X.; Yang, C. An improved photovoltaic power forecasting model with the assistance of aerosol index data. *IEEE Trans. Sustain. Energy* **2015**, *6*, 434–442. [CrossRef]

34. Ehsan, R.M.; Simon, S.P.; Venkateswaran, P.R. Day-ahead forecasting of solar photovoltaic output power using multilayer perceptron. *Neural Comput. Appl.* **2017**, *28.12*, 3981–3992. [CrossRef]

35. Wojtkiewicz, J.; Hosseini, M.; Gottumukkala, R.; Chambers, T.L. Hour-Ahead Solar Irradiance Forecasting Using Multivariate Gated Recurrent Units. *Energies* **2019**, *12*, 4055. [CrossRef]

36. Abdel-Nasser, M.; Mahmoud, K. Accurate photovoltaic power forecasting models using deep LSTM-RNN. *Neural Comput. Appl.* **2019**, *31*, 2727–2740. [CrossRef]

37. Lai, G.; Chang, W.; Yang, Y.; Liu, H. Modeling long-and short-term temporal patterns with deep neural networks. In Proceedings of the 41st International Conference on Research and Development in Information Retrieval, Ann Arbor, MI, USA, 8–12 July 2018; pp. 95–104.

38. Maitanova, N.; Telle, J.-S.; Hanke, B.; Grottke, M.; Schmidt, T.; Maydell, K.; Agert, C. A Machine Learning Approach to Low-Cost Photovoltaic Power Prediction Based on Publicly Available Weather Reports. *Energies* **2020**, *13*, 735. [CrossRef]

39. Fouilloy, A.; Voyant, C.; Notton, G.; Motte, F.; Paoli, C.; Nivet, M.L.; Guillot, E.; Duchaud, J.L. Solar irradiation prediction with machine learning: Forecasting models selection method depending on weather variability. *Energy* **2020**, *165*, 620–629. [CrossRef]

40. Notton, G.; Voyant, C.; Fouilloy, A.; Duchaud, J.L.; Nivet, M.L. Some applications of ANN to solar radiation estimation and forecasting for energy applications. *Appl. Sci.* **2019**, *9*, 209. [CrossRef]

41. Husein, M.; Chung, I.-Y. Day-Ahead Solar Irradiance Forecasting for Microgrids Using a Long Short-Term Memory Recurrent Neural Network: A Deep Learning Approach. *Energies* **2019**, *12*, 1856. [CrossRef]

42. Aslam, M.; Lee, J.-M.; Kim, H.-S.; Lee, S.-J.; Hong, S. Deep Learning Models for Long-Term Solar Radiation Forecasting Considering Microgrid Installation: A Comparative Study. *Energies* **2020**, *13*, 147. [CrossRef]

43. Dercole, F.; Sangiorgio, M.; Schmirander, Y. An empirical assessment of the universality of ANNs to predict oscillatory time series. *IFAC-PapersOnLine* **2020**, in press.

44. Sangiorgio, M.; Dercole, F. Robustness of LSTM Neural Networks for Multi-step Forecasting of Chaotic Time Series. *Chaos Solitons Fractals* **2020**, *139*, 110045. [CrossRef]
45. Hochreiter, S.; Schmidhuber, J. Long short-term memory. *Neural Comput.* **1997**, *9*, 1735–1780. [CrossRef]
46. Schmidhuber, J. Deep learning in neural networks: An overview. *Neural Netw.* **2015**, *61*, 85–117. [CrossRef] [PubMed]
47. Goodfellow, I.; Bengio, Y.; Courville, A. *Deep Learning*; MIT Press: Cambridge, MA, USA, 2016.
48. Chollet, F. Keras: The python deep learning library. Astrophysics Source Code Library. Available online: https://keras.io (accessed on 25 June 2020).
49. Paszke, A.; Gross, S.; Massa, F.; Lerer, A.; Bradbury, J.; Chanan, G.; Killeen, T.; Lin, Z.; Gimelshein, N.; Antiga, L.; et al. PyTorch: An imperative style, high-performance deep learning library. *Adv. Neural Inf. Process. Syst.* **2019**, 8024–8035.
50. Ineichen, P.; Perez, R. A new airmass independent formulation for the Linke turbidity coefficient. *Solar Energy* **2012**, *73*, 151–157. [CrossRef]
51. Perez, R.; Ineichen, P.; Moore, K.; Kmiecik, M.; Chain, C.; George, R.; Vignola, F. A new operational model for satellite-derived irradiances: Description and validation. *Solar Energy* **2002**, *73*, 307–317. [CrossRef]
52. PV Performance Modeling Collaborative. Available online: https://pvpmc.sandia.gov (accessed on 2 May 2020).
53. Miramontes, O.; Rohani, P. Estimating 1/fα scaling exponents from short time-series. *Phys. D Nonlinear Phenom.* **2002**, *166*, 147–154. [CrossRef]
54. Fortuna, L.; Nunnari, G.; Nunnari, S. *Nonlinear Modeling of Solar Radiation and Wind Speed Time Series*; Springer International Publishing: Cham, Switzerland, 2016.
55. Jiang, A.H.; Huang, X.C.; Zhang, Z.H.; Li, J.; Zhang, Z.Y.; Hua, H.X. Mutual information algorithms. *Mech. Syst. Signal Process.* **2010**, *24*, 2947–2960. [CrossRef]
56. Zhang, J.; Florita, A.; Hodge, B.M.; Lu, S.; Hamann, H.F.; Banunarayanan, V.; Brockway, A.M. A suite of metrics for assessing the performance of solar power forecasting. *Solar Energy* **2015**, *111*, 157–175. [CrossRef]
57. Nash, J.E.; Sutcliffe, J.V. River flow forecasting through conceptual models part I—A discussion of principles. *J. Hydrol.* **1970**, *10*, 282–290. [CrossRef]
58. Glorot, X.; Bordes, A.; Bengio, Y. Domain adaptation for large-scale sentiment classification: A deep learning approach. In Proceedings of the 28th international conference on machine learning (ICML), Bellevue, DC, USA, 28 June–2 July 2011; pp. 513–520.
59. Yosinski, J.; Clune, J.; Bengio, Y.; Lipson, H. How transferable are features in deep neural networks? *Adv. Neural Inf. Process. Syst.* **2014**, 3320–3328.
60. Cuéllar, M.P.; Delgado, M.; Pegalajar, M.C. An application of non-linear programming to train recurrent neural networks in time series prediction problems. In Proceedings of the 7th International Conference on Enterprise Information Systems (ICEIS), Miami, FL, USA, 24–28 May 2005; pp. 95–102.

Article

A Machine Learning Approach to Low-Cost Photovoltaic Power Prediction Based on Publicly Available Weather Reports

Nailya Maitanova [1,*], Jan-Simon Telle [1], Benedikt Hanke [1], Matthias Grottke [2], Thomas Schmidt [1], Karsten von Maydell [1] and Carsten Agert [1]

1 DLR Institute of Networked Energy Systems, Carl-von-Ossietzky-Str. 15, 26129 Oldenburg, Germany; Jan-Simon.Telle@dlr.de (J.-S.T.); Benedikt.Hanke@dlr.de (B.H.); Th.Schmidt@dlr.de (T.S.); Karsten.Maydell@dlr.de (K.v.M.); Carsten.Agert@dlr.de (C.A.)
2 Hammer Real GmbH, Sylvensteinstr. 2, 81369 Munich, Germany; Matthias.Grottke@hammer.ag
* Correspondence: Nailya.Maitanova@dlr.de

Received: 20 December 2019; Accepted: 28 January 2020; Published: 7 February 2020

Abstract: A fully automated transferable predictive approach was developed to predict photovoltaic (PV) power output for a forecasting horizon of 24 h. The prediction of PV power output was made with the help of a long short-term memory machine learning algorithm. The main challenge of the approach was using (1) publicly available weather reports without solar irradiance values and (2) measured PV power without any technical information about the PV system. Using this input data, the developed model can predict the power output of the investigated PV systems with adequate accuracy. The lowest seasonal mean absolute scaled error of the prediction was reached by maximum size of the training set. Transferability of the developed approach was proven by making predictions of the PV power for warm and cold periods and for two different PV systems located in Oldenburg and Munich, Germany. The PV power prediction made with publicly available weather data was compared to the predictions made with fee-based solar irradiance data. The usage of the solar irradiance data led to more accurate predictions even with a much smaller training set. Although the model with publicly available weather data needed greater training sets, it could still make adequate predictions.

Keywords: photovoltaic power prediction; publicly available weather reports; machine learning; long short-term memory; integrated energy systems; smart energy management

1. Introduction

The building sector consumed one-third of global final energy use in 2016. About 80% of this final energy consumption was supplied by fossil fuels. The combustion of this fossil fuel amount caused 28% of global energy-related CO_2 emissions, which represent one of the main reasons for the greenhouse effect in the atmosphere and global warming. A plan to limit global warming is described in the Paris Agreement, which entered into force on 4 November 2016. The first point of this agreement defines that the increase in global average temperature should not exceed 2 °C in comparison to the pre-industrial level [1].

Many countries included the aims of the Paris Agreement in their National Climate Action Plans, which, among other things, outline the national policies for reduction of the greenhouse gas emissions in the building sector through to 2050. The Climate Action Plan of Germany aims to make the building stock virtually climate neutral by reducing the primary energy demand of buildings by at least 80% compared to 2008 levels by 2050. This ambitious goal can be achieved by a combination of increasing efficiency, using renewable energy, and sector coupling with the energy and transport sectors [2].

These solutions, particularly the coupling between different sectors, face big challenges. One of these challenges is the development and integration of energy management systems for the buildings. Such smart energy management system can have a significant impact, especially on the energy consumption of non-residential commercial buildings, because this type of building has an advantage of simultaneity between load behavior and locally generated energy from photovoltaic (PV) systems [3,4]. However, the fluctuating electricity generation from PV systems challenges the energy management system to use a prediction of PV power output. This type of power prediction leads to increasing the self-consumption, avoiding higher grid fees, and efficient controlling of temporal coincidence between integrated energy systems such as PV systems, battery electric vehicles (BEV), heat pumps, and other flexible loads.

As a continuation of a previous study Hanke et al. [5], a predictive model based on a machine learning approach was developed within this study to forecast the PV power output for the next 24 h. This model had the same conditions as before [5]; the input dataset for the predictive approach consists of only measured power values of PV systems and free publicly available weather reports. The prediction of PV power is made without any technical information about the PV system (e.g., without installed capacity, efficiency, inclination of the PV modules, etc.). Wang et al. [1] used also only historical PV power output and numerical weather prediction for the short-term forecasting of PV power. One of the main differences between the study [6] and this study is the type of forecasting method, whereby Wang et al. [6] proposed interval forecasting of PV power with lower and upper boundaries, while the predictive model of this study makes deterministic point forecasting.

The study [5] and this study are not the first surveys, which used publicly available data for making predictions. For example, Kwon et al [7] used publicly available weather data (temperature, humidity, dew point, and sky coverage) to make predictions of global horizontal irradiance (GHI), which can theoretically be used for PV power prediction. The given study made direct predictions of PV power using publicly available weather reports as input data. However, these reports do not provide measurements and predictions of GHI. For this reason, the input dataset was extended with an additional descriptive feature. Two possible additional features were investigated within this study: PV power under clear-sky conditions and maximum PV power, calculated from power measurements of the last five days.

The requirement to use free publicly available weather data can be explained by the assumption that most commercial building integrated and grid-connected PV systems do not have any profitable business model. These systems do not generate enough revenue from feed-in tariffs and self-consumption. Without enough revenue, the buildings with PV systems cannot have an energy management system based on cost-intensive forecast data. Using publicly available weather reports is necessary to ensure that energy management systems with an integrated developed predictive approach can operate at a low-cost level.

In addition to using publicly available weather reports, the predictive model should also satisfy other requirements; it should operate fully automatically, be continuously learning, be transferable to other PV systems, and adapt to changes in weather conditions and the PV system, such as degradation of the solar modules.

With regard to these requirements, different studies were investigated in order to find an appropriate predictive approach. In recent years, more and more machine learning algorithms were developed and applied for time series predictions, as References [8–11] showed in their reviews about PV power forecast techniques. Mohammed et al. [12] investigated different machine learning models for forecasting PV power for the next 24 h: seven individual machine learning methods (k-nearest neighbors, decision tree, gradient boosting, etc.) and three ensemble models. Both individual and ensemble models were compared to the benchmark models (autoregressive integrated moving average model, naïve, and seasonal naïve). Das et al. [13] used support vector regression to forecast PV power, and they compared the prediction to a persistence and conventional artificial neural network (ANN). In particular, ANNs are used widely for PV power prediction. Rosato et al. [14] proposed three predictive approaches based on ANN, which were used to predict power output for a large-scale PV plant in Italy. An ANN model was also used by Khandakar et al. [15] to predict the PV power output in Qatar, but the authors of this study additionally investigated two different feature selection techniques. In References [16–18], it was discussed that the application of long short-term memory (LSTM) neural networks provides particularly good results in PV power forecasting.

Taking into account the conducted literature research and the defined requirements for the predictive model, it was not suitable for this study to use classical splitting of the dataset into training and test sets and training the model only once. The dataset with the measured values used in this study is regularly updated with current weather data and PV measurements. Therefore, a re-training of the model with new data occurs at regular time intervals (every 12 h). Together with the current data, the weather forecast is also updated regularly (every 3 h). After each update of the weather forecast the developed model makes updated predictions of the PV power output for the next 24 h. These procedures can ensure continuous learning of the predictive model and addressing concept drift, i.e., adaptation to possible changes in data, completely new data, and new relationships between input and output.

The developed predictive approach was tested for two PV systems, which have different installed capacity, age, solar cell types, and location. This test checked whether the predictive approach can be transferred to different PV systems, despite having different individual parameters. In addition to this, PV power was predicted for two seasons of the year with different levels of GHI, in order to investigate how seasonal fluctuations of GHI can influence the accuracy of PV power prediction.

After making predictions of PV power for two seasons using two different PV systems, it was necessary to evaluate the quality of the developed predictive approach. For this purpose, the developed predictive approach was compared to a benchmark model (seasonal naïve forecast). This comparison analysis was implemented with the help of seasonal mean absolute scaled error (MASE). Furthermore, the PV power prediction of the model, based on publicly available weather data, was compared to predictions made with fee-based solar irradiance data.

At the end of Section 1, it is important to underline the novelty of this study and its contribution to the knowledge of PV power prediction. The "low-cost" predictive approach, which was developed within this study, can predict PV power output without any knowledge of the PV system, i.e., no technical information of the PV system is required. Only measured PV power values and publicly available weather reports were used as input data for the predictive approach. These publicly available weather reports do not have any values of the GHI; thus, the PV power prediction was made without the prediction of GHI. Moreover, all conducted simulations proved that the developed "low-cost" approach can operate fully automatically, can predict the power output of different PV systems, and can adapt for different seasons of the year.

2. Data

In this section, the origin, main characteristics, and quality of input data are explained. In supervised machine learning, the input data are divided into descriptive and target features [19]. The descriptive features of this study included historical weather measurements and numerical weather predictions for two different locations in Germany: Oldenburg and Munich. These features were used for prediction of PV power output, which was defined as a target feature.

The datasets for Oldenburg and Munich included weather data and measured PV power for different observation periods. The observation period for Oldenburg lasted from 5 May 2017 until 10 April 2018, and the observation period for Munich lasted from 5 March 2019 until 30 June 2019. These datasets were explored extensively to determine possible data quality issues and to estimate correlation between descriptive and target features for feature selection.

2.1. Photovoltaic Power Output

The origins of the PV power measurements used in this study are roof-top PV systems. The first system is in operation at the DLR Institute of Networked Energy Systems in Oldenburg since November 2010. Another system was newly installed on a commercial building in Munich and it generates electricity since November 2018.

The investigated PV systems have not only different locations, but also different installed capacities, solar cell types, etc. The main technical characteristics of these two systems are presented in Table 1.

Table 1. Main technical characteristics of the investigated photovoltaic (PV) systems.

Location	Oldenburg	Munich
Installation year	2010	2018
Total capacity	1.14 kW$_P$	99.9 kW$_P$
Orientation	237°	177.5°
Inclination	7°	10°
Solar cell type	a-Si	mono-Si
Nominal cell efficiency at standard test conditions	6.6%	17.9%

From 5 May 2017 until 10 April 2018, the PV system in Oldenburg generated 675.63 kWh of electricity. The measured values of the PV system in Munich are available for the DLR institute since 5 March 2019 and, from this time until 30 June 2019, it produced 50,740.14 kWh of energy.

The technical characteristics of the PV systems were not considered in the predictive model and they are given here only for better understanding of the prediction results. Only measured values of PV power output were used in the prediction model. These measured values for both PV systems were recorded with a time resolution of 5 min.

2.2. Weather Data

The requirements for the predictive model defined that input data should be open and publicly available, in order to ensure low-cost operation of the approach. Therefore, the input weather data for the predictive model were taken from an online service "OpenWeatherMap" (OWM) of a company Openweather Ltd. The main activity profile of this company includes providing current weather data, historical weather data, and weather forecasts of different locations to developers who want to present meteorological data on their homepages, mobile applications, or other web services. If the developers make fewer than 60 calls per minute, the current weather and five-day weather forecasts are available for free. The data of OWM are licensed by the Open Data Commons Open Database License (ODbL). Because the data from OWM are available to the public, the data are here referred to as "publicly available weather reports" [20].

OWM provides current and forecast values of various weather parameters, including ambient air temperature, pressure, air humidity, cloudiness factor, wind speed, precipitation type, etc. Current and forecast values from OWM have different time resolutions; meteorological parameters of the current weather are given every 30 min, but the forecasted weather data are available only every 3 h. The data receiving time of the meteorological values from OWM is in coordinated universal time (UTC) [20].

The used publicly available weather reports do not contain measured and predicted values of GHI. In order to evaluate how the absence of GHI in input data impacts the accuracy of PV power prediction, further weather data were purchased for use as input data for the same model. Because these data are not publicly available, they are here referred to as "fee-based solar irradiance data". These fee-based data include measured values of GHI, which are collected by pyranometer. They also contain calculated values of GHI, direct normal irradiance (DNI), and diffuse horizontal irradiance (DHI) under clear-sky conditions and angles of solar zenith and azimuth. The GHI prediction is based on an optimized combination of different forecasting methods, including satellite cloud images and different numerical weather prediction models [21]. This dataset does not include other meteorological parameters, such as temperature or humidity. The measured values and predictions of solar irradiance in the fee-based dataset are presented in 1 h and 15 min time intervals, respectively. These data were provided by the DLR Institute of the Networked Energy Systems and the Energy Meteorology Group, Institute of Physics, Oldenburg University. Kalisch et al. [21] described these data and published a dataset for the year 2014. For this study, the solar irradiance data (W/m^2) were taken from the year 2019.

The problem of different time resolutions of the input data was dealt with in the data pre-processing step, as described later.

2.3. Additional Descriptive Feature

The publicly available weather reports from OWM do not provide measured and predicted values of GHI, which are strongly correlated with PV power output. Because of this reason, the input dataset with weather parameters was extended with an additional descriptive feature. Two possible features were investigated: PV power under clear-sky conditions and maximum PV power calculated from the PV power measurements of previous days. Then, these values were inserted in the input dataset and used for training of the model and for making predictions.

2.3.1. PV Power under Clear-Sky Conditions

PV power output under clear-sky conditions was the first additional descriptive feature investigated within this study. The process of this feature generation consisted of three steps.

Step 1: Calculation of the total solar irradiance on a horizontal surface, which is called GHI, under clear-sky conditions. GHI was estimated with the help of pvlib, a special open-source python toolbox for modeling PV system performance [22]. This toolbox provides three different simple clear-sky models in order to estimate solar irradiance on horizontal surface under clear-sky conditions: Ineichen, Haurwitz, and simplified soils [22]. These and other clear-sky models were investigated by Reno et al. [23], and the Ineichen model was ranked among the more accurate approaches. For this study, GHI under clear-sky conditions was estimated in pvlib with the Ineichen model, because this model does not need very specific information and it showed good performance in Reference [23]. The Ineichen model in pvlib needs the following input data for the calculation of GHI under clear-sky conditions [22]:

- Latitude, longitude, altitude, and time zone of PV system location;
- Some meteorological parameters, like air temperature and pressure (if these parameters are not available, pvlib uses default values: T = 12 °C, pressure = none);
- Time period for which the GHI is calculated.

At the end of the first step, a time series with GHI values under clear-sky conditions was generated for the definite location of the PV system.

Step 2: Calculation of PV system-specific clear-sky index $CSI_{PV\ system}$ (m^2). For this step, the maximum value of the PV power measurements $P_{PV,\ max}$ (W) was divided by the maximum value of the GHI under clear-sky conditions $GHI_{clear\ sky,max}$ (W/m^2) (see Equation (1)).

$$CSI_{PV\ system} = \frac{P_{PV,\ max}}{GHI_{clear\ sky,max}}. \tag{1}$$

According to the motivation of this study, the prediction of the PV power output should be made without any technical information of the PV system. The calculated index is a parameter for initial assessment of the size, installed capacity, and efficiency of the PV system.

Step 3: Multiplication of GHI values under clear-sky conditions from the first step with the PV system-specific clear-sky index from the second step. The third step generates a time series of power output, which can theoretically be produced by the PV system under clear-sky conditions. In the study, this additional feature is called "clear-sky PV power".

Figure 1 presents two time series for one summer day (1 June 2017) and one winter day (1 December 2017) for the PV system in Oldenburg (see technical parameters of this PV system in Table 1). The yellow areas in Figure 1 present PV power output under clear-sky conditions. As it can be seen, this feature takes into account seasonal properties such as lower solar irradiance and the shorter daytime in winter.

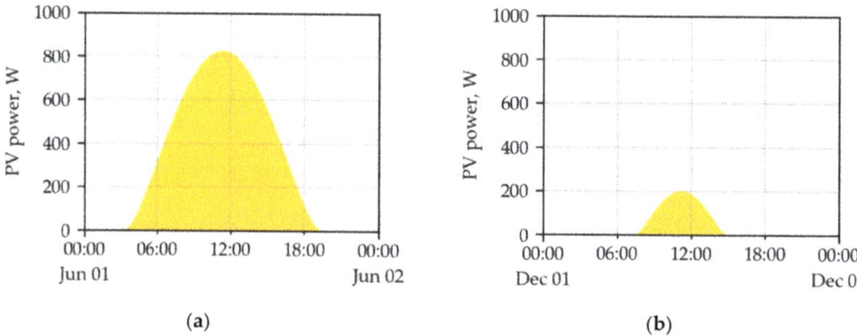

Figure 1. PV power output under clear-sky conditions on (**a**) June 1st and (**b**) December 1st 2017 estimated for the PV system in Oldenburg with an installed capacity of 1.14 kW$_P$.

2.3.2. Maximum PV Power

The maximum PV power curve was the next additional descriptive feature investigated within this work. The calculation methodology of maximum PV power and the optimal number of days to look back were taken from a fully automated model of Hanke et al. [5]. Table 2 presents the working steps for the estimation of maximum power from PV power measurements of the last five days.

Table 2. Working steps for the estimation of the maximum PV power of the last five days.

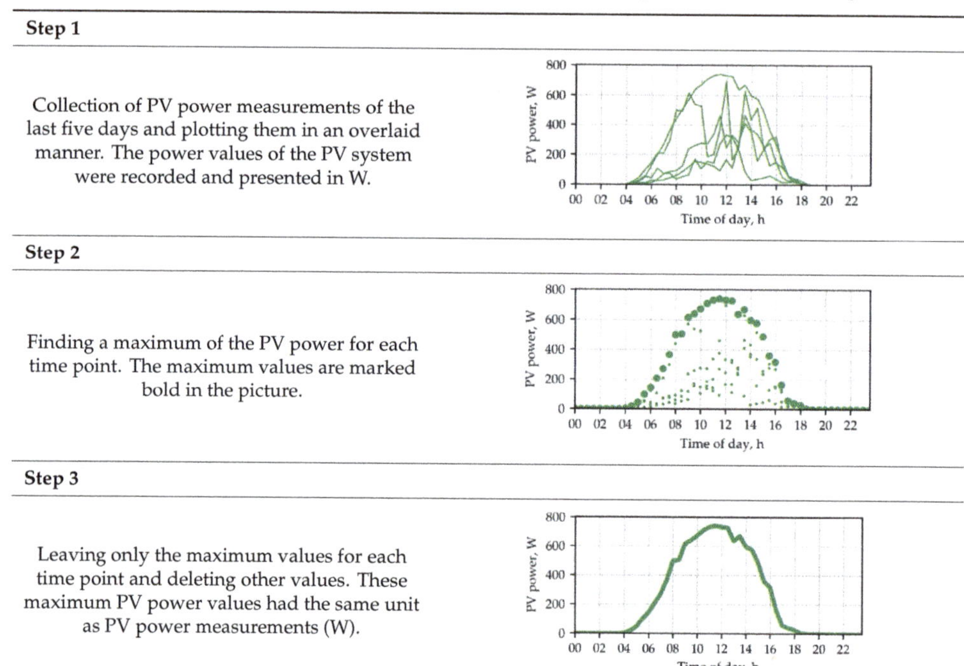

Step 1	
Collection of PV power measurements of the last five days and plotting them in an overlaid manner. The power values of the PV system were recorded and presented in W.	
Step 2	
Finding a maximum of the PV power for each time point. The maximum values are marked bold in the picture.	
Step 3	
Leaving only the maximum values for each time point and deleting other values. These maximum PV power values had the same unit as PV power measurements (W).	

Table 2 explains only the calculation approach for the estimation of the maximum PV power output; it does not explain why this parameter was calculated using power values of the last five days. This number of days was not chosen randomly; rather, Hanke et al. [5] investigated different numbers of days. In this study, the model was consistently trained with data from only yesterday up to data of the last 30 days. Then, the simulation results were compared to each other and to the measured values. Statistical errors, mean absolute error (MAE), and root-mean-square error (RMSE) were selection criteria for choosing the optimal number of days for training of the model. Because the model which was trained with the last five days had the lowest statistical errors in the study [5], this number of days was chosen for generation of the additional feature, i.e., maximum PV power, within this study.

Both maximum PV power and clear-sky PV power were used not only for training the model, but also for verification of the PV power prediction. The prediction of PV power output should not be greater than these additional features. If the predicted PV power was greater than the additional feature, then this predicted value was replaced with the additional feature value. The main advantage of this verification rule was that it prevented the prediction of PV power output overnight, as well as great overestimation by day.

2.4. Data Exploration

The predictive model, which was developed within this study, is a data-driven approach. For this kind of approach, input data and quality of these data play key roles in prediction accuracy. Therefore, exploration of the input data is one of the main steps in data pre-processing and feature selection [19].

Firstly, all input data were explored with the goal of determining whether the data suffered from any data quality issues. Both the publicly available weather reports with current meteorological parameters and the dataset with measured PV power output suffered from missing values. Only 2.9% of values in the publicly available weather dataset were missing. The dataset with the PV measurements had 5.7% missing values. The rule of thumb of Kelleher et al. [19] helped to decide whether these amounts of missing values were critical or not. This rule recommends removing a feature from the input dataset if the proportion of missing values exceeds 60% of the whole dataset. In this case, the amount of relevant information stored in the rest of the data is very low. According to this rule, the proportion of missing values in the datasets with weather parameters and PV measurements was uncritical, and both datasets could be used as input data for the predictive model. The next common data quality issue involves outliers; publicly available weather reports had only one outlier (humidity value) within the investigated period, and the PV dataset did not have any outliers in the same period. According to the OWM homepage, humidity is calculated as a percentage and varies between zero and one hundred. During data exploration, it was determined that the humidity value on 12 March 2018 at 21:00 was 107%.

Secondly, the quality of the numerical weather forecast from publicly available weather reports could also be evaluated, because the weather forecast was available for the whole investigated period of time. The prediction accuracy of publicly available weather reports was evaluated with the help of three statistical metrics.

Mean absolute error (MAE):

$$MAE = \frac{1}{n} \sum_{i=1}^{n} \left| \overline{Y}_i - Y_i \right|. \tag{2}$$

Root-mean-square error (RMSE):

$$RMSE = \sqrt{\frac{1}{n} \sum_{i=1}^{n} \left(\overline{Y}_i - Y_i \right)^2}. \tag{3}$$

Symmetric mean absolute percentage error (sMAPE) [24]:

$$sMAPE = \frac{100\%}{n} \sum_{i=1}^{n} \frac{\left|\overline{Y_i} - Y_i\right|}{\left|\overline{Y_i} + Y_i\right|}.$$

(4)

In Equations (2)–(4), Y_i is the measured value, $\overline{Y_i}$ is the predicted value, and n is the number of values.

The statistical metrics can only be calculated for the numerical variables. The meteorological parameter "precipitation" in publicly available weather reports is a categorical feature, which contains categories such as "rain" and "snow". In order to evaluate the forecast accuracies for this parameter, it was converted into dummy variables. After this procedure, the dataset contained two new columns with discrete variables: "precipitation (rain)" and "precipitation (snow)".

The meteorological parameters cover different ranges; for example, cloudiness and humidity cover the range [0, 100], ambient air temperature covers the range [−10, 29], etc. In order to compare them with each other, all values were converted into the range [0, 1] using range normalization [19]. Afterward, MAE, RMSE, and sMAPE for all meteorological parameters were calculated. The statistical metrics which indicate the quality of weather forecast of publicly available weather reports are presented in the Table 3.

Table 3. Statistical metrics of OpenWeatherMap (OWM) forecast: MAE—mean absolute error; RMSE—root-mean-square error; sMAPE—symmetric mean absolute percentage error.

Investigated Meteorological Parameters from OWM	MAE	RMSE	sMAPE (%)
Temperature	0.023	0.031	2.61
Pressure	0.017	0.022	1.56
Humidity	0.219	0.263	15.43
Cloudiness	0.236	0.339	33.79
Wind speed	0.079	0.100	20.11
Precipitation (rain)	0.415	0.644	41.48
Precipitation (snow)	0.023	0.150	2.31

Among all meteorological parameters, temperature and pressure, with an sMAPE of less than 3%, had the best prediction accuracy. The high statistical errors of cloudiness and precipitation make sense, because these meteorological parameters are the hardest to predict. The accuracy of humidity forecasting lay between the accuracies for temperature and cloudiness.

2.5. Correlation Analysis and Feature Selection

In the previous subsections, the meteorological parameters were investigated separately from the target feature, i.e., the PV power output. The next step was the calculation of the relationship between weather data, additional features (clear-sky PV power and maximum PV power), and measured PV power. The relationship between all these features was estimated with the help of covariance *cov* and correlation *corr* metrics [19]. The formulas for the calculation of these metrics are given below.

$$cov(a, b) = \frac{1}{n-1} \sum_{i=1}^{n} \left((a_i - \overline{a}) \times (b_i - \overline{b}) \right).$$

(5)

$$corr(a, b) = \frac{cov(a, b)}{sd(a) \times sd(b)}.$$

(6)

In Equations (5) and (6), a, b represent features a and b, $\overline{a}, \overline{b}$ represent the means of features a and b, and *sd* is the standard deviation.

The correlation values between meteorological parameters from publicly available weather reports, two additional features, and PV power output are presented as a heat map in Figure 2.

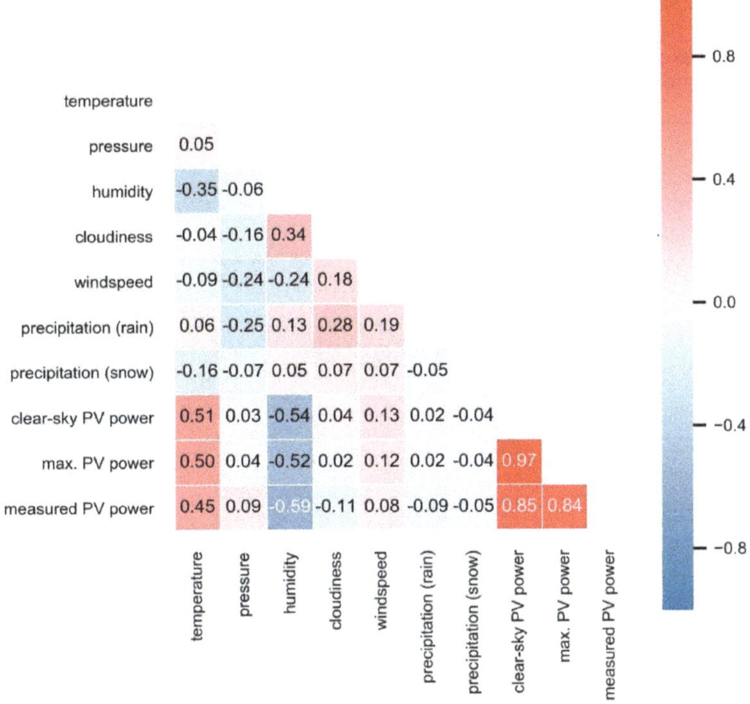

Figure 2. Correlation values between descriptive, additional, and target features. Values close to −1 mean a strong negative linear correlation, values close to 1 mean a strong positive linear correlation, and values around 0 mean no linear correlation [19].

The correlation analysis showed that air temperature and humidity have a stronger relationship with PV power output in comparison with the other meteorological features. The strong positive correlation between PV power and temperature can be explained by the fact that daily curves of the solar irradiation and air temperature were similar to each other. In general, an increase in solar irradiation also causes an increase in air temperature. The same explanation is valid for the strong negative correlation between PV power and humidity. However, in this case, the increase in solar irradiation causes a decrease in humidity.

Despite the commonly perceived fact that the PV generation strongly depends on the current cloud cover, the given probability for cloudiness from the OWM showed a weak correlation with the measured PV power. The calculated maximum PV power values and the expected PV power under clear-sky conditions had the strongest positive relationship with PV power output, as the correlation values were greater than 0.80. Pressure, wind speed, and both precipitation features had the lowest correlation with the measured PV power, or the relationship between these features was strongly non-linear.

The two precipitation features "rain" and "snow" were combined as a common feature, which received the name "precipitation". In the case of rain or snow, the parameter "precipitation" was equal to one; otherwise, it was zero. The correlation value between the new descriptive feature "precipitation" and the target feature "measured PV power" was −0.14.

The conducted correlation analysis helped for a better understanding of the data. Furthermore, the results of this analysis were applied for feature selection. Hall [25] defined the feature selection as a process of identifying and removing features which contain irrelevant or redundant information. The presence of irrelevant information can reduce the prediction accuracy and make the results less understandable. By the same token, the removal of irrelevant features can improve the performance of the machine learning algorithm, whereby the operation time decreases and the algorithm operates more effectively [25].

There are a lot of techniques for feature selection, but this research work used correlation values in order to identify and select the input features with relevant information. Only those descriptive features with correlation values greater than 0.1 or lower than −0.1 were selected for the predictive model. The selected features were air temperature, humidity, cloudiness, precipitation, maximum PV power, and clear-sky PV power.

The heat map in Figure 2 presents the correlation values for the whole observation period (one year). The next step was to investigate whether the correlation values were dependent on seasons or whether they remained constant throughout the whole year. For this purpose, the correlation was calculated for each month in the investigated period. Figure 3 presents the monthly correlation values among the measured PV power, four selected meteorological parameters, and two additional features.

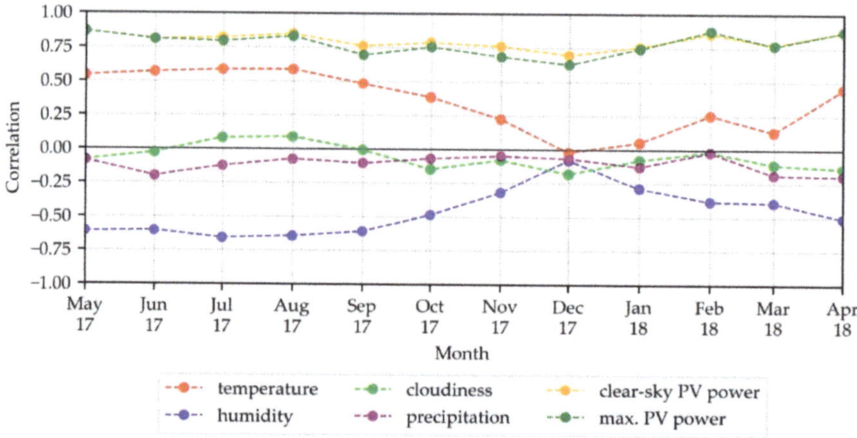

Figure 3. Monthly correlation values among measured PV power, selected meteorological parameters, and additional features.

The correlation between the measured PV power and two additional features remained the strongest over the whole observation period. Moreover, the monthly correlation values between these parameters fluctuated slightly over the year; for example, the correlation values between the measured PV power and calculated maximum PV power ranged between 0.63 in December 2017 and 0.87 in April 2018. The relationship between the PV measurements and air temperature had strong seasonal dependency. The correlation between these features was much stronger in warm seasons. It decreased in autumn and reached its minimum in winter months. In spring, the correlation values began increasing again. The same tendency can be seen by observing the relationship between measured PV power and humidity. The correlation between the measured PV power and two remaining weather parameters did not fluctuate as strongly, and it lay between 0 and −0.25 for almost the whole year.

3. Methodology

3.1. Machine Learning Algorithm

There are two main approaches to PV power output forecasts: performance method and machine learning method. The performance or physical method needs technical specifications of the PV system and prediction of the solar irradiance for this location. However, the main aim of the developed predictive approach is to obtain the PV power output without any information about the PV system (except historical measured values of the generated power). Thus, the performance method could not be used according to the motivation of this study. The machine learning method does not need any information of the system. This was the first reason for choosing the machine learning approach. The second reason was an absence of the solar irradiance in the publicly available weather reports.

The next step was to select a machine learning technique among the many techniques of forecasting the PV power output. The artificial neural network (ANN) is currently the most used machine learning approach for the prediction of PV power. In 24% of all studies in Reference [10], the researchers predicted PV power using ANN models. ANN has many different variants with their own advantages and disadvantages, such as feed-forward, convolutional, recurrent, etc. This study used a special architecture of artificial recurrent neural network (RNN) named the long short-term memory network (LSTM).

LSTM was firstly proposed by Hochreiter and Schmidhuber [26]. This new model was developed in order to overcome a classic problem of the RNN, i.e., that error signals in the RNN blow up or vanish during their backpropagation through time. The LSTM model developed by Hochreiter and Schmidhuber [26] is not affected by this problem. Since then, the classical structure of LSTM was improved and developed further by many different scientists. One of the most relevant improvements was the addition of a component called a "forget gate", which was developed and described by Gers et al. [27]. This additional component and all standard components of the LSTM module are presented in Figure 4.

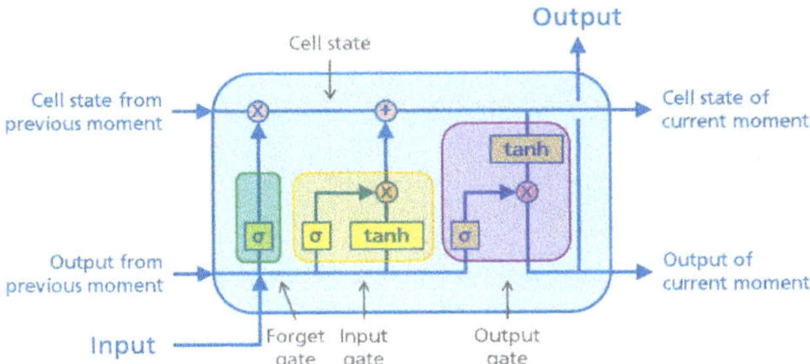

Figure 4. Internal structure of a single long short-term memory (LSTM) block (following the description of Hua et al. [28]).

The key components of the LSTM block are the cell state and gates (input, output, and forget). The cell state is displayed as a horizontal line running through the top of the LSTM module in Figure 4. This component is responsible for remembering the current state of the neural network. The gates are also marked and labeled in the figure. They consist of different units, such as a sigmoid layer (σ), tangent hyperbolic layer (tanh), and multiplication operator ($\times$). These components help to control the information flow to the memory cell. The forget gate looks at the input of the current cell and the output of the previous cell and decides how much information remains in the current cell state. This gate returns a number between 0 and 1. The next step is to control what new information is stored in the current cell state. This is the task of the input gate, which combines sigmoid and tanh layers. The last gate, i.e., the output gate, decides what information flows to the next memory block [28].

The following characteristics of the LSTM network were the main reasons for choosing it for the predictive model of this study [26]:

- LSTM is able to learn long-term dependencies that are typically found in time series. All used input datasets were time series (weather data and measurements of PV power).
- This structure of the ANN can remember relevant information for long periods of time.
- In the case of great time lags, the special structure of the LSTM prevents the error signals from increasing or vanishing.

The detailed mathematical explanation of the LSTM, as well as application cases, can be found in References [26–28].

3.2. Description of Developed Predictive Model

After the neural network structure was chosen, it was decided which programming language and framework would be used to develop the predictive model. The predictive model was designed and trained with machine learning library Keras, which was written in Python. This open-source framework was developed as part of the research project ONEIROS (open-ended neuro-electronic intelligent robot operating system). Keras is intended for the quick implementation of neural networks, both convolutional and recurrent, for different experimental purposes [29].

However, before training the model with the chosen machine learning technique, i.e., LSTM, the input data were prepared. The data preparation, model training, and other main steps of PV power forecasting are presented in a simplified flow chart in Figure 5.

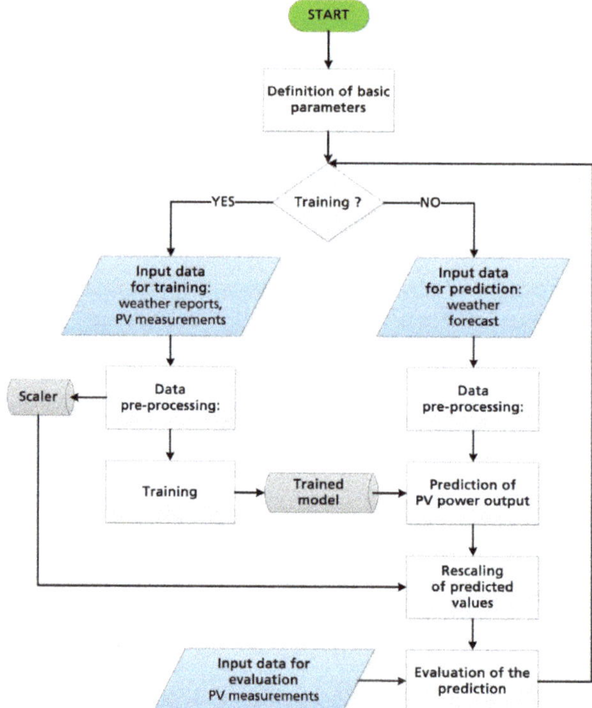

Figure 5. Simplified flow chart of the predictive model.

The developed predictive tool is initialized at the very beginning of the operation. In this case, initialization means a definition of parameters such as geographical coordinates, location, and name of the additional descriptive feature. These parameters are necessary to generate the additional feature, i.e., clear-sky PV power or maximum PV power. The reasons and procedures of additional feature generation were described in Section 2.3. If maximum PV power was chosen as the additional input feature, the predictive tool waited five days to collect enough data to calculate the maximum power output (see Section 2.3.2).

After the definition of the additional feature, the main process of PV power prediction could start. This process occurred continuously in an endless loop. Each iteration of this loop started by making a decision about the execution of the training process (see rhombus in Figure 5). In the case of a positive decision, the predictive model was trained with updated weather data and PV measurements. A positive decision occurred twice per day, at 00:00 and at 12:00. As known from the previous chapter, all input data used in this study had data quality issues, e.g., missing values and different time resolutions. Therefore, data pre-processing was an unavoidable step, which included the following functions: timestamp transformation of PV measurements from local time UTC + 01:00 to UTC, detection of the double or completely inconsistent timestamps, imputation of missing values by linear interpolation, and transformation of data time resolutions into the pre-defined unified time resolution. This pre-defined unified time resolution for all used input data was 30 min. The last step of data pre-processing was the normalization of values. The normalization involved a scaling of values into the range [0, 1] using range normalization. The descriptive and target features were scaled separately from each other, and the scalers were saved for future use in the forecasting process, i.e., for scaling of the input weather forecast values and rescaling of the predicted PV power values.

After the normalization step, the scaled values of the weather reports and the PV measurements were used to train the predictive model. The training process consisted of three steps: architecture definition, compilation (configuration of the learning process), and training. The architecture of the model had the following characteristics:

- First layer with five input parameters;
- Two hidden LSTM layers with 64 and 32 neurons;
- Output dense layer;
- The total number of trainable parameters was 30,369.

Then, the learning process was configured and the model was trained for 100 epochs. After this process, the model architecture and the weights were saved in order to use them later for the prediction of PV power output.

As mentioned above, the model was trained only at midnight and at noon. At other times, the prediction of PV power output was made with the help of the model, which was saved after the previous training step. The last steps in the loop included rescaling of the predicted values and evaluation of the forecast accuracy.

3.3. Content and Sizes of the Training and Test Sets

Because the PV power output depends strongly on the weather conditions and seasons, it was checked whether the developed predictive model can forecast the PV power output for different seasons. For this purpose, the model was trained with the weather data of warm and cold periods, and then the trained model was used to make predictions for warm and cold periods, respectively.

Both the content and the size of the training set were varied. Figure 6 shows four different sizes of the training set (seven days, 14 days, 30 days, and 90 days), one constant size of the test set (23 days), and a general splitting of the whole dataset into training and test sets.

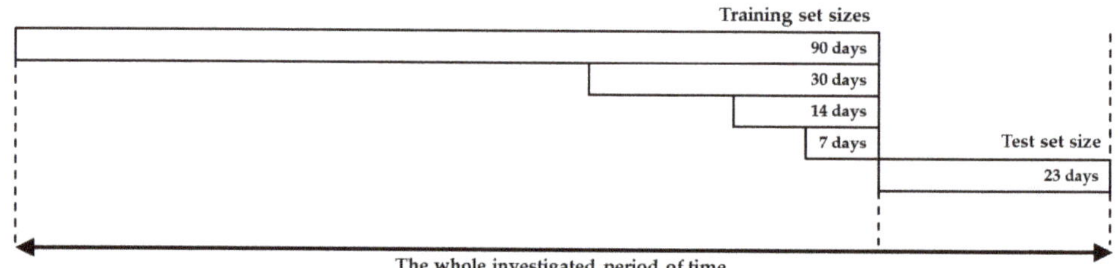

Figure 6. Splitting the dataset into training and test sets.

As seen in Figure 6, the first prediction of PV power output always began at the same time point of the test set, independent of the training set size. One after the other, the pre-defined training set sizes were used to train the model and make predictions of PV power. Then, the impact of the training set sizes on the forecast accuracy was investigated by comparing the predictions with each other.

3.4. Evaluation of Prediction Accuracy

The evaluation of the prediction accuracy was done by calculating standard statistical errors MAE, RMSE, and MAPE (see Equations (2)–(4)). Antonanzas et al. [10] described an adaptation of the classical MAPE for the evaluation of PV power forecasting.

$$MAPE = \frac{100\%}{n} \sum_{i=1}^{n} \frac{\left|P_{pred} - P_{meas}\right|}{P_0},$$

(7)

where P_0 is the installed capacity of the PV system.

Another measure for estimation of the forecast accuracy is the mean absolute scaled error (MASE), which was described in References [10,30].

$$MASE = \frac{MAE}{\frac{1}{n-m}\sum_{i=1}^{n}\left|P_{meas,i} - P_{meas,i-m}\right|}.$$

(8)

This error differs from classical statistical errors in the fact that MASE is independent of the scale of the data. An MASE less than one points to the used predictive method being better than the average naïve forecast [30]. The naïve forecast and persistence forecast represent simple forecasting methods, whereby the forecast value is equal to the last measured value. There is an extension of the classical naïve forecast for seasonal data called the seasonal naïve forecasting method. According to this method, the forecast value is equal to the last measured value of the same season (season can be day, month, year, etc.) [31].

Because PV power output can be defined as seasonal data, it was decided to use the seasonal MASE to evaluate prediction accuracy in this study. Some assumptions which were needed to calculate MASE within this work are listed below.

- The season was taken as one day;
- m in Equation (8) was set to 48, because the time resolution of the data was 30 min;
- $P_{meas,i}$ in Equation (8) was the measured power of the PV system at time i;
- $P_{meas,i-m}$ in Equation (8) was the measured power of the PV system at the same time yesterday.

In addition to statistical errors, the difference between the measured and predicted daily energies plays an important role in energy management systems. This measure is called the energy forecast error, and it was calculated with the following equation [32]:

$$\Delta E = \frac{E_{daily,\ model} - E_{daily,measured}}{E_{daily,measured}}.$$

(9)

The sign of the energy forecast error indicates whether the predictive model overestimates or underestimates the measured energy production of PV system.

4. Results and Discussion

In this chapter, the PV power prediction, made with the publicly available weather reports, is presented, discussed, and compared with the prediction with fee-based solar irradiance data. Moreover, it was also checked whether the developed predictive model meets the defined requirements, such as self-learning ability, transferability, etc.

4.1. Choice of the Additional Descriptive Feature

Section 2.3. described two additional descriptive features: clear-sky PV power and maximum PV power. Here, these two features were compared by making predictions of PV power output and estimating the prediction accuracy. The developed predictive model was trained with the same weather data and the same power measurements of the PV system in Oldenburg (the installed capacity of this PV system is 1.14 W_P). Both training datasets contained 90 days of data. The architecture of the predictive model and the procedure of the predictive process also remained the same. The only difference was that the first input dataset contained maximum PV power and the second input dataset contained clear-sky PV power as the additional descriptive feature.

After training the model and predicting PV power, the MAE of this prediction was calculated for each day in the test set. These MAE values are presented in Figure 7 in the form of boxplots.

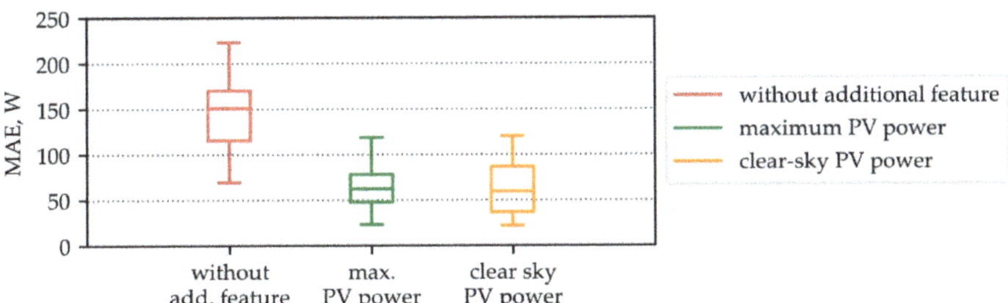

Figure 7. Distribution of daily MAE values of PV power prediction after training the model without any additional feature and with maximum PV power or clear-sky PV power as additional features (training set 90 days; test set 8 August 2017–31 August 2017).

A boxplot is a very representative way of displaying the distribution of values. The main components of boxplots are the box, median, whiskers, and outliers. The box or interquartile range contains the middle 50% of all values. The line in the box indicates the median. The space between the lower whisker and the box covers the range between the minimum and lower quartile. The space above the box includes the values from the upper quartile until the maximum.

The first important conclusion which can be drawn from Figure 7 is that the extension of input data with one of the additional features (maximum PV power or clear-sky PV power) improved the prediction accuracy significantly; the median MAE decreased from 150 W to 67 W. The second important conclusion is that both additional features resulted in a similar accuracy of the PV power prediction. MAE values of these predictions varied between 25 W and 125 W. The only difference between these prediction accuracies was the distribution of the errors between the lower quartile, interquartile range, and upper quartile. Because MAEs of the predictions with both additional features had almost the same range, they could be equally used for PV power prediction.

To generate the maximum PV power, only two parameters are needed for the calculation: measured power of the PV system and pre-defined number of days looking back (see calculation procedure in Section 2.3.2). The feature "clear-sky PV power" required more input data for its generation, including geographical coordinates, location name, and calculated specific clear-sky index for the feature generation. Thus, the additional feature "maximum PV power" was more independent than the feature "clear-sky PV power". For this reason, it was decided to use the maximum PV power as the additional descriptive features for all predictions, as described later.

4.2. Prediction with Publicly Available Weather Reports

The developed predictive model with publicly available weather reports as input data must be able to make reasonably good predictions of PV power output. Moreover, the model must meet the requirements defined in Section 1. One of these requirements was the suitability of the model for different seasons of the year. This requirement was checked using the PV system in Oldenburg. Therefore, the predictive model was tested for two periods of time with different weather and solar irradiance conditions: a warm period from 8 August until 30 August 2017 and a cold period from 19 March until 10 April 2018.

The prediction accuracies for warm and cold periods are displayed in Figure 8 in the form of boxplots with the distribution of MAE, RMSE, MAPE, and MASE values. The predictions for both warm and cold periods were made using the predictive model, which was trained with four training set sizes one after the other. All training and test sets contained the data from publicly available weather reports. The X-axis in the figure presents the four sizes of the training set, which were used to train the model. The Y-axis shows the distribution of the error values for warm and cold periods.

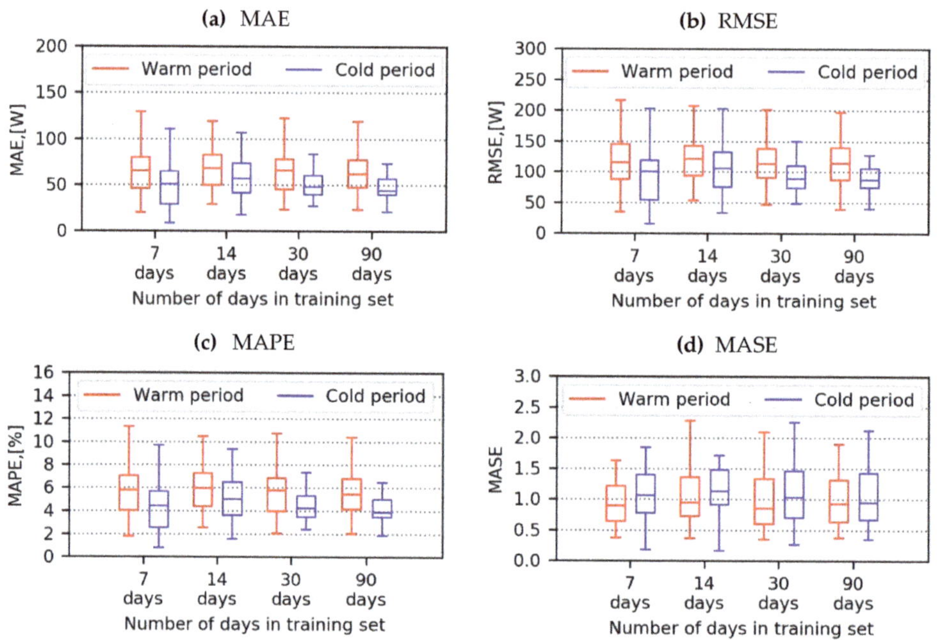

Figure 8. Daily values of MAE, RMSE, MAPE, and mean absolute scaled error (MASE) of PV power prediction for warm and cold seasons, which were made after training with four sizes of the training set containing publicly available weather reports.

The training dataset size was increased step by step in order to investigate and improve the prediction accuracy of the model. The prediction accuracy improved with the increase in size of the training set. This was especially significant in the cold period; for example, if the training set was increased from seven days to 90 days, the median MAE of the cold period decreased from 51 W to 45 W (almost 12%).

The next important point, which can be seen in Figure 8, is that the scale-dependent errors MAE and RMSE in the cold period were lower than in the warm period (see Figure 8a,b). These results can be explained by the fact that the measured values of PV power in winter were much lower than in summer. The percentage error MAPE had the same disadvantage. For this reason, it was important to calculate the scale-independent metric MASE and to use it for comparison of the prediction accuracy across different seasons.

The distribution of the daily MASE values for the two seasons and four training sets are also presented as boxplots in Figure 8d. It is obvious from this figure that the prediction of the PV power in the cold period was less accurate than that in the warm period. The MASE medians of the cold period lay above 1.0 for almost all training sets, except for the set with 90 days, and the MASE medians in the warm period were about 0.90 for all training sets. One of the possible reasons for the better prediction in the warm period is that solar irradiance and, consequently, PV power output in the warm season was more stable, whereas the cold season had a lot of days with strongly fluctuating solar irradiance during the day. However, the developed predictive model should forecast the PV power output for all seasons of the year equally well. In this case, only the training set with 90 days could ensure appropriate prediction accuracy for both warm and cold seasons; the MASE medians of the PV power prediction for this training set were about 0.90, regardless of the season.

After testing whether the predictive model could accurately predict PV power for different seasons, the same model was also tested with regard to forecasting PV power for a completely different PV system. This system is located in Munich, Germany, and its installed capacity is almost one hundred times greater than the PV system in Oldenburg. The technical parameters of this PV system were presented in Table 1. The power output prediction for the PV system in Munich was also made for 23 days from 8 June until 30 June 2019. The dataset of the Munich PV system was split in the same way as the dataset of the Oldenburg PV system (see Figure 6).

Because the installed capacity of the Munich PV systems was much greater than the capacity of the Oldenburg PV system, only the scale-independent statistical metric MASE could be used for comparison of the prediction accuracy of these two systems. The average values of MASE for the whole test periods of the two PV systems are displayed as a dot chart in Figure 9.

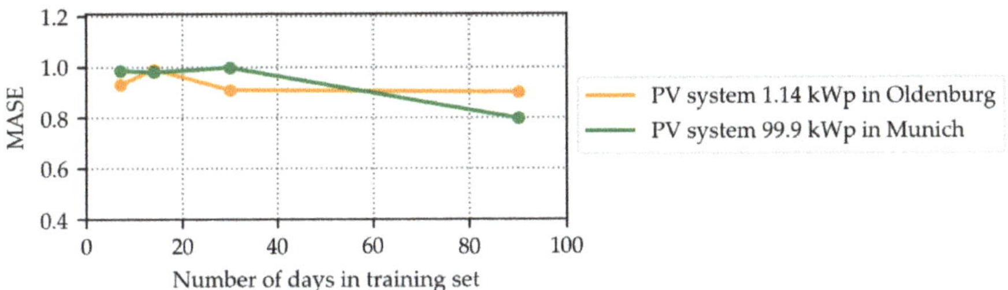

Figure 9. Average of MASE values for two different PV systems in Oldenburg and Munich.

The predicted values of the Munich PV system had similar average values of MASE to the predicted values of the Oldenburg's system. The most accurate prediction occurred again after 90 days of training, but the extension of the Munich training set to 90 days led to a greater improvement of the prediction accuracy. Figure 9 not only shows the forecasting quality; it also verifies that the developed predictive model was able to forecast the power output for the different PV systems without any technical information about these systems, except for the measured power values, using only publicly available weather reports without direct prediction of GHI.

4.3. Prediction with Fee-Based Solar Irradiance Data

In this section, the PV power prediction with publicly available weather data was compared to the prediction with fee-based solar irradiance data. The main difference between these two data sources lies in the fact that publicly available weather data contain only indirect values of the GHI such as cloudiness and precipitation, while fee-based data contain measurements and predictions of the GHI, which are highly correlated with PV power output. The prediction of PV power output presented in this section was also made using the same predictive model described in Section 3.

In the previous subsection, it was proven that the developed predictive approach could be used to make predictions for two different PV systems. This is why fee-based solar irradiance data were used to make predictions for only one PV system. The PV system in Munich was chosen for this purpose.

Firstly, the MASE values of the PV power predictions with publicly available weather reports and fee-based solar irradiance data for all training sets were compared with each other, in order to determine the optimal sizes of the training set for each data origin. Figure 10 shows these values.

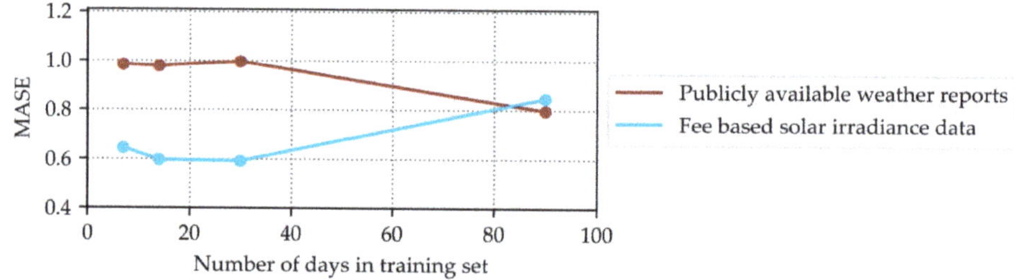

Figure 10. MASE of PV power predictions with publicly available data and fee-based data. The prediction was made for the Munich PV system (installed capacity of 99.9 kW$_P$).

Figure 10 shows that the prediction accuracy of the model with the solar irradiance data was much better when using shorter training datasets. The extension of the training set with solar irradiance data from 30 days to 90 days led to an increase in MASE. The predictive model made the most accurate prediction if the training set contained 14 days of solar irradiance data or 90 days of publicly available data. The PV power predictions with these exact training set sizes are compared below.

The next measure for the comparison between predictions, made with two data origins, was the error between the predicted and measured daily energies, calculated using Equation (9). The normalized distribution of energy forecast errors of the predictive model, which was trained with 90 days of publicly available weather data and 14 days of fee-based data, is displayed in Figure 11.

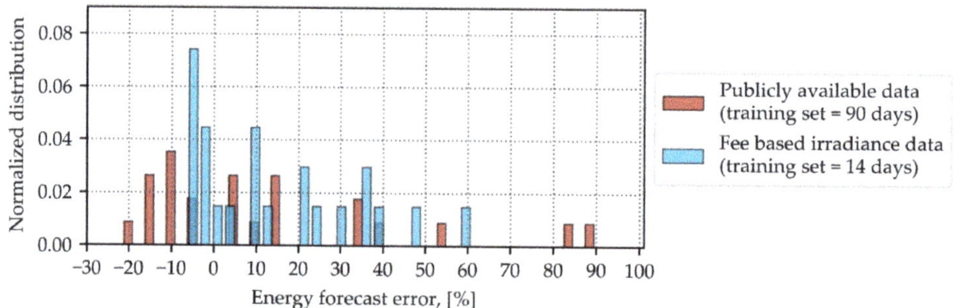

Figure 11. Normalized distribution of the energy forecast errors of predictions with publicly available weather reports (training set with 90 days) and fee-based data (training set with 14 days). The prediction was made for the Munich PV system (installed capacity of 99.9 kW$_P$).

Two main conclusions can be drawn from Figure 11. The first conclusion is that the developed predictive model was slightly inclined to overestimate the measured values regardless of the input data used; about 60% of the energy forecast errors had a positive sign The second conclusion is that the predictive model with fee-based solar irradiance data could predict daily energy more accurately than the same model trained with publicly available weather reports. The energy forecast errors of the prediction with solar irradiance data varied between −10% and 60%, but the usage of publicly available weather reports for the input data led to an increase in forecast errors in both directions, i.e., overestimating and underestimating.

It is relevant to consider the predicted PV power output not only for the whole test period, but also for single days. This is why two days with different solar irradiance were selected from the test set: one with clear-sky conditions during the whole day (29 June 2019) and another with strongly fluctuating solar irradiance (20 June 2019). The developed model was used to make predictions for these days with publicly available weather reports and fee-based solar irradiance data. The results were compared to each other and to the measured values. The predicted and measured power curves for the selected days are presented in Figure 12.

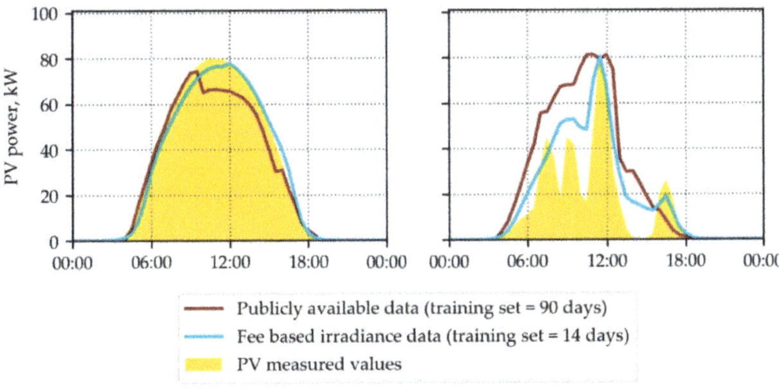

Figure 12. Measured PV power of the Munich PV system in comparison to the PV power predictions made by the model with publicly available weather reports (training set with 90 days) and fee-based solar irradiance data (training set with 14 days) on 20 and 29 June 2019.

Despite the smaller training set, the model trained with fee-based irradiance data could predict PV power output more accurately for both selected days than the model trained with publicly available weather reports. Moreover, the training with solar irradiance data led to the model predicting single peaks and drops of the PV system accurately even with fluctuating PV power production on 20 June 2019. Despite the fact that publicly available weather reports do not have direct predictions of the GHI, the model trained for 90 days with this dataset could forecast the main trends of the PV power production for both days. Furthermore, this model was also able to predict the rapid power drop on the day with strongly fluctuating solar irradiance (20 June 2019 at 12:00).

5. Conclusions

The developed predictive approach is a data-driven method, where the quality of input data plays a key role. Therefore, the accuracy of the publicly available weather reports was investigated at the very beginning of the study. The accuracy of the PV power output prediction cannot be better than the accuracy of the used input data.

The evaluation of the prediction accuracy indicated that the machine learning approach provided adequate results of PV power prediction for the next 24 h even with publicly available weather reports. Although the publicly available weather data from OWM are to be used mainly on websites and mobile applications, they can be also used for the purposes of PV power prediction. This study also proved that it is possible to predict the PV power output without solar irradiance measurements and forecasts, and without any technical information about the PV system, except the measured power values. Because publicly available weather reports do not provide GHI values, this input dataset was extended with an additional descriptive feature, i.e., the maximum PV power of the last five days. The addition of this feature improved the prediction accuracy, prevented the prediction of PV power overnight, and avoided overestimated predictions of PV power by day.

The requirements of the predictive approach were defined in the motivation of the study. The first requirement was a fully automated online operation of the day-ahead PV power forecasting. This requirement was proven during the simulation, where the weather forecast was updated every 3 h. In the same time interval, the PV power was predicted for the next 24 h. The constant updating of the training set with current weather data and PV measurements resulted in a periodic re-training of the predictive model every 12 h. The second requirement was the transferability of the model for different seasons and PV systems. The suitability for different seasons was tested using a simulation with the dataset from Oldenburg, which contains weather data for warm and cold periods. The comparison of the simulation results with these two datasets pointed to an influence of seasons on prediction accuracy, as well as the ability of the model to adapt to seasonal weather changes. The transferability of the model to the PV systems with different locations, sizes, and technical parameters was proven by the prediction of the PV power for two completely different PV systems. During the transferability check, the main disadvantage of MAE, RMSE, and MAPE was detected. Therefore, the seasonal MASE was selected to compare the forecasting accuracies across different seasons and PV systems.

Afterward, the PV power predictions with publicly available weather reports were compared to the predictions with fee-based solar irradiance data. The predictive model, which was fitted with publicly available weather data, needed more training data in order to make more accurate predictions of the PV power. The best accuracy of the prediction with publicly available weather reports occurred using the training set with data from the last 90 days (the time resolution of all data was 30 min). If fee-based solar irradiance data were used, the training set with the last two weeks of data led to the most accurate prediction. The predictive model which used solar irradiance data not only had better prediction accuracy, but it also forecasted single power peaks and drops of the PV system more accurately than the model which used publicly available data.

The developed predictive approach with publicly available weather data is not suitable for applications which require higher accuracy and finer resolution, e.g., grid stabilization. However, the accuracy of the developed predictive approach with publicly available weather reports is suitable for other applications, such as forecast-based energy management system for buildings with PV systems. An energy management system based on PV power prediction can increase the self-consumption of the PV system and optimize its operation and flexible loads, such as BEVs and heat pumps. Moreover, if the PV power prediction is considered for the distribution of energy demand over the day, it can support a reduction in peak load, which prevents power limits on the house connection point from being exceeded and avoids high grid fees.

The publicly available weather reports from OWM and the PV power measurements are continuously recorded in order to collect more input data for training and validation of the developed predictive approach with a greater input dataset. This greater input dataset contains more information about seasonal circumstances, rapid weather changes, and relationships between weather data and PV power, which can improve the accuracy of the predictive approach.

In this study, the accuracy of the developed predictive model with publicly available weather reports was improved in different ways, such as selection of appropriate input features and machine learning algorithm, optimal configuration of the LSTM network, increase in training set size, etc. However, the prediction accuracy may also be improved in the future if publicly available weather data sources provide measurements and prediction of solar irradiance.

Future investigations should not only deal with the improvement of the developed predictive approach, but also investigate many other questions. One of these refers to the optimal size of the training set. The accuracy of PV power prediction strongly depends on the type of solar irradiance for the next 24 h; in other words, the prediction accuracy depends on whether the prediction is made for a day with constant solar irradiance or for a day with strongly fluctuating solar irradiance. Therefore, it will be interesting to investigate dynamic sizes of the training set, i.e., the training set size can be adapted automatically depending on the weather forecast and type of solar irradiance. Dynamic training set sizes can also be more efficient in cases of disruptive changes like snow cover on PV modules or the failure of strings.

The focus for future investigation can be on an economic analysis of the PV power prediction. This study used only statistical metrics (MAE, RMSE, MAPE, and MASE) to evaluate the performance of the developed predictive approach and to estimate influences of the data origin on the prediction accuracy. Additionally, predictions made with different data origins can also be evaluated with economic metrics, e.g., whether paying for solar irradiance data can provide considerable economic benefit. Another goal for further investigations is a combination of the developed machine learning approach for PV power prediction with another approach for load prediction. These two approaches use different input features, predict different target features, and can use different measures for prediction accuracy evaluation. However, in order to evaluate the combination of these predictions uncertainty quantification needs to be conducted for both approaches. Afterward, these predictive tools can be integrated into the energy management system of the buildings.

Author Contributions: Conceptualization, N.M., J.-S.T. and B.H.; Data curation, N.M., B.H., T.S. and M.G.; Formal analysis, N.M.; Funding acquisition, B.H. and K.v.M.; Investigation, N.M. and J.-S.T.; Methodology, N.M., J.-S.T. and B.H.; Project administration, J.-S.T.; Resources, C.A., K.v.M. and M.G.; Software, N.M. and J.-S.T.; Supervision, B.H. and M.G.; Visualization, N.M.; Writing—original draft, N.M; Writing—review & editing, J.-S.T. All authors have read and agreed to the published version of the manuscript.

Funding: The authors would like to acknowledge the financial support of the "Federal Ministry for Economic Affairs and Energy" of the Federal Republic of Germany for the project "EG2050: EMGIMO: Neue Energieversorgungskonzepte für Mehr-Mieter-Gewerbeimmobilien" (03EGB0004G and 03EGB0004A). For more details, visit www.emgimo.eu. The presented study was conducted as part of this project.

Conflicts of Interest: The authors declare no conflicts of interest.

References

1. Global Alliance for Buildings and Construction. *Towards a Zero-Emission, Efficient, and Resilient Buildings and Construction Sector*; Global Status Report 2017; Global Alliance for Buildings and Construction, International Energy Agency, UN Environment: Paris, France, 2017.
2. Federal Ministry for the Environment, Nature Conservation, Building and Nuclear Safety (BMUB). *Climate Action Plan 2050. Principles and Goals of the German Government's Climate Policy*; Federal Ministry for the Environment, Nature Conservation, Building and Nuclear Safety (BMUB): Berlin, Germany, 2016.
3. National Renewable Energy Laboratory; National Center for Photovoltaics. *Photovoltaics and Commercial Buildings—A Natural Match*; National Renewable Energy Laboratory: Golden, CO, USA, 1998.
4. National Renewable Energy Laboratory, Office of Energy Efficiency & Renewable Energy. *Nationwide Analysis of U.S. Commercial Building Solar Photovoltaic (PV) Breakeven Conditions*; National Renewable Energy Laboratory: Golden, CO, USA, 2015.
5. Hanke, B.; Bottega, M.; Peters, D.; Maitanova, N.; Telle, J.-S.; Grottke, M.; von Maydell, K.; Agert, C. Fully Automated Photovoltaic System Modelling for Low Cost Energy Management Applications Based on Power Measurement Data. In Proceedings of the 35th European Photovoltaic Solar Energy Conference and Exhibition, Brussels, Belgium, 24–28 September 2018; pp. 1588–1593.
6. Wang, S.; Sun, Y.; Zhou, Y.; Mahfoud, R.J.; Hou, D. A New Hybrid Short-Term Interval Forecasting of PV Output Power Based on EEMD-SE-RVM. *Energies* **2019**, *13*, 87. [CrossRef]
7. Kwon, Y.; Kwasinski, A.; Kwasinski, A. Solar Irradiance Forecast Using Naïve Bayes Classifier Based on Publicly Available Weather Forecasting Variables. *Energies* **2019**, *12*, 1529. [CrossRef]

8. Raza, M.Q.; Nadarajah, M.N.; Ekanayake, C. On recent advances in PV output power forecast. *Sol. Energy* **2016**, *136*, 125–144. [CrossRef]

9. Mosavi, A.; Salimi, M.; Faizollahzadeh Ardabili, S.; Rabczuk, T.; Shamshirband, S.; Varkonyi-Koczy, A.R. State of the Art of Machine Learning Models in Energy Systems, a Systematic Review. *Energies* **2019**, *12*, 1301. [CrossRef]

10. Antonanzas, J.; Osorio, N.; Escobar, R.; Urraca, R.; Martinez-de-Pison, F.; Antonanzas-Torres, F. Review of photovoltaic power forecasting. *Sol. Energy* **2016**, *136*, 78–111. [CrossRef]

11. Das, U.K.; Tey, K.S.; Seyedmahmoudian, M.; Mekhilef, S.; Idris, M.Y.I.; Van Deventer, W.; Horan, B.; Stojcevski, A. Forecasting of photovoltaic power generation and model optimization: A review. *Renew. Sustain. Energy Rev.* **2018**, *81*, 912–928. [CrossRef]

12. Mohammed, A.; Aung, Z. Ensemble Learning Approach for Probabilistic Forecasting of Solar Power Generation. *Energies* **2016**, *9*, 1017. [CrossRef]

13. Das, U.K.; Tey, K.S.; Seyedmahmoudian, M.; Idris, M.Y.I.; Mekhilef, S.; Horan, B.; Stojcevski, A. SVR-Based Model to Forecast PV Power Generation under Different Weather Conditions. *Energies* **2017**, *10*, 876. [CrossRef]

14. Rosato, A.; Altilio, R.; Araneo, R.; Panella, M. Prediction in Photovoltaic Power by Neural Networks. *Energies* **2017**, *10*, 1003. [CrossRef]

15. Khandakar, A.; Chowdhury, M.E.H.; Kazi, M.-K.; Benhmed, K.; Touati, F.; Al-Hitmi, M.; Gonzales, A.J.S.P. Machine Learning Based Photovoltaics (PV) Power Prediction Using Different Environmental Parameters of Qatar. *Energies* **2019**, *12*, 2782. [CrossRef]

16. Kuzmakova, A.; Colas, G.; McKeehan, A. *Short-term Memory Solar Energy Forecasting at University of Illinois*; University of Illinois: Champaign, IL, USA, 2017.

17. Gensler, A.; Henze, J.S.B.; Raabe, N. Deep Learning for Solar Power Forecasting—An Approach Using Autoencoder and LSTM Neural Networks. In Proceedings of the 2016 IEEE International Conference on Systems, Man, and Cybernetics (SMC 2016), Budapest, Hungary, 9–12 October 2016; pp. 2858–2865.

18. Abdel-Nasser, M.; Mahmoud, K. Accurate photovoltaic power forecasting models using deep LSTM-RNN. *Neural Comput. Appl.* **2017**, *31*, 2727–2740. [CrossRef]

19. Kelleher, J.D.; Namee, B.M.; D'Arcy, A. *Fundamentals of Machine Learning for Predictive Data Analytics: Algorithms, Worked Examples, and Case Studies*; The MIT Press: Cambridge, MA, USA, 2015.

20. Openweather Ltd. OpenWeatherMap. Available online: https://openweathermap.org (accessed on 7 January 2019).

21. Kalisch, J.; Schmidt, T.; Heinemann, D.; Lorenz, E. *Continuous Meteorological Observations in High-Resolution (1Hz) at University of Oldenburg*; PANGAEA. Data Publisher for Earth & Environmental Science: Bremerhaven, Germany, 2015. [CrossRef]

22. Holmgren, W.F.; Hansen, C.W.; Mikofski, M.A. pvlib python: A python package for modeling solar energy systems. *J. Open Source Softw.* **2018**, *3*, 884. [CrossRef]

23. Reno, M.J.; Hansen, C.W.; Stein, J.S. *Global Horizontal Irradiance Clear Sky Models: Implementation and Analysis*; Sandia National Laboratories: Albuquerque, NM, USA, 2012.

24. Flores, E. A pragmatic view of accuracy measurement in forecasting. *Omega* **1986**, *14*, 93–98. [CrossRef]

25. Hall, M.A. *Correlation-based Feature Selection for Machine Learning*; University of Waikato: Hamilton, New Zealand, 1999.

26. Hochreiter, S.; Schmidhuber, J. Long short-term memory. *Neural Comput.* **1997**, *9*, 1735–1780. [CrossRef] [PubMed]

27. Gers, F.A.; Schmidhuber, J.; Cummins, F. Learning to Forget: Continual Prediction with LSTM. In Proceedings of the 9th International Conference on Artificial Neural Networks: ICANN'99, Edinburgh, UK, 7–10 September 1999; pp. 850–855.

28. Hua, Y.; Zhao, Z.; Li, R.; Chen, X.; Liu, Z.; Zhang, H. Deep Learning with Long Short-Term Memory for Time Series Prediction. *IEEE Commun. Mag.* **2019**, *57*, 114–119. [CrossRef]

29. Chollet, F. Keras. 2015. Available online: https://keras.io (accessed on 22 October 2019).

30. Hyndman, R.J.; Koehler, A.B. Another look at measures of forecast accuracy. *Int. J. Forecast.* **2006**, *22*, 679–688. [CrossRef]

31. Hyndman, R.; Athanasopoulos, G. *Forecasting: Principles and Practice*, 2nd ed.; OTexts: Melbourne, Australia, 2018; Available online: https://www.otexts.com/fpp2 (accessed on 4 November 2019).
32. Zinsser, B. Jahresenergieerträge unterschiedlicher Photovoltaik-Technologien bei verschiedenen klimatischen Bedingungen. Ph.D. Thesis, University Stuttgart, Stuttgart, Germany, 2010.

Article

Machine Learning Based Photovoltaics (PV) Power Prediction Using Different Environmental Parameters of Qatar

Amith Khandakar [1,*], Muhammad E. H. Chowdhury [1], Monzure- Khoda Kazi [2], Kamel Benhmed [1], Farid Touati [1], Mohammed Al-Hitmi [1] and Antonio Jr S. P. Gonzales [1]

[1] Electrical Engineering Department, College of Engineering, Qatar University, Doha 2713, Qatar
[2] Chemical Engineering Department, College of Engineering, Qatar University, Doha 2713, Qatar
* Correspondence: amitk@qu.edu.qa; Tel.: +974-4403-4235

Received: 27 May 2019; Accepted: 14 July 2019; Published: 19 July 2019

Abstract: Photovoltaics (PV) output power is highly sensitive to many environmental parameters and the power produced by the PV systems is significantly affected by the harsh environments. The annual PV power density of around 2000 kWh/m^2 in the Arabian Peninsula is an exploitable wealth of energy source. These countries plan to increase the contribution of power from renewable energy (RE) over the years. Due to its abundance, the focus of RE is on solar energy. Evaluation and analysis of PV performance in terms of predicting the output PV power with less error demands investigation of the effects of relevant environmental parameters on its performance. In this paper, the authors have studied the effects of the relevant environmental parameters, such as irradiance, relative humidity, ambient temperature, wind speed, PV surface temperature and accumulated dust on the output power of the PV panel. Calibration of several sensors for an in-house built PV system was described. Several multiple regression models and artificial neural network (ANN)-based prediction models were trained and tested to forecast the hourly power output of the PV system. The ANN models with all the features and features selected using correlation feature selection (CFS) and relief feature selection (ReliefF) techniques were found to successfully predict PV output power with Root Mean Square Error (RMSE) of 2.1436, 6.1555, and 5.5351, respectively. Two different bias calculation techniques were used to evaluate the instances of biased prediction, which can be utilized to reduce bias to improve accuracy. The ANN model outperforms other regression models, such as a linear regression model, M5P decision tree and gaussian process regression (GPR) model. This will have a noteworthy contribution in scaling the PV deployment in countries like Qatar and increase the share of PV power in the national power production.

Keywords: PV power prediction; artificial neural network; renewable energy; environmental parameters; multiple regression model

---

1. Introduction

Due to global warming and climate change concerns, many pieces of energy legislation and incentives to promote the use of renewable energy have been established worldwide. Among renewable energy resources, photovoltaics (PV) energy is one of the most-promising supplements for fossil fuel-generated electricity, and has received a lot of attention recently it is abundant, inexhaustible, and clean. Arabian Peninsula is blessed with solar irradiance of more than 2000 kWh/m^2 annually [1]. Due to this high amount of solar irradiance in this region, PV technology has potential in comparison to other renewable energy sources (e.g., wind energy or tidal energy). Solar energy is gaining popularity day-by-day, due to some other salient features like noise and pollution free technology with low maintenance cost. Together with the ever-decreasing prices of PV modules and continuous depletion of fossil fuel, it is expected that the penetration level of PV energy into modern electric power and energy systems will further increase. However, due to the chaotic and erratic nature of the weather systems, the power output of the PV energy system exhibits strong uncertainties regarding its intermittency, volatility, and randomness. These uncertainties may potentially degrade the real-time control performance, reduce system economics, and, thus, pose a great challenge for the management and operation of electric power and energy system. Predicting the power efficiency of a PV power plant is very crucial in making the best economic benefit out of it. The PV output power is directly related to the solar irradiance on the PV panels, which is a well-known fact. However, other meteorological parameters (e.g., ambient temperature, relative humidity, wind speed and dust accumulation) have been reported to influence the PV efficiency as well [2–5]. This association substantially increases in the harsh environment of the Gulf region. There are several recent works that showed the negative influence of dust accumulation on the PV panel on PV output power prediction [6,7]. The authors hypothesize that suitable weather parameters at a specific geographic location can be an important aspect for PV power forecasting. Moreover, PV power prediction can be particularly useful when multiple energy sources are combined to produce a hybrid energy matrix. Since solar energy source is highly intermittent, it is difficult to maintain system stability with an intolerable proportion of renewable energy injection. Solar power forecasting can be used to improve system stability by providing approximated future power generation to system control engineers. This will help the utility companies to devise a mechanism to design a switching controller to switch between the energy sources in a hybrid energy source [8]. It can be hypothesized that the key design parameters for the switching controller will be linked to the environmental parameters due to its potential effect in PV power generation.

Several recent works reported different approaches for PV output power forecasting and estimation. In detail, the specific literature on PV plant power production estimation presents three different types of models: Phenomenological, stochastic/statistical learning and hybrid ones. Deterministic approaches, based on physical phenomena, try to predict PV plant output by considering the electrical model of the PV devices constituting the plant using software like PVSyst, System Advisor Model (SAM). A deterministic approach was used to model electrical, thermal, and optical characteristics of PV modules [9]. Most of the published researches for PV power forecasting concentrate only on the deterministic forecast, i.e., point forecast. Deterministic forecasting methods sometimes fail to evaluate the uncertainties exhibited in PV power data. Probabilistic PV power forecasting models that can statistically describe these uncertainties have received much attention recently. One of the mainstreams for generating probabilistic uncertainty is to use an ensemble of deterministic forecaster. The main shortcoming of ensemble-based PV power forecasting model is their high computational cost—which may cause a real-time problem for practical implementation. Another demerit with respect to the methodologies used in deterministic and probabilistic PV power forecasting is their shallow learning models. Because of the complicated nature of the weather system, these shallow models may be inadequate to fully extract the corresponding nonlinear features and static traits in PV power data. Therefore, more investigation on the deterministic technique to provide high accuracy by optimization of artificial neural network (ANN) can lead to better performance.

By using the hourly solar resource and meteorological data, the model has been validated for different modules types. Statistical and machine learning ones, such as: Artificial neural network (ANN), support vector machine (SVM), multiple linear regression (MLR), adaptive neuro-fuzzy inference system (ANFIS) operate without any a priori knowledge of the system under consideration. They try to "understand" the relation between inputs and outputs by adequately analyzing a dataset containing acquired input and output variables collection. Statistical learning algorithms have many advantages. Firstly, they are able to learn from them, and they can also work in the presence of incomplete data. Secondly, once trained, they are able to generalize and to provide predictions. Their features make them suitable to be used in different contexts. Different machine learning (ML) algorithms to predict output power have been investigated for other renewable energy sources rather than solar energy [10,11]. ML gives insights into the properties of data dependencies and shows the significance of individual characteristics in datasets [10]. Jawaid et al. [12], compared different ANN algorithms without showing the details of the prediction model and their comparative performance numerically. Several other works predicted the solar irradiance using machine learning techniques, rather than the PV power itself [13–15]. Some of the researches were only focusing on training and testing of one machine learning algorithm for PV power prediction [16]. An adaptive ANN was used to model and size a stand-alone PV plant, using a minimum input dataset [17,18]. An ANFIS was applied to model the different devices constituting a PV power system and its output signals [19]. A linear regression model and an ANN were applied to estimate daily global solar radiation [20,21]. Thirdly, a hybrid model can combine different models to overcome limitations characterizing one single technique [22]. In addition, "ensemble" methods [23] build predictive models by integrating multiple strategies in order to improve the overall prediction performance.

It can be noted that the previous works extensively explore different ML-based prediction models; however, there is still scope for improvement in prediction accuracy. In this manuscript, authors have reported the following new contributions:

1. Development of a PV and weather system testbed with the continuous calibration of sensors. This continuous calibration of the sensors ensures that the weather data is measured accurately.
2. A moderately sized dataset, with several PV and environmental parameters, was acquired during the two years' deployment period of the PV system. The prediction model reported in this manuscript were trained and tested on the actual dataset, which can be shared publicly upon request.
3. Comparison of several multiple regression models and ANN-based prediction models.
4. Exploring the ANN extensively and finding the best prediction model using an ANN that can provide the lowest Root Mean Squared Error (RMSE). It was also evaluated that the prediction is a biased prediction or not.
5. Exploring the best set of features that could predict the PV power accurately.

The rest of the paper is organized as follows: Section 2 describes the materials and methodologies used in the paper; Section 3 describes the analysis techniques used in this work, Section 4 discusses the results, and finally, Section 5 concludes the work.

2. Materials and Methodology

To analyze the effect of PV performance due to the PV and environmental parameters, an in-house PV setup was designed and implemented, which acquired and recorded the PV performance and environmental parameters. The experimental setup comprised of two sub-systems (as shown in Figure 1: One at the rooftop which was equipped with sensors for acquiring the data and the other sub-system in the laboratory which was used for archiving the data and plotting them in real time.

The system on the terrace included PV modules (characteristics are shown in Table 1), signal conditioning circuits for all sensors of weather parameters, an Arduino Mega 2560 microcontroller, a wireless transceiver (XBee/Wifi) and a controllable electronic resistive load along with a DC-DC converter. Maximum power point tracking (MPPT) algorithm was implemented to produce pulse width modulation (PWM) signals to drive the controllable electronic load, which was emulating different levels of currents and voltage across the load without varying the actual load resistance itself. Power-voltage (P-V) and current-voltage (I-V) curves were plotted using the calibrated voltage and current sensors' data across the emulated electronic load. The MPPT was used to adjust the orientation of the PV panels to optimize solar irradiance, while achieving maximum PV output power yield. The sub-system at research laboratory consisted of XBee/Wifi adapters, connected to a workstation, for receiving and logging data from the rooftop subsystem wirelessly. All measurements from the rooftop sub-system sensors were received on demand or periodically at a specified time interval by these wireless adapters. Received data were recorded as a LabVIEW measurement file and displayed on the LabVIEW front panel numerically on the workstation screen. The recorded parameters were also uploaded to an open Internet of things (IoT) data platform called Thingspeak [24] for widespread access. Both sub-systems communicate through an Xbee/EtherMega shield connected to the Arduino Mega 2560 microcontroller. The overall block diagram of the PV system is depicted in Figure 1.

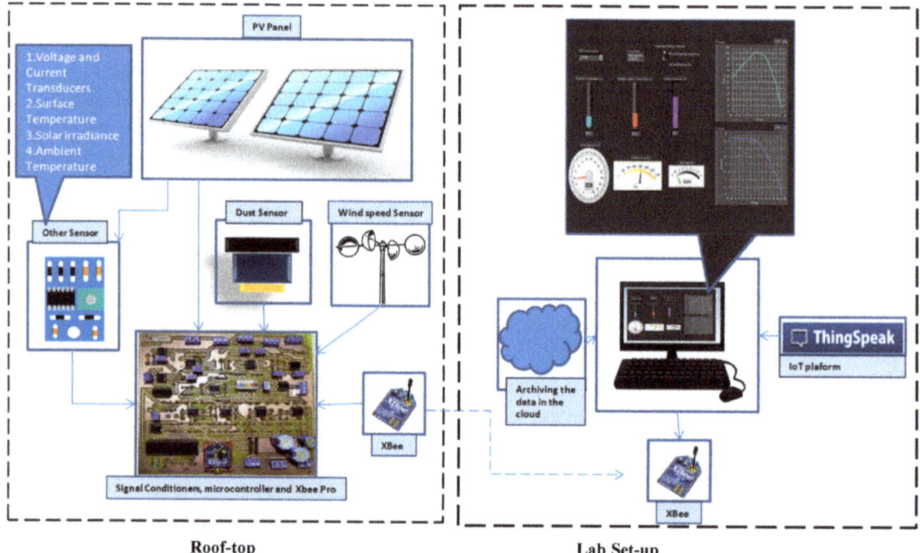

Figure 1. Block diagram of experimental photovoltaics (PV) system set-up: Roof-top sub-system (left) and lab sub-system (right).

Table 1. Poly-crystalline silicon PV module characteristics (manufacturer: PTL Solar, UAE).

Max Power at STC [1]	Area (m^2)	Voc (V)	Isc (A)
80	0.6426	21	5.24

[1] Standard Test Conditions (STC) −1000 W/m^2 and 25 °C.

2.1. Sensor Calibration

The hardware components consist of a microcontroller (Atmega32), six sensors with signal conditioning circuits, a DC-DC Buck-boost converter and long-range XBee Pro wireless modules. The six sensors read the ambient temperature LM35DT (http://www.ti.com/product/LM35/datasheet/pin_configuration_and_functions#SNIS1593406), solar irradiance SP110 (http://www.apogeeinstruments.co.uk/pyranometer-sp-110/), humidity HSM-20G (http://www.geeetech.com/wiki/index.php/Humidity_/Temperature_Sensor_Module_HSM-20G), dust GP2Y1010AU0F(https://digitalmeans.co.uk/shop/compact_optical_dust_sensorgp2y1010au0f), wind speed anemometer (https://www.adafruit.com/products/1733) and the PV module surface temperature sensor PT100 (http://export.farnell.com/labfacility/rtf4-3/sensorpt100-patch-3m/dp/1633500). The PT100 was fixed at the backside of the PV module using a highly thermally conductive adhesive. Also, the voltage and current transducers are used to sense voltages and currents from the PV module in order to plot the P-V and I-V curves. Before installing the overall system, all sensors were tested and calibrated methodically. The BK PRECISION-720 humidity and temperature meter are used as a reference when calibrating the humidity, surface and ambient temperatures' sensors. The temperature of a heating element is controlled to generate various ambient and surface temperatures for sensors' calibration. The HSM-20G sensor was calibrated using steam generated inside an encapsulated box where both the HSM-20G humidity sensor along with the BK-PRECISION-720 m was placed. Simultaneous measurements were performed by taking the readings from both the BK PRECISION meter and the humidity and temperature sensors. The commercially available INSPEED VORTEX wind speed sensor, using a CATEYE VELO8 display, was used to calibrate the anemometer (wind speed sensor with analog output). The voltage and current sensors were calibrated using Yokogawa GS510 SMU (source measurement unit) with standard procedures. For the dust sensor, we used the firm calibration curve. In the laboratory, different dust levels were deposited on the sensing element of the GP2Y1010AU0F sensor, which were found to be within the operating range of this sensor. All the calibration results were repeatable. The output voltages of the various sensors were amplified in order to match the full-scale analog range of the microcontroller's analog-to-digital converter (ADC) without causing ADC saturation errors. However, the dust and PV surface temperature sensors do not directly provide analog signals, and a circuit was developed so that they can be interfaced to the microcontroller. For the PV surface temperature sensor, which is a resistive type (resistance temperature dependent, RTD), a constant current source circuit is devised to provide an output voltage that is linearly dependent to the variation of resistance. For the dust sensor, its output pulse lies on a 0.32 ms pulse width that needs to be acquired correctly by sampling at 0.28 ms of the pulse. All the sensors conditioning circuits are integrated into a single printed circuit board (PCB).

The maximum power point tracking (MPPT) used a back-boost converter, which serves as a direct load to the PV modules. Through a gate driver circuit, by adjusting the firing angles of the insulated gate bipolar transistor (IGBT) switch of the back-boost converter, the microcontroller keeps adjusting the output voltage of the PV modules until reaching the maximum power point. Then, the microcontroller reads the corresponding voltages and currents of the PV module through the voltage and current transducers above discussed. Furthermore, two XBee Pro transceivers were used to transfer the measured data wirelessly from the rooftop sensors and electronics modules to a LABVIEW-based monitoring station (Figure 1), which plots I-V curves, P-V curves and also save the measured data for future analysis.

2.2. Machine Learning-Based Prediction

The process of applying ML on any dataset to predict unknown output values consists of three general phases (Figure 2): Pre-processing of data to extract features, training the prediction models and observing validation accuracy on training dataset and evaluation of the pre-trained model for the test dataset. Firstly, the acquired dataset was pre-processed to make it suitable in format, free of anomalies, such as missing, outliers and erroneous data values. Most importantly, then the relevant features were extracted. We have used the collected parameters, e.g., Temperature, Relative Humidity, PV surface Temperature, Irradiance, Dust accumulation and Wind Speed as features for the training and testing; which eased this sub-task. Training and testing dataset was created using the cvpartition function in Matlab, which allows to randomly partition the training and testing data into 80% and 20% respectively. In this study, 380 instances were used for training and validation, whereas 95 instances were used for testing. In the prediction phase, data with known output response values were used for training several ML algorithms using Regression Learner from Statistical and Machine Learning Toolbox and Neural Network toolbox of Matlab.

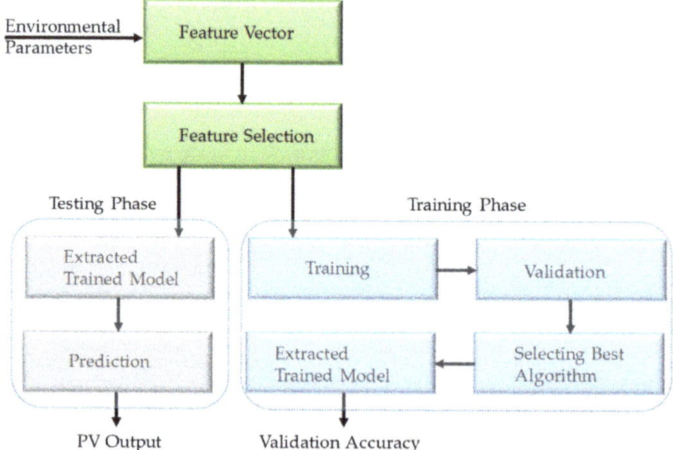

Figure 2. Block diagram of the machine learning-based training and prediction stages.

An additional step of feature selection can be used to optimize the trained algorithm. Selection of features is the process of selecting a subset of relevant, high-quality and non-redundant features to create learning models with better accuracy [25,26]. Several feature reduction techniques were tested to obtain the optimized prediction model with the selected subset of features. In the testing phase, the best performing pre-trained models were evaluated for test dataset, and the evaluation parameters were computed to perform the reliable statistical evaluation. In addition, this process can be made adaptive and can be accomplished to improve model quality as historical data gradually becomes available.

2.3. Features Selection

After processing the data acquired from the data acquisition system, as shown in Figure 1, the acquired PV and environmental parameters were used as features for the training, validation and testing purpose. However, it is important to evaluate whether the complete set of environmental parameters are necessary for the prediction or the feature number can be reduced. Correlation feature selection (CFS) and Relief feature selection (ReliefF) techniques were used to select most contributory features. CFS technique selects feature sub-sets, based on correlation-based heuristic evaluation function, and uses a sub-set search method and calculates the level of redundancy between features in all sub-sets created. It then evaluates the importance of sub-sets, where the low inter-correlation, but high-correlation to the target result are selected. ReliefF is an instance-based algorithm that assigns a relevance weight to each feature that reflects its ability to differentiate class values. Because of sufficient data, ReliefF has the potential to detect interactions higher than pairwise. In order to select the best subset with ReliefF from the ranked features, the lowest ranked features were iteratively removed until the best result was achieved.

2.4. Prediction Models

There are many predictive methods, based on ML, and they can perform differently for the given datasets. Several simple and popular regression and prediction models were attempted in this work to estimate the PV output power. These are namely simple linear regression [27], gaussian process regression (GPR) [28] from the regression learner, M5P regression tree [29]. The simple linear regression model has a linear relationship between the output response and the input parameters. GPR involves a Gaussian process using lazy learning and a measure of the point similarity (kernel function) to predict the value from the training data for an unseen point. The prediction is not only an estimate for that point, but also information about uncertainty. It is a one-dimensional Gaussian distribution (which at that point is the marginal distribution) [28]. In the M5P regression tree model, a tree-based model with an M5 algorithm developed by Quinlan, 1992 [29], combines a conventional decision tree with the linear regression functions at each branch end of the tree; it creates a model that predicts the target's value by learning simple decision rules [30]. In other words, predicted power would be the result of "if... then... else..." statements [8].

In this work, the ANN was also used to predict daily PV output power, which is a very popular machine learning tool for classification and regression application [31]. Figure 3 provides the layered structure of the ANN along with the detailed depiction of forward propagation and weight adjustment. The ANN tries to replicate the machine learning in the similar nature of the human brain with a layered structure (input, hidden and output layers) (Figure 3). Models of ANNs take the form of artificial neurons where a number of inputs are given to each neuron. The activation function is applied to these inputs resulting in neuron activation level (neuron output value) and learning knowledge is provided in the form of training inputs and output pairs (Figure 3). More details on the various training functions/algorithms are listed in Table 2. Each training function has their own advantages and disadvantages, and they work differently with different datasets. It was necessary to explore all the training functions to check which of them works the best for the dataset developed and used in this work.

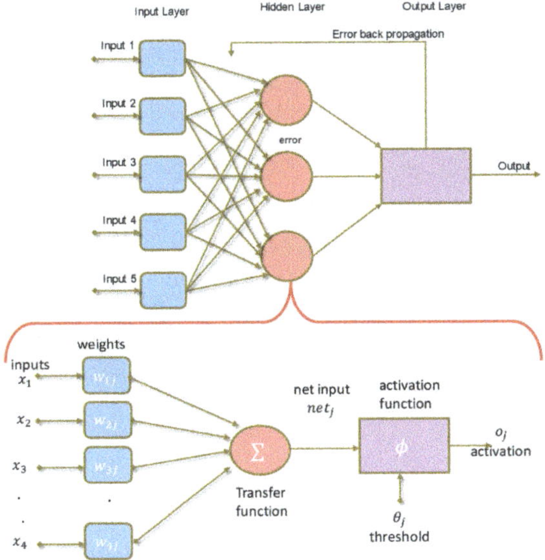

Figure 3. The layered structure of the artificial neural network (ANN).

Table 2. Training algorithms of a ANN in Matlab and their description.

Matlab Algorithm Functions	Description
trainlm	Levenberg-Marquardt
trainbfg	BGFS Quasi-Newton
trainrp	Resilient Back propagation
trainscg	Scaled Conjugate Gradient
traincgf	Conjugate Gradient with Powell/Beale Restarts
traincgp	Polak-Ribiere Conjugate Gradient
trainoss	One Step Secant
traingdx	Variable Learning Rate Back propagation
trainbr	Bayesian Regularization back propagation

The artificial neural network and other regression-based predictors were implemented on Matlab 2017b version on a workstation with the below specification:

Processor: Intel® Core™ i7-7500U CPU @ 2.70 GHz

Installed memory (RAM): 16.0 GB

System type: 64-bit operating system, x64 based processor

In this work, various combinations of the number of hidden layers and training functions were explored to find the best combination that predicts the PV power most accurately, as shown in Figure 4. An in-house written Matlab script was used to train automatically 10 different training functions (Table 2). The script was written to change the number of hidden layers from 10 to 300 in increments of 10 and each training was performed using a particular training function and a specific number of hidden layers. This is due to the fact that each run provides different network and the best network out of that 10 tries is selected for that specific combination of training function and number of hidden layers. Later, a comparison was carried out between the best set of networks with different training functions, and the number of hidden layers and the best network for each function amongst all the combination was selected. A final comparison was made with the best network of different training functions, and the best network among all functions was selected. Figure 4 shows the flowchart of how the best model selection was carried out.

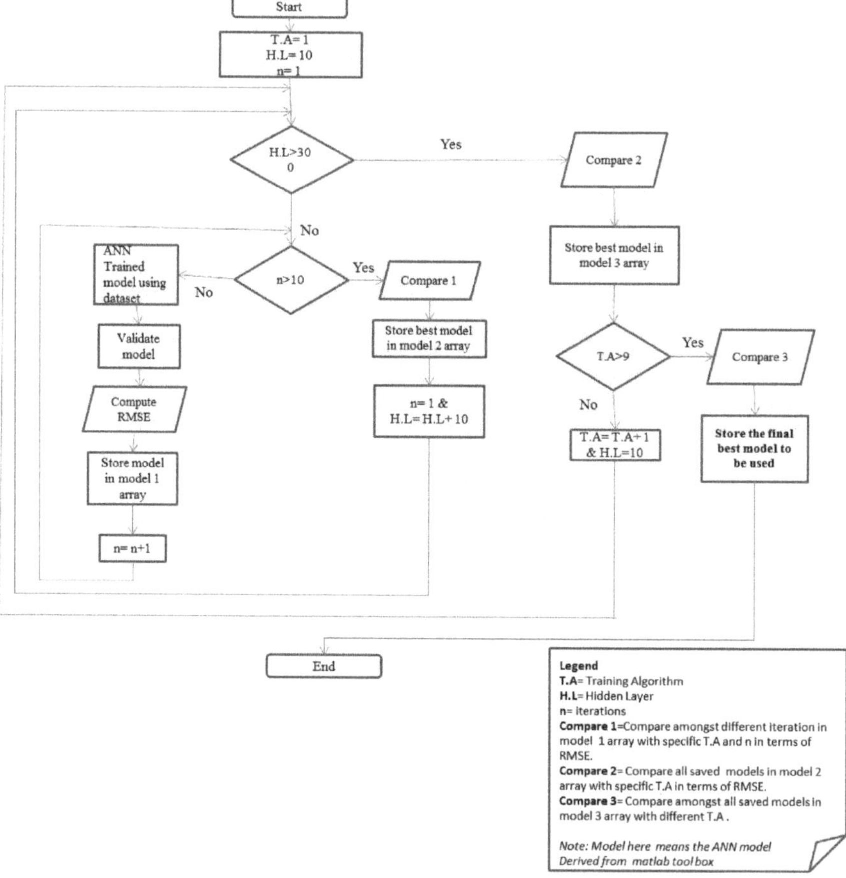

Figure 4. Method to find the best ANN to predict PV power.

Table 3 summarizes the network settings for the ANN-based PV power prediction. It can be seen in Table 3 that the optimum number of hidden layers providing the best model were different for all features, CFS technique and ReliefF technique.

Table 3. Neural network parameters. CFS, correlation feature selection.

Parameters		Values
Number of hidden layer	when all features were used	60
	when CFS filtering was used	260
	when ReliefF filtering was used	180
Training data		70%
Validation data		15%
Testing data		15%
Number of Folding in Cross validation		5

2.5. Bias Calculation and Correction in Prediction

The biased forecast is described as a tendency to either over-forecast (the forecast is more than the actual), or under-forecast (the forecast is less than the actual). To improve the forecast accuracy in the presence of bias is possible if the bias is correctly identified. The correction of the forecast error can be achieved by adjusting the forecast by the appropriate amount in the appropriate direction, i.e., increase it in the case of under-forecast bias, and decrease it in the case of over-forecast bias.

Two different techniques are used in this work to calculate the bias in the forecast:

(i) Tracking signal-based technique
(ii) NFM technique

Tracking Signal-Based Technique

The other common metric used to measure forecast accuracy is the tracking signal. The "Tracking Signal" quantifies "Bias" in a forecast. No product can be planned from a badly biased forecast. Tracking Signal is the gateway test for evaluating forecast accuracy. The tracking signal in each period is calculated using the formula as follows:

$$\text{Tracking signal} = \frac{\text{Actual} - \text{Forecast}}{\text{ABS}(\text{Actual} - \text{Forecast})}$$

Once this is calculated, for each period, the numbers are added to calculate the overall tracking signal. A forecast history totally void of bias is returned a value of zero, with 12 observations, the worst possible result would return either +12 (under-forecast) or −12 (over-forecast). Such a forecast history generally returns a value greater than 4.5 or less than negative 4.5 would be considered out of control. NFM Technique

Normalized Forecast Metric (NFM) can be used to measure the bias. The formula of NFM to calculate bias is:

$$\text{NFM} = \frac{(\text{Forecast} - \text{Actual})}{(\text{Forecast} + \text{Actual})}$$

As can be seen, this metric stays between −1 and 1, with 0 indicating the absence of bias. Consistent negative values indicate a tendency to under-forecast, whereas consistent positive values indicate a tendency to over-forecast. Over a 12 period window, if the added values are more than 2, we consider the forecast to be biased towards over-forecast. Likewise, if the added values are less than −2, we consider the forecast to be biased towards under-forecast. A forecasting process with a bias eventually get off-rails unless steps are taken to correct the course from time to time.

The bias correction and change factor methods work well for bias correcting non-stochastic variables. The quantile mapping (QM) technique removes the systematic bias in the predicted output and has the benefit of accounting for biases in statistical downscaling approaches.

3. Analysis

Several statistical analyses were carried out to evaluate the performance of machine learning algorithms for PV output power prediction. To compare models' performances, various evaluation metrics are commonly used: (i) Correlation coefficient, which measures the linear dependency between two variables; (ii) mean absolute error (MAE), which takes the average of the absolute difference between the real and predicted values; (iii) mean square error (MSE) measures the average squared error and the square difference between target and predicted values were calculated and averaged; (iv) root mean square error (RMSE) is the square root of MSE and similar to MAE, but it averages the squares of the difference and then finds the square root where it actually puts a heavier weight on larger errors; and (v) coefficient of determination (R^2) always lies between $-\infty$ to 1 and is the ratio between how well the prediction model in comparison to naive mean model. These parameters provide better descriptions of predictor performance [32].

$$\text{Correlation Coefficient}, r = \frac{\text{Con}(X, Y)}{\sigma_x \sigma_y} \tag{i}$$

$$\text{Mean absolute error, MAE} = \frac{1}{n} \sum_n |X - Y| \tag{ii}$$

$$\text{Mean Squared Error, MSE} = \frac{\sum |X - Y|^2}{n} \tag{iii}$$

$$\text{Root mean square error, RMSE} = \sqrt{\frac{\sum |X - Y|^2}{n}} = \sqrt{\text{MSE}} \tag{iv}$$

$$\text{coefficient of determination, or } R^2 = 1 - \frac{\text{MSE (Model)}}{\text{MSE (Baseline)}} \tag{v}$$

$$\text{MSE(Baseline) is calculated by } \frac{\sum |X - \overline{Y}|^2}{n} \tag{vi}$$

where X is the actual data vector, Y and $\overline{Y}$ are the predicted data vector and mean of the predicted data vector.

Different ANNs and regression models were compared using MAE, MSE, RMSE, r-value and R^2 value. After extensively exploring the ANN training functions that provide a better prediction of the PV, Bayesian regularization backpropagation algorithm was used from the neural network toolbox of Matlab. A built-in Matlab function for Bayesian regularization backpropagation minimizes a linear combination of squared errors and weights and then determines the correct combination so as to produce a network that generalizes well. It updates the weight and bias values according to Levenberg-Marquardt optimization [33]. The best ANN selection technique (as shown in Figure 4) was repeated for three different scenarios: (i) When all the features were used; (ii) when features selected using CFS technique are used; and (iii) when features selected using ReliefF technique.

4. Results and Discussion

The prototype system (setup shown in Figure 1) was used for collecting the PV and environmental parameters and PV power output data from the period of November 2014 until October 2016. Summary of the PV and environmental parameters and the data used for deriving the predictive model of the PV power is shown in Table 4.

Table 4. Details of the environment parameters used for the predictive model.

Environmental Parameters	Max	Min	Unit
Temperature	61.0000	14.6365	Degree Celsius
Relative Humidity	90.76345	27.8157	%
PV surface Temperature	74.3968	9.3037	Degree Celsius
Irradiance	1033.5290	38.0076	W/m^2
Dust Accumulation	1.1142	0.0553	mg/m^3
Wind Speed	34.2437	0.5893	km/h
Power	114.2017	0.0368	W

Table 5 shows that Temperature, PV Temperature, Irradiance and Accumulated dust were the selected feature using CFS algorithm, whereas the irradiance, wind speed, PV temperature, and environmental temperature were selected as highly ranked features by ReliefF technique.

Table 5. Selected features vector.

Selection Technique		Selected Features
Filters	CFS	Temperature PV Temperature Irradiance Dust Accumulation
	ReliefF (Ranked Features)	Irradiance Wind Speed PV Temperature Temperature

Table 6. summarizes the evaluation matrix for different regression techniques evaluated in this study. Linear regression, M5F tree and GPR were implemented using MATLAB with all the features, and also with reduced features using CFS and ReliefF. The reason for selecting these regression techniques, because they provided the best performance compared to other regression techniques commonly used. Table 6A shows the performance matrix of the different regression techniques in the validation phase, whereas Table 6B shows their performance matrix for the unseen test dataset.

Table 6. (A) Performance comparison between the various regression techniques (validation phase). (B) Performance comparison between the various regression techniques (testing phase).

(A)

Selection Criterion	Features	Linear Regression Model		M5P Tree Model		GPR	
Without feature Selection (all features are used)	Temperature Relative Humidity PV Temperature Irradiance Dust Accumulation Wind Speed	r MAE MSE RMSE R^2	00.9853 05.3592 59.1440 07.6905 00.8100	r MAE MSE RMSE R^2	00.8908 06.7867 80.1867 08.9547 00.9167	R MAE MSE RMSE R^2	00.9833 04.2601 44.8460 06.6967 00.8600
CFS	Temperature PV Temperature Irradiance Dust Accumulation	r MAE MSE RMSE R^2	00.9814 05.7514 66.7100 08.1680 00.7900	r MAE MSE RMSE R^2	00.9727 06.6892 92.8209 09.6344 00.7048	R MAE MSE RMSE R^2	01.0000 05.1657 61.4970 05.1657 00.8100
ReliefF	Irradiance Wind speed PV Temperature Temperature	r MAE MSE RMSE R^2	00.9837 05.3727 60.0210 07.7473 00.8100	r MAE MSE RMSE R^2	00.9694 06.9110 103.9819 10.1972 00.6693	R MAE MSE RMSE R^2	01.0000 04.2257 46.4300 06.8139 00.8500

Table 6. *Cont.*

(B)							
Selection Criterion	**Features**	**Linear Regression Model**		**M5P Tree Model**		**GPR**	
Without feature Selection (all features are used)	Temperature Relative Humidity PV Temperature Irradiance Dust Accumulation Wind Speed	r MAE MSE RMSE R^2	00.8550 13.1340 212.1000 14.5639 00.7310	r MAE MSE RMSE R^2	00.7139 16.7506 418.5574 20.4587 00.5096	R MAE MSE RMSE R^2	00.8662 12.6338 195.7792 13.9921 00.7502
CFS	Temperature PV Temperature Irradiance Dust Accumulation	r MAE MSE RMSE R^2	00.8672 12.4394 194.3095 13.9395 00.7520	r MAE MSE RMSE R^2	00.6809 27.3516 1050.19 32.4066 00.4636	R MAE MSE RMSE R^2	00.8895 11.5247 161.6557 12.7144 00.7912
ReliefF	Irradiance Wind speed PV Temperature Temperature	r MAE MSE RMSE R^2	00.8650 12.7339 197.5275 14.0544 00.7482	r MAE MSE RMSE R^2	00.5797 18.8818 614.7668 24.7945 00.3361	R MAE MSE RMSE R^2	00.8815 11.5482 173.3196 13.1651 00.7771

This is clearly revealed from the tables in Table 6, the overall performance of the evaluated regression techniques for the testing dataset was not similar to that of the training dataset. For the testing dataset, CFS-based feature selection technique outperforms all features and ReliefF-based techniques for Linear and GPR regression techniques; however, M5P outperforms for all features-based technique.

Training and validation performance of the ANN-based prediction models were shown in Figures 5 and 6 for the three different techniques. The validation performance, as shown in Figure 5, the best training performance was observed at different epochs for different techniques. Out of the numerous models developed by the ANN using the different set of features, the best epochs were obtained 707, 91 and 598 respectively. This could be used in future to derive the model for predicting the PV power. Figure 6 shows the error histogram with 20 bins, where bins represent the vertical bars in the graph. Total error from each neural network ranges from (−25.84 to 16.88), (−49.08 to 26.43) and (−43.83 to 20.03) respectively. Each vertical bar represents the number of samples from corresponding dataset, which lies in a particular bin. There is a zero-error line in the graph, and more than 80% of the errors lie within +10 Watt. It is typically assumed that any algorithm which could predict the output where 80% of the error lying within 10% (i.e., approximately 10 W) of the target value, is a very good predictive model [34].

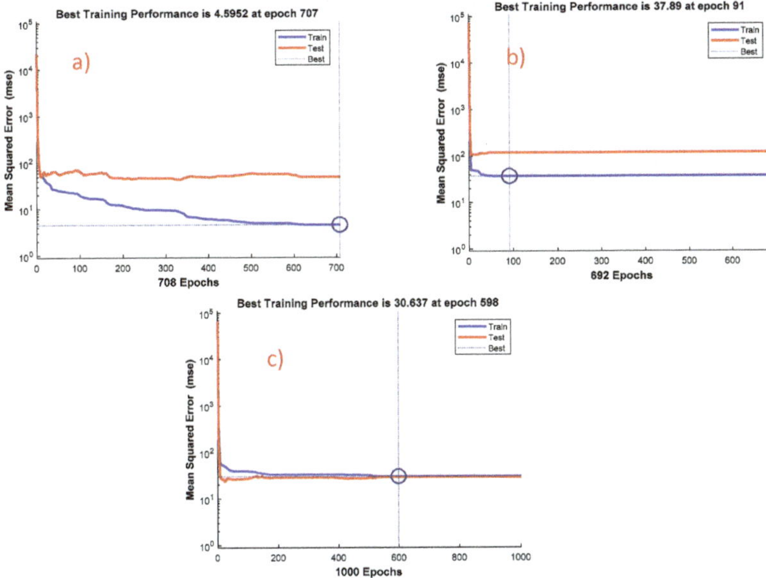

Figure 5. Comparison of the MSE for different techniques: (**a**) With all features; (**b**) CFS feature selection technique; (**c**) ReliefF feature selection technique for training and validation.

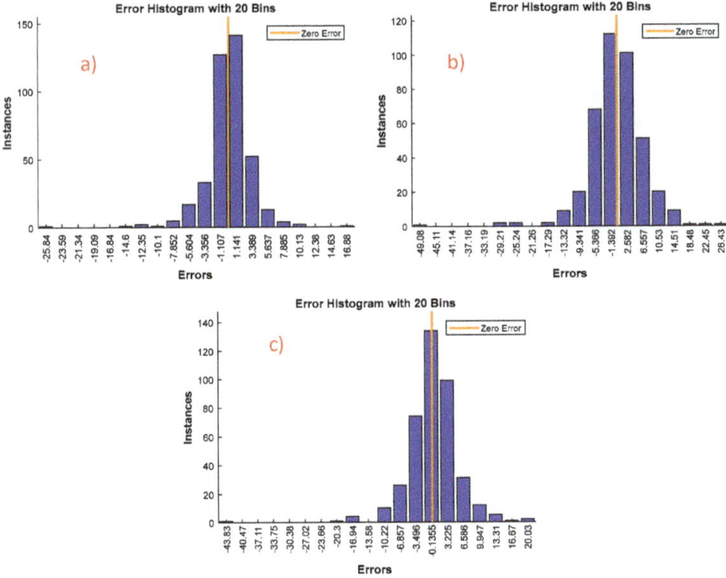

Figure 6. Error Histogram (**a**) with all features; (**b**) CFS feature selection technique; (**c**) ReliefF feature selection technique for training and validation.

Figure 7 shows the relation between the original power output and the predicted power output using the best epochs from the ANN. The dots represent the original power output, the blue line is the best linearized predictive model derived from ANN, and the dotted line represents the best linear relation for the true target. The difference between the dotted line and the blue line is represented by the correlation coefficient, i.e., in Figure 7, r represents the successful linearized model developed by the ANN using three different techniques. However, the difference between the predictive model trend-line and the true trend-line was noticed minimum for all features, which is evident in Figure 6a and Table 7.

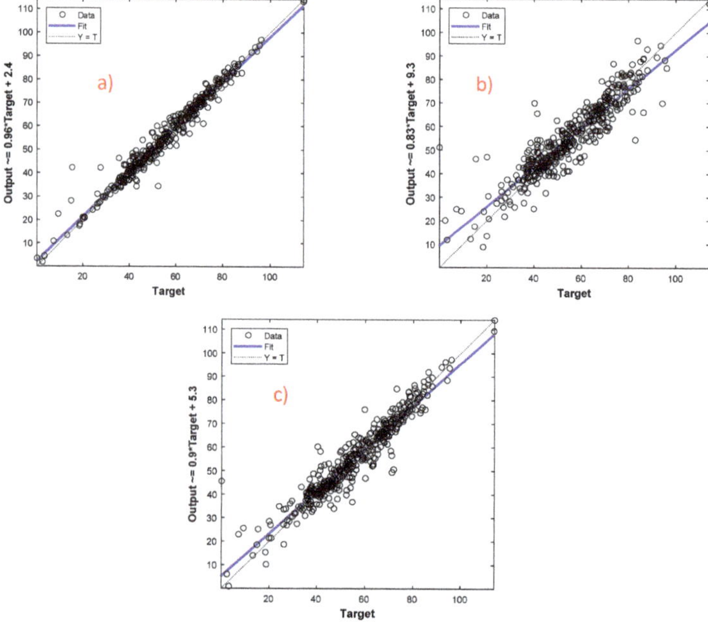

Figure 7. Relation between the original power output and the predicted power output in training and validation: (**a**) With all features; (**b**) CFS feature selection technique; (**c**) ReliefF feature selection technique.

The best ANN model found using the Figure 4 approach, was validated during training and validation process and results were reported in Table 7A. The testing dataset was used to validate the trained model, and the results were shown in Table 7B. As seen in Table 7A, the ANN provides really good prediction compared to the other regression techniques. By using the various feature selection techniques, it was found that the ANN provides the lowest RMSE, i.e., 2.1436 for all feature set; whereas ReliefF feature selection technique provides the second-best performance in terms of RMSE, i.e., 5.5351 in the validation phase. Similar performance was observed for the testing dataset as well. All features technique outperforms others, and the best RMSE was observed to be 5.4784. Figure 8 shows the comparison of the ANN predicted power with the actual power using test dataset for different sets of features. It is apparent from Figure 9 that there is a consistent over-forecasting or under-forecasting in the predicted output, i.e., the predicted output was biased in prediction for some instances. Figure 9 shows the biased forecasting for all the features-based predictions, as shown in Figure 8a. It can be seen that tracking signal-based bias calculation can identify bias more accurately than the NFM technique.

Table 7. (A) Performance comparison between the various ANN techniques (validation phase). (B) Performance comparison between the various ANN techniques (testing phase).

(A)			
Selection Method	**Features**	**ANN**	
Without features Selection (all features are used)	Temperature Relative Humidity PV Temperature Irradiance Dust Accumulation Wind Speed	r MAE MSE RMSE R^2	00.9967 02.1275 04.5952 02.1436 00.9641
CFS	Temperature PV Temperature Irradiance Dust Accumulation	r MAE MSE RMSE R^2	00.9852 04.8239 37.8900 06.1555 00.8396
ReliefF	Irradiance Wind speed PV Temperature Temperature	r MAE MSE RMSE R^2	00.9910 03.7225 30.6370 05.5351 00.9032

(B)			
Selection Method	**Features**	**ANN**	
Without features Selection (all features are used)	Temperature Relative Humidity PV Temperature Irradiance Dust Accumulation Wind Speed	r MAE MSE RMSE R^2	00.9856 03.2945 30.0134 05.4784 00.9538
CFS	Temperature PV Temperature Irradiance Dust Accumulation	r MAE MSE RMSE R^2	00.9713 07.7453 130.3845 11.4186 00.8213
ReliefF	Irradiance Wind speedPV Temperature Temperature	r MAE MSE RMSE R^2	00.9804 05.0234 36.2345 06.0145 00.9013

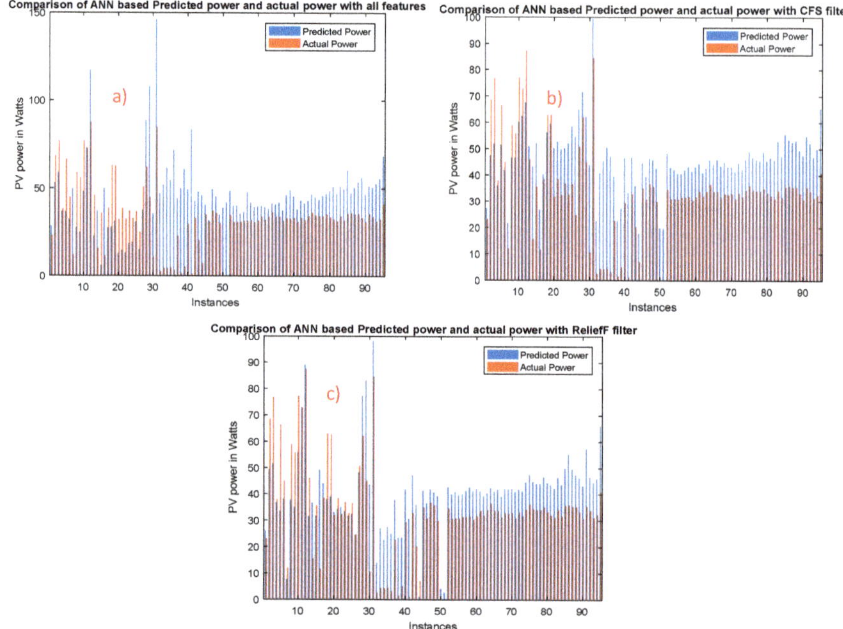

Figure 8. Comparison of the ANN predicted power with the actual power using test dataset: (**a**) With all features; (**b**) CFS feature selection technique; (**c**) ReliefF feature selection technique.

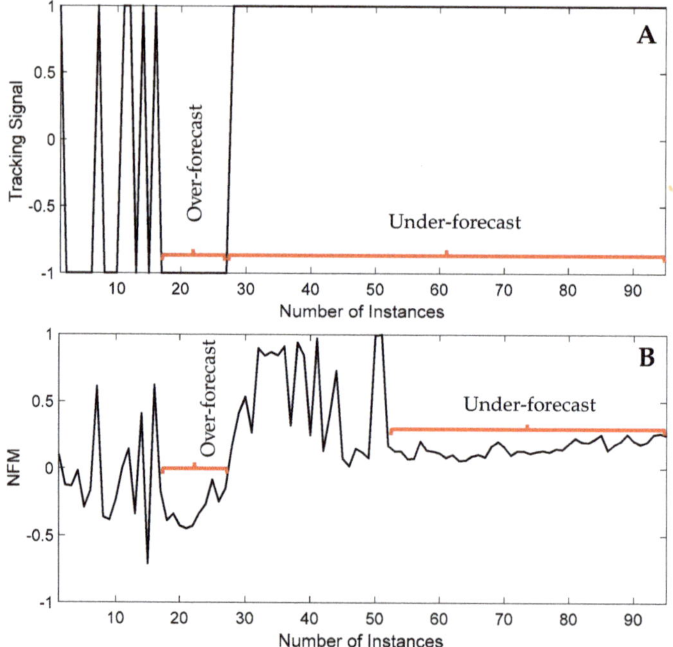

Figure 9. Bias calculation for a biased prediction for all features based PV power prediction: (**A**) Using tracking signal technique; (**B**) using NFM technique.

In order to confirm if the ANN derived models are efficient in predicting PV power, test dataset (not included in training) was used for testing the performance of the ANN algorithms. From Figures 8 and 10, it is evident that using all the features, it is closer to predict the actual power than CFS and ReliefF filtering. It should be noted that more data should be included in the training dataset to increase the accuracy of prediction. It has been mentioned in the literature by several groups that the bias correction can improve accuracy; however, some other article showed a contrary performance. Most importantly, by incorporating bias correction in the prediction algorithm, overall computational complexity and cost will significantly increase. Moreover, shallow Convolutional Neural Network (CNN) can be used for PV power prediction to remove the biasing problem. The ANN's computational complexity could be less than CNN or deep learning approach and more flexible for real-time prediction.

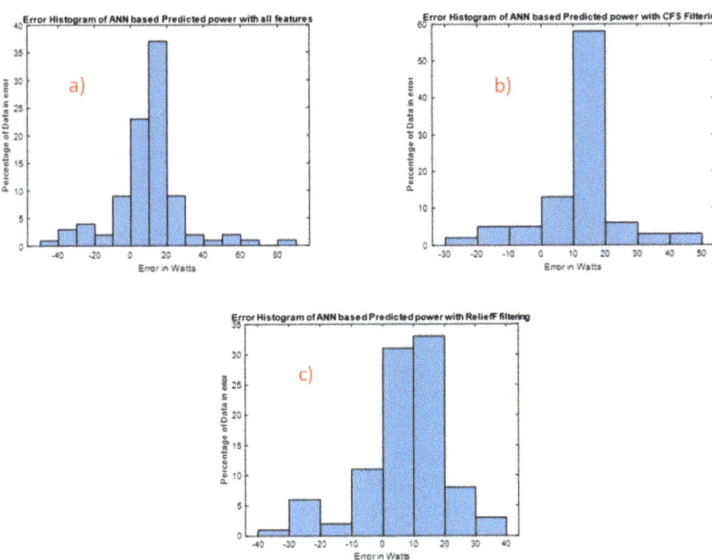

Figure 10. Error Histogram between the Predicted and actual PV with test dataset: (**a**) With all features; (**b**) CFS feature selection technique; (**c**) ReliefF feature selection technique.

5. Conclusions

An in-house PV system was developed at Qatar University to monitor, analyze and evaluate the performance of PV using various weather factors. The PV and environmental data collected from the system was used to develop a prediction model that can be used to predict the PV power in advance. To conclude, the prediction model was developed using several regressions—and ANN-based networks using the data collected by the PV system. Two feature selection techniques (CFS and ReliefF) were used to select subsets of applicable, high-quality and non-redundant characteristics. Compared to the three best regression models (simple linear regression model, M5P decision tree model and GPR), the ANN was more accurate in predicting the output power with RMSE of 2.1436. It was found that using feature selection techniques along with the ANN can predict the PV power with RMSE of 6.1555 and 5.5351, respectively. The trained ANN models are simpler and can be used to accurately predict the output power of PV systems with minimal computational complexity. Since the PV system was designed and tested in Qatar, this work can help the researchers in the Gulf to utilize the optimized algorithm and its performance in prediction for this region. We believe this would help the solar industry of this region in a great deal for optimizing the overall PV output. More PV and environmental data are being acquired for training a more accurate predictive model using the approach described in this work and will be compared with CNN-based approach in the future. Moreover, the concept of cooling the PV using the techniques discussed in Reference [5] are going to combine with the existing PV system, which can potentially help in increasing the efficiency and also avoid the accumulation of dust, which can affect the PV performance. Future works can open new horizons in this domain regarding different bias correction algorithm, with five years' system data, and more parameters for prediction—including cooling and cleaning effect on the PV system.

Author Contributions: Experiments were designed by A.K., and M.E.H.C.; Experiments were performed by F.T., A.J.S.P.G., K.B., M.-K.K.; Results were analyzed by A.K., M.E.H.C., F.T. and M.H.; All authors were involved in the interpretation of data and writing the paper.

Funding: The publication of this article was funded by the Qatar National Library.

Acknowledgments: The authors would like to thank the Electrical Engineering Department of Qatar University for providing the opportunity to conduct the experiments at Qatar University.

Conflicts of Interest: The authors certify that they have no affiliations with or involvement in any organization or entity with any financial interest or non-financial interest in the subject matter or materials discussed in this manuscript.

References

1. Sayyah, A.; Horenstein, M.N.; Mazumder, M.K. Energy yield loss caused by dust deposition on photovoltaic panels. *Sol. Energy* **2014**, *107*, 576–604. [CrossRef]
2. Almasoud, A.; Gandayh, H.M. Future of solar energy in Saudi Arabia. *J. King Saud Univ. Eng. Sci.* **2015**, *27*, 153–157. [CrossRef]
3. Ennaoui, A.; Figgis, B.; Plaza, D.M. Outdoor Testing in Qatar of PV Performance, Reliability and Safety. In *Qatar Foundation Annual Research Conference Proceedings*; HBKU Press Qatar: Doha, Qatar, 2016.
4. Touati, F.; Al-Hitmi, M.; Alam Chowdhury, N.; Abu Hamad, J.; Gonzales, A.J.S.P. Investigation of solar PV performance under Doha weather using a customized measurement and monitoring system. *Renew. Energy* **2016**, *89*, 564–577. [CrossRef]
5. Ahmad, N.; Khandakar, A.; El-Tayeb, A.; Benhmed, K.; Iqbal, A.; Touati, F. Novel Design for Thermal Management of PV Cells in Harsh Environmental Conditions. *Energies* **2018**, *11*, 3231. [CrossRef]
6. Benghanem, M.; Almohammedi, A.; Khan, M.T.; Al-Masraqi, A. Effect of dust accumulation on the performance of photovoltaic panels in desert countries: A case study for Madinah, Saudi Arabia. *Int. J. Power Electron. Drive Syst.* **2018**, *9*, 1356–1366. [CrossRef]
7. Darwish, Z.A.; Kazem, H.A.; Sopian, K.; Alghoul, M.A.; Alawadhi, H. Experimental investigation of dust pollutants and the impact of environmental parameters on PV performance: An experimental study. *Environ. Dev. Sustain.* **2018**, *20*, 155–174. [CrossRef]
8. Touati, F.; Chowdhury, N.A.; Benhmed, K.; Gonzales, A.J.S.P.; Al-Hitmi, M.A.; Benammar, M.; Gastli, A.; Ben-Brahim, L. Long-term performance analysis and power prediction of PV technology in the State of Qatar. *Renew. Energy* **2017**, *113*, 952–965. [CrossRef]
9. King, D.L.; Kratochvil, J.A.; Boyson, W.E. *Photovoltaic Array Performance Model*; US-Department of Energy: Washington, DC, USA, 2004; pp. 1–43.
10. Mishra, S.; Dash, P. Short term wind power forecasting using Chebyshev polynomial trained by ridge extreme learning machine. In Proceedings of the 2015 IEEE Power, Communication and Information Technology Conference (PCITC), Bhubaneswar, Odisha, India, 15–17 October 2015.
11. Netsanet, S.; Zhang, J.; Zheng, D.; Agrawal, R.K.; Muchahary, F. An aggregative machine learning approach for output power prediction of wind turbines. In Proceedings of the 2018 IEEE Texas Power and Energy Conference (TPEC), College Station, TX, USA, 8–9 February 2018.
12. Jawaid, F.; NazirJunejo, K. Predicting daily mean solar power using machine learning regression techniques. In Proceedings of the 2016 Sixth International Conference on Innovative Computing Technology (INTECH), Dublin, Ireland, 24–26 August 2016.
13. Li, J.; Ward, J.K.; Tong, J.; Collins, L.; Platt, G. Machine learning for solar irradiance forecasting of photovoltaic system. *Renew. Energy* **2016**, *90*, 542–553. [CrossRef]

14. Moosa, A.; Shabir, H.; Ali, H.; Darwade, R.; Gite, B. Predicting Solar Radiation Using Machine Learning Techniques. In Proceedings of the 2018 Second International Conference on Intelligent Computing and Control Systems (ICICCS), Madurai, India, 14–15 June 2018.
15. Khosravi, A.; Koury, R.N.N.; Machado, L.; Pabon, J.J.G. Prediction of hourly solar radiation in abu musa island using machine learning algorithms. *J. Clean. Prod.* **2018**, *176*, 63–75. [CrossRef]
16. Sheng, H.; Xiao, J.; Cheng, Y.; Ni, Q.; Wang, S. Short-term solar power forecasting based on weighted Gaussian process regression. *IEEE Trans. Ind. Electron.* **2018**, *65*, 300–308. [CrossRef]
17. Hiyama, T.; Karatepe, E. Investigation of ANN performance for tracking the optimum points of PV module under partially shaded conditions. In Proceedings of the 2010 Conference Proceedings IPEC, Singapore, 27–29 October 2010; pp. 1186–1191.
18. O'Leary, D.; Kubby, J. Feature Selection and ANN Solar Power Prediction. *J. Renew. Energy* **2017**, *2017*, 7. [CrossRef]
19. Mellit, A. Artificial intelligence based-modeling for sizing of a stand-alone photovoltaic power system: proposition for a new model using neuro-fuzzy system (ANFIS). In Proceedings of the 2006 3rd International IEEE Conference Intelligent Systems, Piscataway, NJ, USA, 1 March 2007; pp. 606–611.
20. Bocco, M.; Willington, E.; Arias, M. Comparison of regression and neural networks models to estimate solar radiation. *Chil. J. Agric. Res.* **2010**, *70*, 428–435. [CrossRef]
21. Nikhil, P.G.; Subhakar, D. Approaches for developing a regression model for sizing a stand-alone photovoltaic system. *IEEE J. Photovolt.* **2014**, *5*, 250–257. [CrossRef]
22. Wu, Y.K.; Chen, C.R.; Abdul Rahman, H. A novel hybrid model for short-term forecasting in PV power generation. *Int. J. Photoenergy* **2014**, *2014*, 9. [CrossRef]
23. Yokoyama, J. Short term load forecasting improved by ensemble and its variations. In Proceedings of the 2012 IEEE Power and Energy Society General Meeting, San Diego, CA, USA, 22–26 July 2012; pp. 1–6.
24. Ray, P.P. A survey of IoT cloud platforms. *Future Comput. Inform. J.* **2016**, *1*, 35–46. [CrossRef]
25. Guyon, I.; Elisseeff, A. An introduction to variable and feature selection. *J. Mach. Learn. Res.* **2003**, *3*, 1157–1182.
26. Wang, H.; Khoshgoftaar, T.M.; Gao, K.; Seliya, N. High-dimensional software engineering data and feature selection. In Proceedings of the 2009 21st IEEE International Conference on Tools with Artificial Intelligence, Newark, NJ, USA, 2–4 November 2009.
27. Yan, X.; Su, X. *Linear Regression Analysis: Theory and Computing*; World Scientific: Singapore, 2009.
28. MacKay, D.J.; Mac Kay, D.J. *Information Theory, Inference and Learning Algorithms*; Cambridge University Press: Cambridge, UK, 2003.
29. Quinlan, J.R. Learning with continuous classes. In Proceedings of the 5th Australian Joint Conference on Artificial Intelligence, Singapore, 16–18 November 1992.
30. Wang, Y.; Witten, I.H. Induction of model trees for predicting continuous classes. Available online: https://researchcommons.waikato.ac.nz/handle/10289/1183 (accessed on 17 July 2019).
31. Mellit, A.; Pavan, A.M. A 24-h forecast of solar irradiance using artificial neural network: Application for performance prediction of a grid-connected PV plant at Trieste, Italy. *Sol. Energy* **2010**, *84*, 807–821. [CrossRef]
32. Sheiner, L.B.; Beal, S.L. Some suggestions for measuring predictive performance. *J. Pharmacokinet. Biopharm.* **1981**, *9*, 503–512. [CrossRef] [PubMed]
33. MacKay, D.J. Bayesian interpolation. *Neural Comput.* **1992**, *4*, 415–447. [CrossRef]
34. Bevington, P.R.; Robinson, D.K.; Blair, J.M.; Mallinckrodt, A.J.; McKay, S. Data reduction and error analysis for the physical sciences. *Comput. Phys.* **1993**, *7*, 415–416. [CrossRef]

Communication

Simulating Power Generation from Photovoltaics in the Polish Power System Based on Ground Meteorological Measurements—First Tests Based on Transmission System Operator Data

Jakub Jurasz [1,2,*], **Marcin Wdowikowski** [3] **and Mariusz Figurski** [3,4]

[1] Department of Engineering Management, Faculty of Management, AGH University, 30-059 Cracow, Poland
[2] School of Business, Society and Engineering, MDH University, 722-20 Västerås, Sweden
[3] Institute of Meteorology and Water Management-National Research Institute, 01-673 Warsaw, Poland; marcin.wdowikowski@imgw.pl (M.W.); mariusz.figurski@pg.edu.pl (M.F.)
[4] Faculty of Civil and Environmental Engineering, Gdansk University of Technology, 80-233 Gdansk, Poland
* Correspondence: jurasz@agh.edu.pl

Received: 5 July 2020; Accepted: 12 August 2020; Published: 17 August 2020

Abstract: The Polish power system is undergoing a slow process of transformation from coal to one that is renewables dominated. Although coal will remain a fundamental fuel in the coming years, the recent upsurge in installed capacity of photovoltaic (PV) systems should draw significant attention. Owning to the fact that the Polish Transmission System Operator recently published the PV hourly generation time series in this article, we aim to explore how well those can be modeled based on the meteorological measurements provided by the Institute of Meteorology and Water Management. The hourly time series of PV generation on a country level and irradiation, wind speed, and temperature measurements from 23 meteorological stations covering one month are used as inputs to create an artificial neural network. The analysis indicates that available measurements combined with artificial neural networks can simulate PV generation on a national level with a mean percentage error of 3.2%.

Keywords: photovoltaics; artificial neural networks; national power system

---

1. Introduction

The transformation of the power system is a continuous process, and to fully realize this we will need years of ongoing commitment and well-thought decisions on country and regional levels. In the past two years in the Polish power system, we could observe a significant increase in the installed capacity in both residential and commercial photovoltaic (PV) systems. To a set of interesting investments in PV capacity one could potentially include a 600 kWp system located near Porąbka-Żar pumped-storage hydropower station, 2.5 MW installation for waterworks in Szczecin, and 739 kW for 35 buildings belonging to a housing cooperative in Wrocław. The growing interest in PV systems can be linked to (a) increasing electricity prices, (b) the decreasing cost of PV systems, and (c) growing awareness of the impact of the energy generation sector on the natural environment. (Impact of PV systems from the perspective of Life Cycle Assessment should not be neglected. Readers are referred to other works strictly dedicated to this topic.)

The Polish Transmission System Operator (PSE) has been publishing wind generation data on an aggregated level for quite some time. These time series with an hourly time step are of great importance to visualize and analyze the variability [1] of wind generation with different time horizons. They can also be used to create forecasting or simulation models [2–5]. Recently, the PSE has also made publicly available aggregated generation time series from photovoltaic installations located in Poland.

These data create new opportunities for further research, including an analysis of the complementarity between renewable energy sources in Poland [6] or the use of available ground measurements to simulate PV generation on a country level. The second research direction can be later used to simulate and predict how the growing installed capacity in PV systems in Poland will affect power system operations. Based on available measurement data, and knowing the transmission system constraints, decisions can be made regarding the optimal distribution of renewable generators in the power system.

Considering the information above, the objective of this study is to conduct first tests with regard to the possibility of using ground measurements from meteorological stations to simulate the power generation from PV systems on a country level. Such research results can be found in the international literature both for PV and wind generation, using different kinds of inputs and simulation tools. For example, for Sweden reanalysis based on MERRA (Modern Era Retrospective-Analysis for Research and Applications), data sets have been used to effectively model a national fleet of wind turbines' power output [7]. Recently, Olauson [8] found that ERA5 data sets performed much better than the MERRA reanalysis sets mentioned previously [7]. Similar to the analysis presented in this work, Black et al. [9] used meteorological data and regression techniques to simulate a fleet of PV systems.

2. Materials and Methods

To simulate the energy generation from PV systems on a country level, the method based on an artificial neural network (ANN) was applied. ANNs are a computational system loosely inspired by biological neural networks. They belong to a group of artificial intelligence information processing paradigms that has gained immense popularity in recent years. Simulation or forecasting models based on ANNs have been successfully applied in various areas of cognitive and application research, including water-demand forecasting [10], lake water-level forecasting [11], renewables integration studies [12], in the area of color image identification and reconstruction [13], multi-core optic fibers [14], wind speed prediction [15], or, most importantly from this paper's perspective, in the areas of direct and global radiation prediction [16] and PV energy yield forecasting [17].

A model of feedforward neural network (NN) has been developed in Matlab 2019a software. The Levenberg–Marquardt method was used to optimize the weights and bias values [18]. The input data were divided into training, validation, and testing subsets in proportions of 70, 15, and 15. Since the length of the available time series was limited to one month, the data were divided so that the first 70% of hours was used to teach the neural network, followed by 15% to validate/supervise the teaching process and the remaining 15% to test the NN performance. The number of neurons in the hidden layer was selected following a brute-force approach, namely NNs with the number of hidden neurons k ranging from 1 to *n*, where *n* is the number of input neurons being tested. In the literature, various approaches to solving this problem can be found [18,19]. However, considering the low computing effort to create an ANN, a brute-force approach in this particular case seems to be a justified choice.

As inputs, the PV generation time series covering the month of May (available at https://www.pse.pl/) and time series of wind speed, temperature, and irradiation for 23 meteorological stations located in Poland, obtained from the Institute of Meteorology and Water Management—National Research Institute (IMWM-NRI), were used. The locations of meteorological stations along with the equipment used are presented in Figure 1 and in Table 1. The data have an hourly time step. The nighttime hours and hours when the energy generation from the PV system was less than 5 MW were removed from the input data set. Such low values were removed because hourly temporal resolution does not take into account the spatial distribution of PV systems in Poland, and low generation periods occur during sunset and sunrise. The final data set consisted of 473 hourly records. The data were normalized to a range of 0–1 before the ANN creation procedure. The performance of ANN has been assessed based on two commonly applied metrics, namely (MAPE) mean absolute percentage error (Equation (1)) and (RMSE) root-mean-square error (Equation (2)).

$$MAPE = \frac{1}{n}\sum_{i=1}^{n}\left|\frac{o_i - s_i}{o_i}\right| \tag{1}$$

$$RMSE = \left[\sum_{i=1}^{n}(o_i - s_i)^2 / n\right]^{1/2} \tag{2}$$

where n—sample size, o—observed value, and s—simulated value.

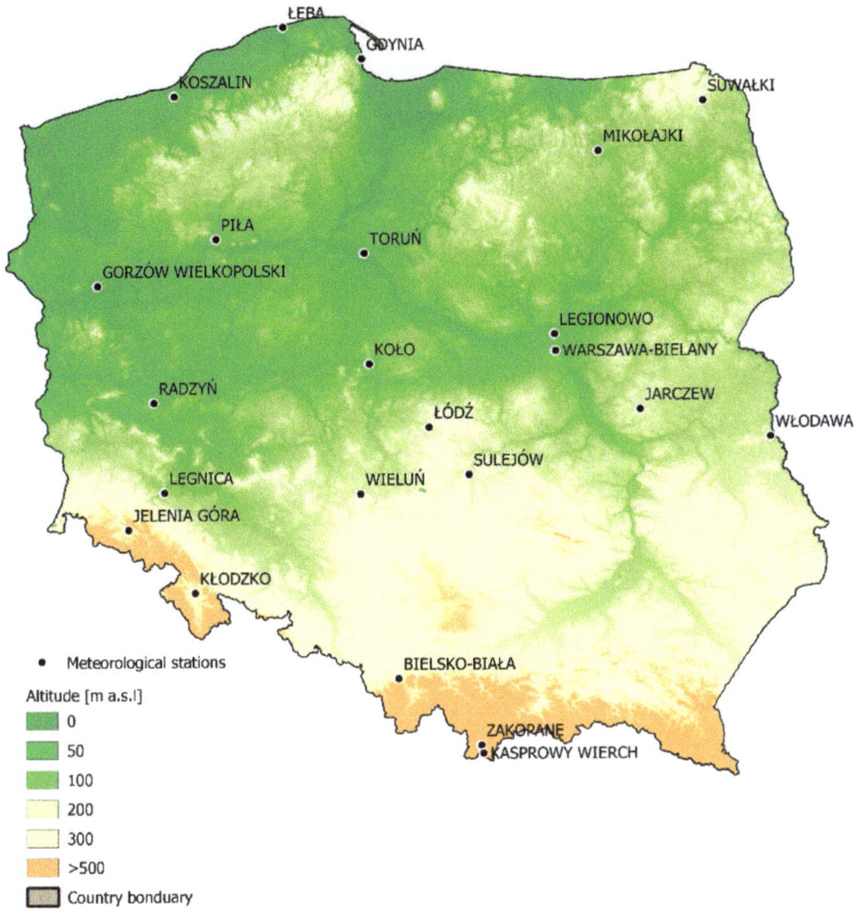

Figure 1. Locations of 23 IMWM-NRI meteorological stations.

Table 1. Meteorological stations measurement sensor description.

No.	Meteorological Element	Sensor Type			
1	Irradiation	KIPP & ZONEN CNR4 Net Radiometer	KIPP & ZONEN CMP11 Pyranometer	X	X
	Location	Piła, Łeba	All 21 others	X	X
2	Air temperature	Pt 100 PVC	Vaisala QMT 103	Vaisala QMT 110	Vaisala HMP45A/45D
	Location	Radzyń, Warszawa-B., Legionowo, Jarczew, Kasprowy W., Wieluń, Zakopane, Łódź, Włodawa, Suwałki	Łeba, Koło, Bielsko-B., Piła, Gorzów W., Sulejów, Mikołajki	Toruń, Koszalin, Legnica, Jelenia G., Kłodzko	Gdynia
3	Wind speed	Vaisala Ultrasonic WS 425	Vaisala Cup Anemometer WAA 151	Gill Instruments WindSonicM	Others (Vaisala WMS 302, WMT 700, 702, 703 and G. Lufft WS 200-UMB
	Location	Gorzów W., Jelenia G., Kłodzko, Koszalin, Legnica Łódź, Mikołajki, Piła, Sulejów, Suwałki, Wieluń, Włodawa, Zakopane	Gdynia, Radzyń, Warszawa-B.	Toruń, Legionowo	Jarczew, Kasprowy, Łeba, Koło, and Bielsko-B.

3. Results and Discussion

The energy yield from photovoltaic modules is a function of irradiation falling on the module area, the modules' efficiency, and their temperature. Since detailed information about the PV systems' location is currently not available to the authors, nor is the specification of the modules used, it is justified to use a black-box approach where the input data are transferred into desired output. Because the temperature of the PV modules is determined by irradiation and ambient air temperature as well as wind speed, which has a cooling effect, the above-mentioned meteorological parameters have been considered as explanatory variables. To simulate an hourly power generation from PV systems in Poland, a set of 69 (3 meteorological parameters from 23 stations: Bielsko-Biała, Gdynia, Gorzów Wielkopolski, Jarczew, Jelenia Góra, Kasprowy Wierch, Kłodzko, Koło, Koszalin, Legionowo, Legnica, Łeba, Łódź, Mikołajki, Piła, Radzyń, Sulejów, Suwałki, Toruń, Warszawa-Bielany, Wieluń, Włodawa, and Zakopane – please see Figure 1) explanatory variables was used. These explanatory variables exhibited various correlation coefficient values with the response variable. For the irradiation, it was on average 0.777, whereas the air temperature and wind speed were, respectively, 0.439 and 0.299. Figure 2 shows the observed hourly irradiation on the 1st of May 2020 and the production of energy from PV systems at the national level. During that day, the observed irradiation in individual hours varied significantly among considered locations (meteorological stations), while the PV systems maintained a relatively smooth energy generation pattern. This phenomenon can be attributed to the spatial smoothing of power generation due to the geographical dispersion of the PV systems [20]. This situation is beneficial from the perspective of variable renewable energy sources (VRES) integration to the power system, although constraints such as transmission network capacity may limit the benefits resulting from spatial smoothing.

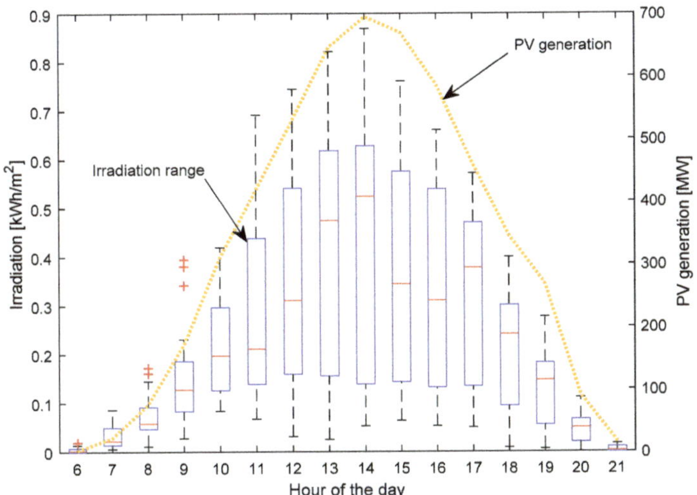

Figure 2. Observed irradiation in 23 meteorological stations along with photovoltaic (PV) production on a country level on 1st of May 2020.

The generation from PV systems during the whole period is visualized in Figure 3. Significant variability between individual daily yield sums can be observed. On a side note, accordingly to the PSE data in May, the PV systems generated 223 GWh, whereas wind turbines generated almost 1.072 TWh, which contributed to covering, respectively, 1.8% and 8.6% of the national demand in this month.

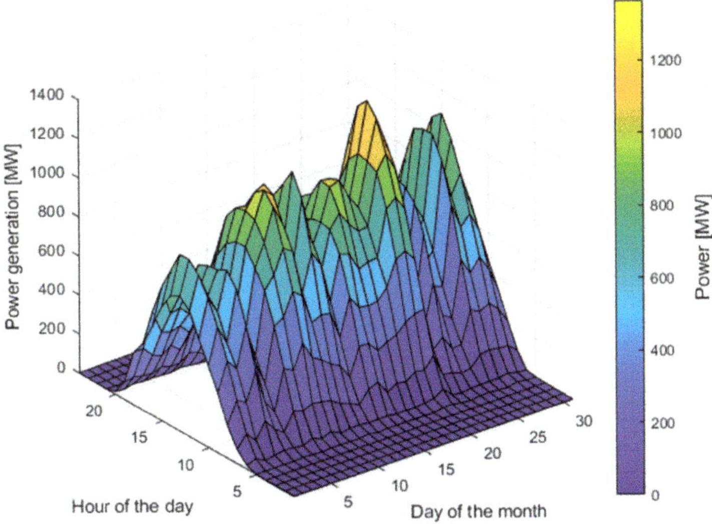

Figure 3. PV generation on a country level during May 2020.

As mentioned in the Methods and Data section, in total 69 potential configurations of the ANNs were tested. For those, the one with the lowest MAPE was selected for further analysis. The performance of the selected and the remaining ANNs as a function of the number of neurons in the hidden layer is presented in Figure 4. The best performing ANN had 15 neurons in the hidden layer and MAPE of 3.2%.

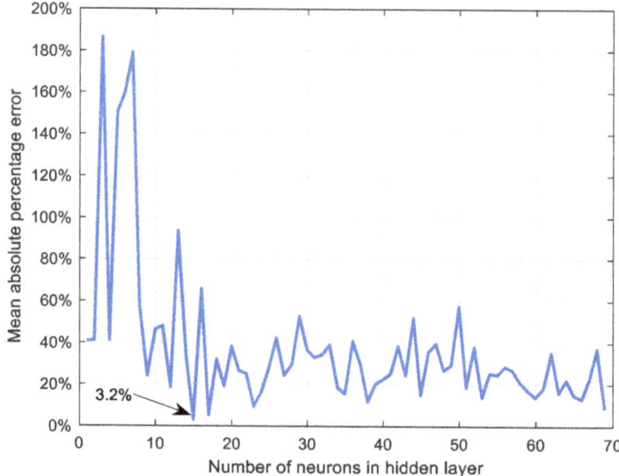

Figure 4. Performance of neural networks for the testing subset based on mean absolute percentage error (MAPE) criteria.

The ANN also performed well in terms of RMSE, which was found to be 39.2 MW. In Figure 5, the performance of the ANN was visualized for the testing subset only. This set is the final verification if the neural network is capable of obtaining good-quality results for input data that remained unknown during the training phase. In Figure 5 it can be noted that the ANN performed very well for the extreme values (very low and high generation), whereas some systematic errors with an unknown source occurred for the mid-range values. On average, it was found that the ANN tended to overestimate the generation by 6 MW. The highest overestimation error was found to be 133 MW, whereas the highest underestimation was 150 MW.

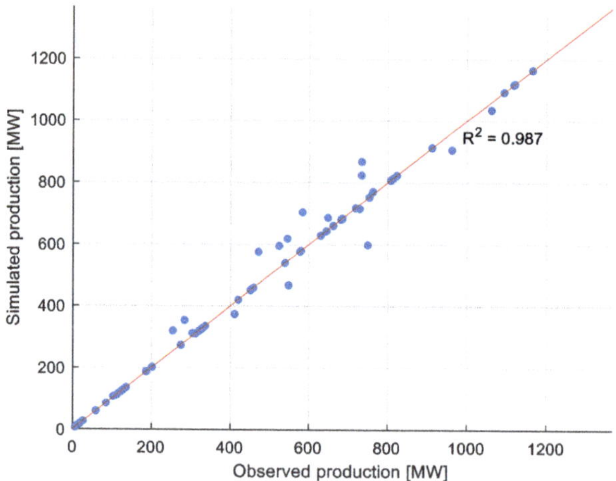

Figure 5. Performance of the neural network for the testing subset. R^2 refers to linear regression. Red line indicates a theoretical perfect match.

Figure 6 visualizes the performance of the neural network over a period of 4.5 days by the end of May 2020. The night hours are excluded from the analysis. As shown in the figure, the values simulated by the ANN followed the real PV systems generation well. During the second day, the ANN wrongly simulated a sudden drop in PV generation, increasing the variability of the modeled time series (in terms of ramp rates). During the fourth day, one can observe that in the midday hours the simulated generation was slightly greater than the observed one. In general, the absolute errors did not exceed 150 MW.

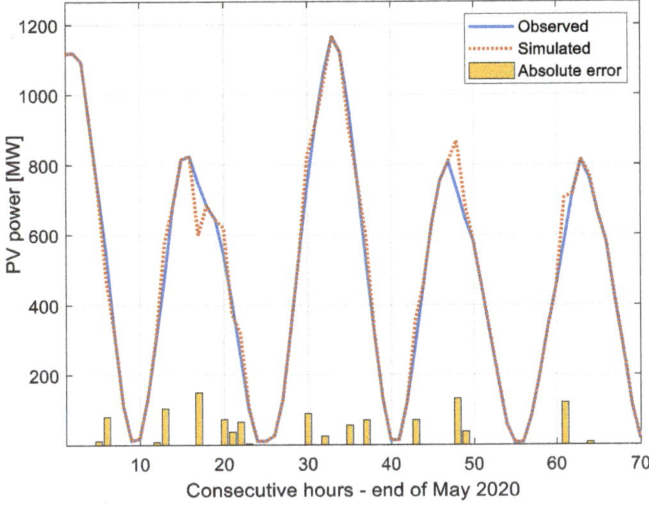

Figure 6. Performance of the neural network for the testing subset. Please note that night hours (no irradiation) are excluded.

4. Conclusions

The presented short communication archives the first results regarding the potential of simulating the generation from the PV systems on a national level based on ground measurements provided by the Institute of Meteorology and Water Management. The conducted analysis based on artificial neural networks revealed that ground measurements consisting of irradiation, wind speed, and air temperature can be used to correctly model the power generation from PV systems on a country level. Despite some limitations, such as neglecting the technical specification of the PV systems, the spatial distribution of PV farms across the country, taking input irradiation on a horizontal rather than an inclined surface, and finally representing the meteorological conditions in Poland based on a sample of 23 locations, it was possible to model the PV generation with a mean absolute percentage error of roughly 3.2%.

Most importantly, the obtained results have some practical implications. In the research and reality of the operation of present energy systems, meteorology starts to play a very important and, in many instances, crucial role in enabling and securing an efficient and reliable operation of the power system. The need for including meteorology-based studies in energy research comes directly from the unprecedented increase in the installed capacity of renewable energy sources, especially the ones in which variability/availability is driven by climate and weather [21]. The Polish power sector is starting a process of transformation. The increasing share of renewables such as wind and, in particular, solar energy (as observed in 2019/2020) is driven by an increasing awareness of the energy sector's impact on the natural environment, the growing cost of electricity generation from conventional fuels, the decreasing cost of renewables, and national/international policy. The power system expansion/development studies [22] call for data with both higher temporal and spatial resolution. This can be provided by either ground measurements, satellite measurements, numerical weather prediction models, or reanalysis data sets. In this paper we have investigated whether the in-house data available from a governmental institution (Institute of Meteorology and Water Management) can be used to simulate aggregated PV generation on a country level. The results obtained proved the high value of the already available data and its promising applications in power system expansion studies as well as research dedicated, for example, to optimal location of PV systems from the perspective of grid topology or the impact of PV systems on the residual load curve.

This study intended to present the results of the first tests on the freshly published data by the Polish Transmission System Operator—PSE. Therefore, we did not go into detail comparing different statistical or machine learning techniques. Clearly, for now, the data sample is relatively short and, at the time of writing, was limited to one month. In the future, we plan to extend this research by investigating (a) in detail the impact of the input set on the model performance, (b) comparing different simulation tools, (c) the impact of a long series of meteorological parameters on the quality of the model, and finally (d) selecting a minimal set of representative meteorological stations sufficient for simulations. The results presented here should be taken with caution since solar irradiation has a high annual variability, and model performance might, therefore, vary depending on the part of the year.

Author Contributions: Conceptualization, J.J. and M.W.; methodology, J.J.; software, J.J.; validation, M.W., M.F.; resources, M.W.; data curation, M.W.; writing—original draft preparation, J.J.; writing—review and editing, M.W. and M.F.; visualization, J.J.; supervision, M.F.; project administration, M.F. All authors have read and agreed to the published version of the manuscript.

Funding: This research received no external funding.

Conflicts of Interest: The authors declare no conflict of interest.

Nomenclature

ANN	Artificial Neural Network
ECMWF	European Centre for Medium-Range Weather Forecasts
ERA5	ECMWF Reanalysis 5th Generation
IMWM-NRI	Institute of Meteorology and Water Management—National Research Institute
MAPE	Mean Absolute Percentage Error
MERRA	Modern Era Retrospective-Analysis for Research and Applications
NN	Neural Network
PS	Power System
PSE	pol. Polskie Sieci Elektroenergetyczne—Polish Transmission System Operator
PV	Photovoltaic
RMSE	Root-Mean-Squared Error
VRES	Variable Renewable Energy Sources

References

1. Augustyn, A.; Kamiński, J. Rola generacji wiatrowej w pokryciu zapotrzebowania na moc w krajowym systemie elektroenergetycznym [The role of wind power generation in the load carrying capability]. *Rynek Energii* **2018**, *1*, 47–52.
2. Kacejko, P.; Wydra, M. Energetyka wiatrowa w Polsce-analiza potencjalnych ograniczeń bilansowych i oddziaływania na warunki pracy jednostek konwencjonalnych [Wind energy in Poland-Analysis of potential power system balance limitations and influence on conventional power units operational conditions]. *Rynek Energii* **2011**, *93*, 25–30.
3. Kacejko, P.; Wydra, M. Nowa metoda szacowania możliwości przyłączeniowych generacji wiatrowej w KSE z uwzględnieniem ograniczeń bilansowych [A new method for estimating connection potential of wind power into National Power System with power balancing constraints]. *Przegląd Elektrotechniczny* **2014**, *90*, 32–35.
4. Paska, J.; Kłos, M. Elektrownie wiatrowe w systemie elektroenergetycznym-przyłączanie, wpływ na system i ekonomika [Wind power plants in electric power system–Connecting, influence on the system and economics]. *Rynek Energii* **2010**, *1*, 3–10.
5. Paska, J. Elektrownie wiatrowe w systemie elektroenergetycznym i ich zdolność do pokrywania obciążenia [Wind power plants in electric power system and their load carrying capability]. *Przegląd Elektrotechniczny* **2009**, *85*, 224–230.

6.   Jurasz, J.; Canales, F.A.; Kies, A.; Guezgouz, M.; Beluco, A. A review on the complementarity of renewable energy sources: Concept, metrics, application and future research directions. *Sol. Energy* **2020**, *195*, 703–724. [CrossRef]

7.   Olauson, J.; Bergkvist, M. Modelling the Swedish wind power production using MERRA reanalysis data. *Renew. Energy* **2015**, *76*, 717–725. [CrossRef]

8.   Olauson, J. ERA5: The new champion of wind power modelling? *Renew. Energy* **2018**, *126*, 322–331. [CrossRef]

9.   Black, J.; Hoffman, A.; Hong, T.; Roberts, J.; Wang, P. Weather data for energy analytics: From modeling outages and reliability indices to simulating distributed photovoltaic fleets. *IEEE Power Energy Mag.* **2018**, *16*, 43–53. [CrossRef]

10.   Adamowski, J.; Karapataki, C. Comparison of multivariate regression and artificial neural networks for peak urban water-demand forecasting: Evaluation of different ANN learning algorithms. *J. Hydrol. Eng.* **2010**, *15*, 729–743. [CrossRef]

11.   Piasecki, A.; Jurasz, J.; Adamowski, J.F. Forecasting surface water-level fluctuations of a small glacial lake in Poland using a wavelet-based artificial intelligence method. *Acta Geophys.* **2018**, *66*, 1093–1107. [CrossRef]

12.   Sharifzadeh, M.; Sikinioti-Lock, A.; Shah, N. Machine-learning methods for integrated renewable power generation: A comparative study of artificial neural networks, support vector regression, and Gaussian Process Regression. *Renew. Sustain. Energy Rev.* **2019**, *108*, 513–538. [CrossRef]

13.   Shabairou, N.; Cohen, E.; Wagner, O.; Malka, D.; Zalevsky, Z. Color image identification and reconstruction using artificial neural networks on multimode fiber images: Towards an all-optical design. *Opt. Lett.* **2018**, *43*, 5603–5606. [CrossRef] [PubMed]

14.   Cohen, E.; Malka, D.; Shemer, A.; Shahmoon, A.; Zalevsky, Z.; London, M. Neural networks within multi-core optic fibers. *Sci. Rep.* **2016**, *6*, 1–14. [CrossRef] [PubMed]

15.   Li, F.; Ren, G.; Lee, J. Multi-step wind speed prediction based on turbulence intensity and hybrid deep neural networks. *Energy Convers. Manag.* **2019**, *186*, 306–322. [CrossRef]

16.   Renno, C.; Petito, F.; Gatto, A. ANN model for predicting the direct normal irradiance and the global radiation for a solar application to a residential building. *J. Clean. Prod.* **2016**, *135*, 1298–1316. [CrossRef]

17.   Al-Dahidi, S.; Ayadi, O.; Alrbai, M.; Adeeb, J. Ensemble approach of optimized artificial neural networks for solar photovoltaic power prediction. *IEEE Access* **2019**, *7*, 81741–81758. [CrossRef]

18.   Çelik, Ö.; Teke, A.; Yildirim, H.B. The optimized artificial neural network model with Levenberg–Marquardt algorithm for global solar radiation estimation in Eastern Mediterranean Region of Turkey. *Clean. Prod.* **2018**, *116*, 1–12. [CrossRef]

19.   Sheela, K.G.; Deepa, S.N. Review on methods to fix number of hidden neurons in neural networks. *Math. Problems Eng.* **2013**, *2013*. [CrossRef]

20.   Marcos, J.; Marroyo, L.; Lorenzo, E.; García, M. Smoothing of PV power fluctuations by geographical dispersion. *Prog. Photovolt. Res. Appl.* **2012**, *20*, 226–237. [CrossRef]

21.   Engeland, K.; Borga, M.; Creutin, J.D.; François, B.; Ramos, M.H.; Vidal, J.P. Space-time variability of climate variables and intermittent renewable electricity production–A review. *Renew. Sustain. Energy Rev.* **2017**, *79*, 600–617. [CrossRef]

22.   Schyska, B.U.; Kies, A. How regional differences in cost of capital influence the optimal design of power systems. *Appl. Energy* **2020**, *262*, 114523. [CrossRef]

Article

Complex Network Analysis of Photovoltaic Plant Operations and Failure Modes

Fabrizio Bonacina [1,*]**, Alessandro Corsini** [1]**, Lucio Cardillo** [2] **and Francesca Lucchetta** [1]

[1]   Department of Mechanical and Aerospace Engineering, Sapienza University of Rome, 00184 Rome, Italy; alessandro.corsini@uniroma1.it (A.C.); francesca.lucchetta@uniroma1.it (F.L.)
[2]   SED Solutions, 03013 Ferentino, Italy; lucio.cardillo@sedsoluzioni.com
*   Correspondence: fabrizio.bonacina@uniroma1.it

Received: 23 April 2019; Accepted: 22 May 2019; Published: 24 May 2019

Abstract: This paper presents a novel data-driven approach, based on sensor network analysis in Photovoltaic (PV) power plants, to unveil hidden precursors in failure modes. The method is based on the analysis of signals from PV plant monitoring, and advocates the use of graph modeling techniques to reconstruct and investigate the connectivity among PV field sensors, as is customary for Complex Network Analysis (CNA) approaches. Five month operation data are used in the present study. The results showed that the proposed methodology is able to discover specific hidden dynamics, also referred to as emerging properties in a Complexity Science perspective, which are not visible in the observation of individual sensor signal but are closely linked to the relationships occurring at the system level. The application of exploratory data analysis techniques on those properties demonstrated, for the specific plant under scrutiny, potential for early fault detection.

Keywords: sensor network; data fusion; complex network analysis; fault prognosis; photovoltaic plants

---

1. Introduction

The cumulative global photovoltaic (PV) capacity has been growing exponentially around the world over recent years. In the decade 2005–2015, the solar PV generation capacity in the EU has increased from 1.9 GW to 95.4 GW [1]. Notwithstanding this, the Europe PV market's conditions are still substantially dependent on regional energy policies and public subsidies for renewable energies. As per the Italian market, in June 2013 the Italian public company GSE (Gestore dei Servizi Energetici) officially announced the discontinuation of the last Feed-in-Tariff incentive after its cap of 6700 million euro was reached [2]. The end of such subsidies has led to new attention being focused on PV plant performance management, lifetime and availability, with a view to reduce operating and maintenance costs [3].

Faults in PV plant components (i.e., modules, converters, connection lines), in addition to downtime correlated penalties, could result in an acceleration of system aging and as a consequence in a reduction of power plant reliability [4]. Typically, faults severity depends on various factors. These factors include the time to detect, time to repair or substitute, COE, and occurrence over time, and all of these factors can have a significant impact on profitability [5]. For these reasons, over the last decade Fault Detection and Diagnosis (FDD) in PV systems has been established as a critical field of research [6]. To mention but a few areas, research has addressed key topics like real-time monitoring, partial shading effects analysis, estimation of natural degradation rate over time and residual life for solar panels [7]. In the FDD arena, a number of studies have proposed data-driven fault detection algorithms based on statistical processing of performance parameters (e.g., power loss factor analysis studies, I-V output characteristics [8–11]) or an exponentially weighted moving average control chart [12]. In addition, model-based techniques, implementing dynamic fault trees [13] or combining PV system performance simulation (voltage ratio, performance ratio, &c) and Fuzzy and neuro-Fuzzy logic classification systems [14,15], have also been studied as methods to detect fault occurrence [16]. The open literature already offers several review papers on this field, illustrating the state-of-the-art and opportunities [4,17–19].

With the advent of Internet of Things, technologies for smart monitoring have found widespread applications, including in PV systems, thereby enabling decision-making processes [20]. However, monitoring systems typically employ several distributed sensors and the collection of raw data opening the field to a number of specific challenges is still unresolved, which is in part due to the different communication standards, heterogeneous structure and the huge volume of data [21,22]. As reported in a number of big data studies [23,24], traditional data analysis in statistics, management and visualization fails with sensor network data streams. High-dimensionality and heterogeneity encourage data mining and artificial intelligence solutions, such as feature extraction, classification, clustering and sampling [25–30]. Furthermore, multi-sensor data fusion techniques have been introduced to provide a robust description of an environment or process of interest by combining observations from a number of different sources [31].

In this context, complex network science emerged as an active field, cross-fertilized by natural, social and computer science, exploiting the general idea that artificial systems like biological ones (metabolic network, worldwide web, distribution network, etc.) are complex systems, composed of a large number of interacting parts (components or entities) with an articulated self-organization. The complexity of the systems, then, resides in interaction and loops among its entities [32]. Looking for example at industrial systems, constitutive elements are governed by well-defined physical laws and are included in a pre-determined process scheme. However, surrounding environments and business laws (determining the operations) influence this by-definition deterministic structure. Moving to the component level, multi-physics processes are nonlinear, as is the component matching. In a general view, these systems can evolve surfacing complex structures characterized by diversity, as well as multiple interactions within and between layers. These characteristics can make industrial systems a complex engineering system [33,34].

In this paper CNA has been used to model the sensor network of a 1 MW PV field, advocating a complex system engineering point of view [35,36]. A data fusion criterion has been developed processing the signals from the sensor equipment with a view to detect fault precursors. The sensors include PV field AC and DC electrical parameters, temperature, and solar irradiance measurements.

The paper, first, illustrates the methodology for data processing with a consideration of connectivity analysis to create the digital twin of the sensor network in the form of a functional graph. The result of these analyses is a correlation matrix that defines the graph structure in terms of the weight of the edges connecting the nodes (element/sensor signal) at a given instant from all the pairs of pre-processed sensor time series. By applying the data processing within a fixed time window, sliding over the monitoring interval, the result is a fully-connected weighted dynamic graph. Then, dynamic graph analysis is used to explore the sensor network in order to unveil hidden correlations among signal time series. The results are discussed by using exploratory data analysis to combine the most significant topological graph metrics in the identification of operational patterns of the PV plant.

2. Methodology

The field sensors are modeled using graph theory as a complex network, where the nodes are represented by the signals from sensors and the edge are evaluated with non-linear statistical correlation functions applied to the time series pairs. Specifically, the data flow is structured according to the following steps:

a.   data collection and pre-processing,
b.   connectivity analysis and graph modeling,
c.   pattern recognition.

At the initial stage (a), heterogeneous data acquired by the sensors are synchronized and cleaned by removing outliers. Then for stage (b), a fully-connected graph is created in which each monitored variable represents a node and all the nodes are connected to each other with directed edges. The output of this phase is a complete weighted graph model (also called functional graph) of the sensor network. After, some typical complex network measurements are applied to extrapolate synthetic properties from the functional graphs. Finally (c), these measurements were used as input for exploratory network analysis, with the aim of grouping the data as a function of multiple topological graph metrics.

The implementation of the proposed methodology has been done using Python version 3.5 [37]. In particular, a Python code has been created using Scikit-learn and NetworkX packages, respectively [38,39].

2.1. Complex Network Analysis

The CNA has its origins in graph theory and is used to describe the properties of complex systems through the mathematical study of networks. The key ingredient in the CNA is the study of the correlation among uni-variate signals recorded from different sources, as is customary for biological network analysis. For this purpose, the mutual information (MI) has been widely used [40–42].

In this paper we propose a novel approach based on the study of correlations between signals of m heterogeneous sensors within a fixed window of n samples. In particular, given two signals $y_{n,i}$ and $y_{n,j}$, taken from the i-th and j-th component of the feature matrix $Y_{n,m}$, where (n) is spanning over time steps, the mutual information MI quantifies the level of uncertainty in $y_{n,i}$ removed by the conditional knowledge of $y_{n,j}$. This measure essentially tells us how much extra information one gets from one signal by knowing the outcomes of the other one [43–45]. Once for each sample, a correlation matrix is obtained computing the MI in the m-dimensional feature space, while sliding the window from the beginning of the data set up to the entire monitoring interval. As a consequence, the evolution of correlation matrix is represented in the form of a functional graph consisting of a set of m nodes associated by k weighted edges representing the connecting force between the pairs of nodes.

The result of this phase is therefore a dynamic graph that contains all the information about the evolution of spatial and temporal relationships between all the entities monitored and allows the definition of parameters quantifying the characteristics of the systems [46].

In the context of CNA, the network measurements represent the most used tools for the extrapolation of synthetic information through the analysis of network topology. These measurements can be evaluated over the entire network (e.g., Shortest Path length, Diameter, Global Efficiency, Modularity, &c) or they can locally refer to nodes (e.g., Centrality Measures). This type of metric has been applied in several social networking studies to address the problem of identifying and ranking those people who exert an unequal amount of influence on the decision-making of others (also called influencers or opinion leaders) and to study the diffusion information within the network [47–49].

In the present approach, both the network measurements relating to the whole network (global scale) and those specific to individual nodes (local scale) were considered. Specifically, the study uses Shortest Path Length [50], Diameter and Radius of the graph [51], among the different global-scale measurements, Eccentricity and Weighted Degree Centrality of the nodes [39,52–54], among the local-scale measurements.

2.2. Exploratory Data Analysis

Promoted by Tukey [55], exploratory data analysis (EDA) represents a powerful tool to maximize insight into the underlying structure of a complex dataset. It facilitates the understanding of the distribution of samples and simplifies the data analysis, pointing out to special observations (outliers), clusters of similar observations, groups of related variables, and crossed relationships between specific observations and variables [56,57]. All this information in turn, can be very informative for further data modeling and it is of paramount importance to improve data knowledge. For these purposes, it encompasses a wide range of statistics and graphical tools (e.g., histogram, box plot, Pareto chart, principal component analysis, etc.) which have been commonly used for decades in various research fields such as archeology, biology, anthropology, medicine, chemometrics, &c [58–60].

In this paper, EDA is used to investigate the results of the complex network analysis and compare them with the sensor signals $Y_{n,m}$. In particular, the 3D scatterplot has been used for visual pattern recognition. This tool is based on a features representation in a multivariable space, allowing us to identify the dependencies between different network measurements in order to discriminate specific operational PV plant clusters.

3. Case Study

The data were collected from a PV plant with a power of 1 MWp. The solar field is connected to two inverters, each with three conversion blocks. Both inverters are grid-tied, feeding a medium voltage power distribution network. They are equipped with fully independent monitoring systems and incorporate a solar power controller to regulate the Maximum Power Point Tracking (MPPT).

Measurements were recorded every 5 min and included AC and DC electrical parameters, AC power output, ambient temperature and solar irradiance. The monitoring system is shown in Figure 1. In particular, solar radiation on the plane of the modules and ambient temperature have been measured by a pyranometer and a PT-100 RTD sensor respectively, both installed on a sensor box near the panels. Both inverters are equipped with resistive potential divider voltage sensors for DC and AC parameters measurement for each conversion block. Shunt resistors have been used to measure 12 string currents as a representative sample of the plant.

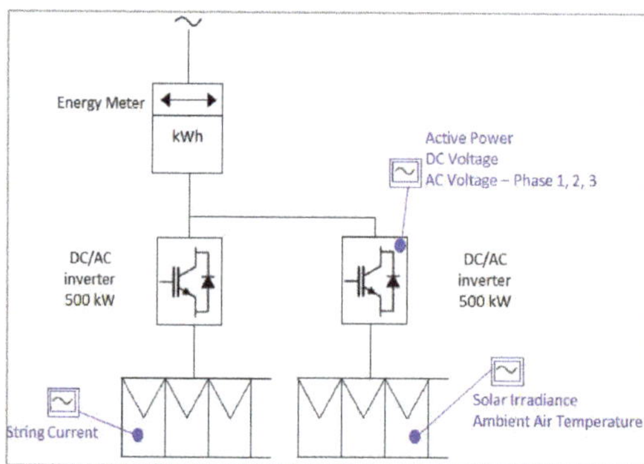

Figure 1. Schematic block circuit diagram of the PV system with sensors and measuring points.

The measurements were collected in the period from May 20th to September 15th 2017. The 47 monitored variables which define the m-dimensional feature matrix $Y_{n,m}$ are listed in Table 1.

Table 1. Monitored PV plant variables.

$Y_{n,1–16}$	Variable	$Y_{n,17–47}$	Variable
1	DC Total Output Power [kW]	17	DC Voltage Inv. 1—block C [V]
2	AC Total Active Output Power [kW]	18,19,20	AC Voltage-phase 1, 2, 3 Inv. 1—block C [V]
3	Total PV Plant Energy Produced [kWh]	21	Active Output Power Inv. 2—block A [kW]
4	Ambient Air Temperature [°C]	22	DC Voltage Inv. 2—block A [V]
5	Solar Irradiance [W/m²]	23,24,25	AC Voltage-phase 1, 2, 3 Inv. 2—block A [V]
6	Active Output Power Inv. 1—block A [kW]	26	Active Output Power Inv. 2—block B [kW]
7	DC Voltage Inv. 1—block A [V]	27	DC Voltage Inv. 2—block B [V]
8,9,10	AC Voltage-phase 1, 2, 3 Inv. 1—block A [V]	28,29,30	AC Voltage-phase 1,2,3 Inv. 2—block B [V]
11	Active Output Power Inv. 1—block B [kW]	31	Active Output Power Inv. 2—block C [kW]
12	DC Voltage Inv. 1—block B [V]	32	DC Voltage Inv. 2—block C [V]
13,14,15	AC Voltage-phase 1, 2, 3 Inv. 1—block B [V]	33,34,35	AC Voltage-phase 1, 2, 3 Inv. 2—block C [V]
16	Active Output Power Inv. 1—block C [kW]	36–47	String currents [A]

Table 2 provides the specification of the sensors used in the monitoring system. As far as the dataset is concerned, the 47 monitored variables $Y_{n,m}$ are used for the connectivity analysis with a sliding window set to 216 data (about 18 h) per variable, along the available 5-months records.

Table 2. Main characteristics of the different sensors involved.

Sensor	Type	Measurement Range	Temperature Range
Solar radiation sensor	Pyranometer	0 to 4000 W/m^2	−20 to + 50 °C
Temperature sensor	RTD (PT-100)	−50 to + 300 °C	−50 to + 300 °C
Current sensors	Shunt resistor	15 A	−40 to + 125 °C
Voltage sensors	Resistive potential divider	3 to 500 V (DC)	−40 to + 85 °C

4. Results

Figure 2 provides the evolution of the solar irradiance and the active power of the inverter 1 (block A) gathered from the measurement system during the period of investigation.

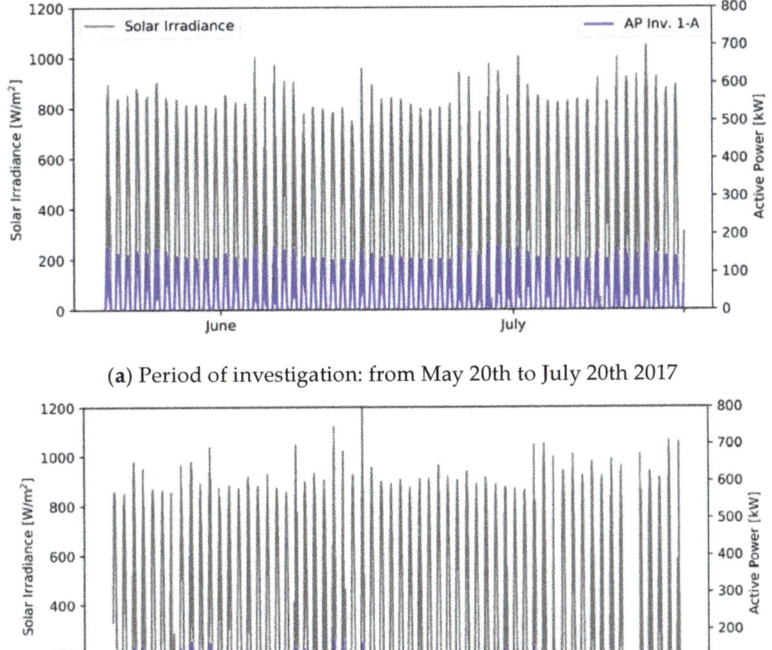

(**a**) Period of investigation: from May 20th to July 20th 2017

(**b**) Period of investigation: from July 21st to September 15th 2017

Figure 2. Solar irradiance and active power of the inverter 1 (block A) during the period of investigation.

The interest on inverter 1 is motivated by the observation of the signals revealing the occurrence of a fault in the period between 31st August and 15th September. In particular, a breakdown of switching devices occurred, which caused the failure of the block A of inverter 1 inductor. In this case the internal protection of the inverter automatically reacted by turning off the block involved in the fault. It is worth noting that the protection reacts within the sampling interval.

Detailing on a three-day period at the end of August, the trends of the active power on the 3 blocks (from A to C) of the two inverters with the irradiation are shown in Figure 3 (inverter 2) and Figure 4 (inverter 1) respectively. The plots refer to 30th August (Figures 3a and 4a), 31st August (Figures 3b and 4b) and September 1st (Figures 3c and 4c).

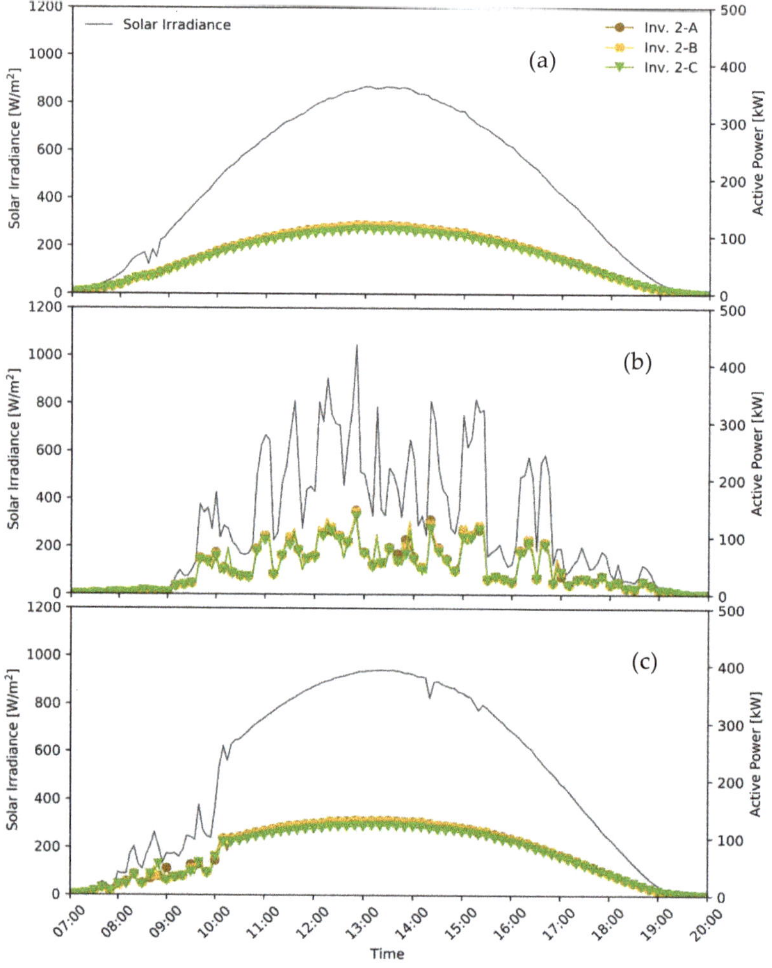

Figure 3. Analysis of sensor signals of the Inverter 2 on 30st August (**a**), on 31th August (**b**) and on 1st September (**c**)—Solar Irradiance over the Active Power of the inverter blocks.

In Figures 3a and 4a, August 30th plots reveals the typical operation of PV plant in sunny condition, where the solar irradiance reaches the peak value of about 950 W/m^2 at mid-day and the active powers of all the inverters blocks follow its trend.

Looking at the monitored data on 31st August (Figures 3b and 4b), it is possible to infer that the PV plant operates under cloudy weather conditions with variations of solar irradiance. In particular, while the inverter 2 (Figure 3b) appears to work in standard conditions, following the solar radiation trend, inverter 1 (Figure 4b) starting from 13:30 features the collapse of active power on block A, which then extends to 1st September (see Figure 4c) and gives evidence of fault occurrence.

To give more hints on this fault event, Figure 5 compares the signals $Y_{n,7-8}$ and $Y_{n,36}$ gathered from inverter 1 block A (Figure 5a) and the corresponding CNA metrics. Specifically, Figure 5b illustrates the behavior of degree centralities of solar irradiance, DC voltage, AC voltage and string currents.

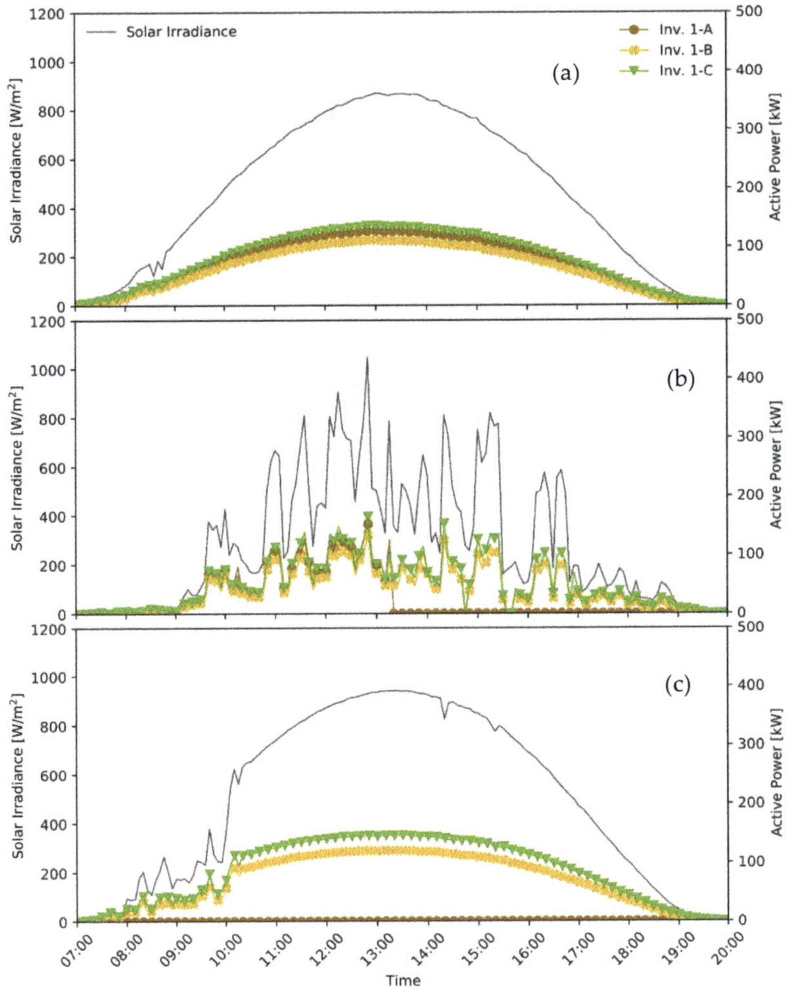

Figure 4. Analysis of sensor signals of the Inverter 1 on 30st August (**a**), on 31th August (**b**) and on 1st September (**c**)—Solar Irradiance over the Active Power of the inverter blocks.

As confirmed by sensor signals (Figure 5a), at 13:30 the fault has an immediate effect on the string currents, while the voltages zeroed around 16:30. When looking at complex network topology measurements (Figure 5b), the fault occurrence correlates with the departure of the degree centrality of solar irradiance from those of voltage-current signals which remain correlated.

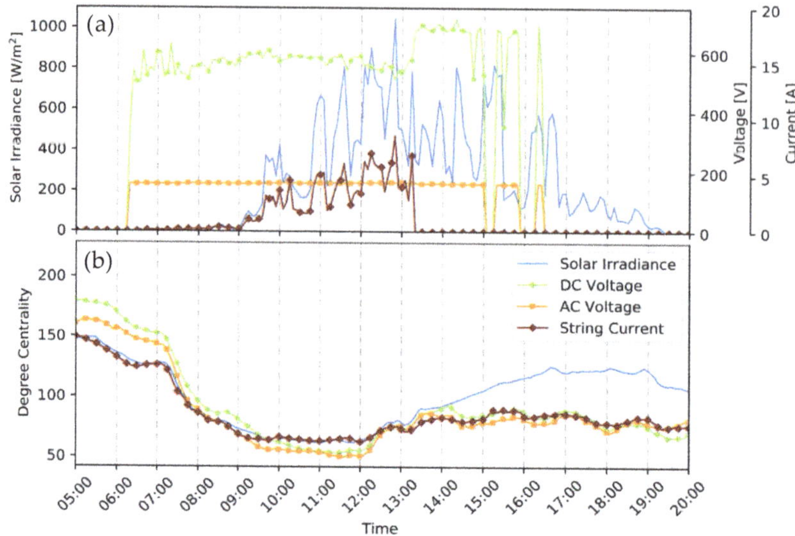

Figure 5. Analysis of (**a**) sensor signals from block A of the inverter 1 compared with (**b**) sensor network CNA metrics on 31st August.

In order to understand the dynamic of the PV field operation, EDA is applied to raw data from the plant monitoring system as well as to the modeled sensor network topology variables. Figure 6, first, shows the results of the three-dimensional scatterplot of sensor signals as a function of time and solar irradiance, focusing on: the active power (Figure 6a), DC voltage (Figure 6b) and AC voltage (Figure 6c) of the inverter 1 block A. Notably, the plots refer to the sampling period from 20th May to 15th September. Data have been colored according to an agglomerative clustering.

Irrespective of the EDA variables, the analysis of the scatter plots in Figure 6a–c identify three patterns. Specifically:

- A to B (from 00:00 of 20 May to 16:00 of 31 August), includes the sub-set of operating conditions of the inverter block with AC and DC voltages, and active power following the trend of actual solar irradiance.
- A′ to B′ (from 00:00 of 20 May to 16:00 of 31 August), includes the sub-set representative of night conditions.
- C to D (from 16:00 of 31 August to 18:00 of 15 September), then, represents the power system shutdown phase after the fault event following which AC and DC voltages, and active power zeroed.

In terms of raw data clustering, it is possible to resolve only the two behaviors which determine the operations before and after the fault event.

Figure 7 illustrates the results of EDA applied to network topology measurements. The scatter plots are created to investigate the correlation in time between a global graph measure (e.g., the graph diameter) and a local node related variable (e.g., sensor signal degree of centrality). In particular, the sensor network graph diameter is plotted against active power degree centrality (Figure 7a), AC voltage degree centrality (Figure 7b), and DC voltage degree centrality (Figure 7c).

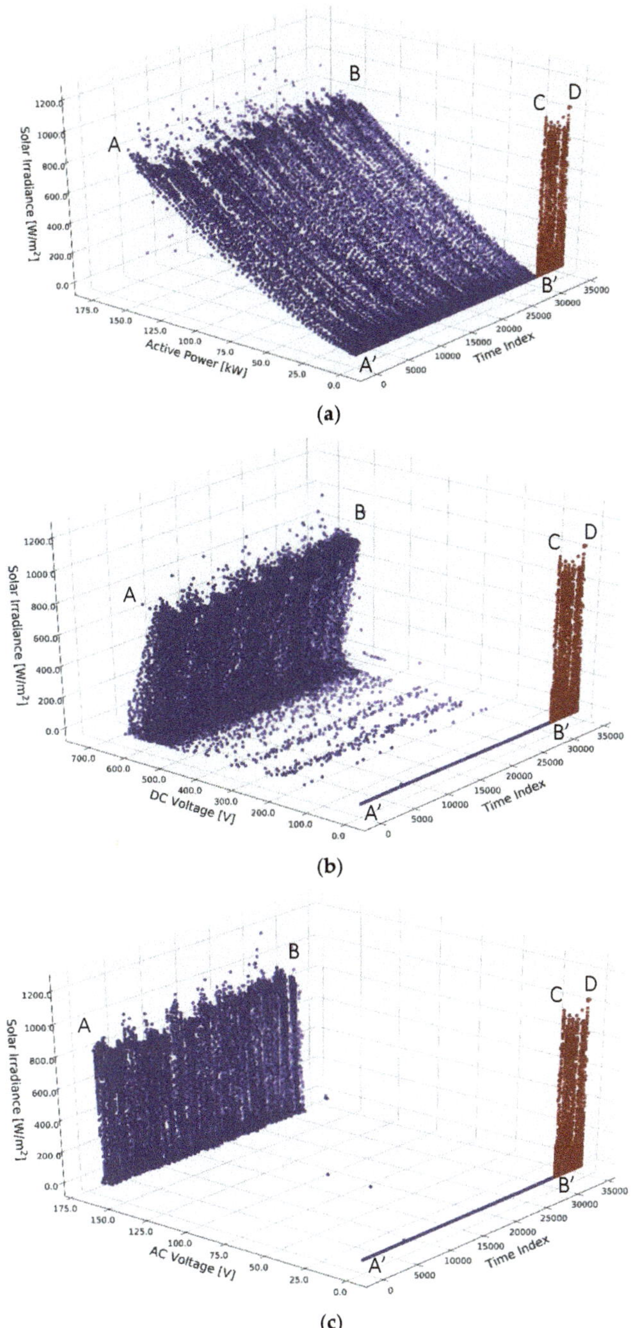

Figure 6. Exploratory data analysis applied to sensor signals from block A of the inverter 1. Trends of (**a**) active power, (**b**) DC voltage and (**c**) AC voltage (x-axis), against solar irradiance (y-axis) and time (z-axis).

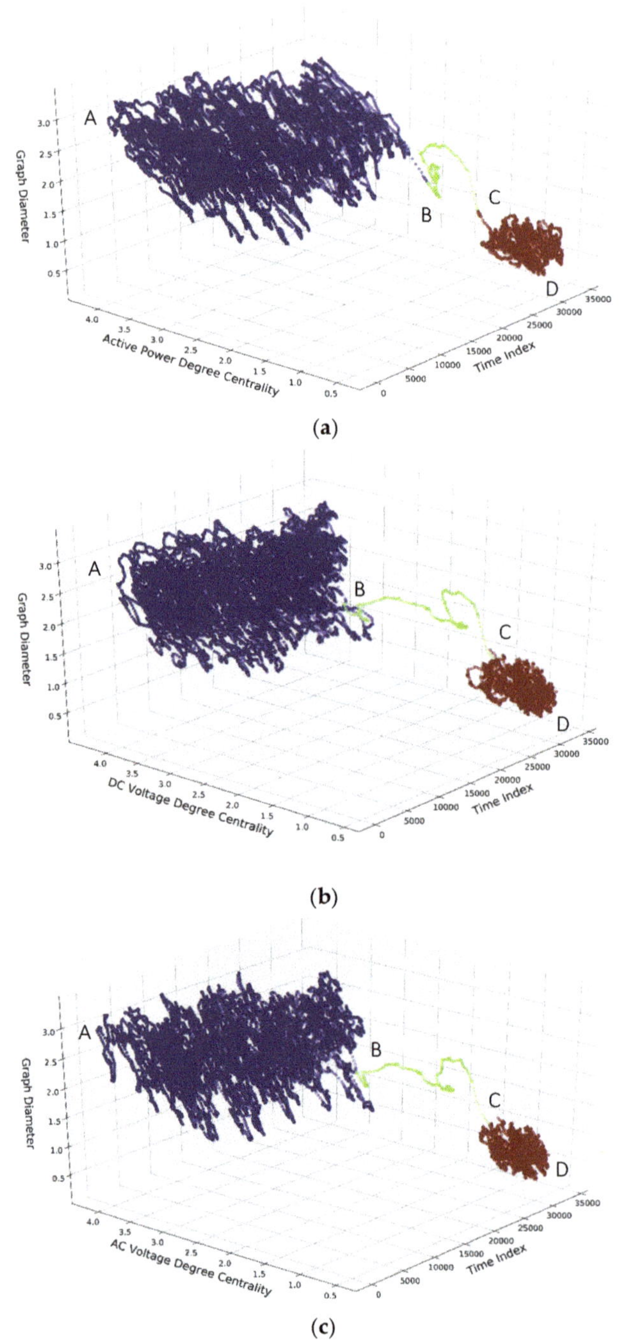

Figure 7. Exploratory data analysis applied to complex network measurements from block A of the inverter 1. Trends of (**a**) active power, (**b**) DC voltage and (**c**) AC voltage normalized weighted degree centralities (x-axis), against network diameter (y-axis) and time (z-axis).

As a first general comment, Figure 7 gives the evidence of a wealth of information emerging from the application of CNA methods. All the scatter plots demonstrate how the combination of global-local network topology measures determine the emergence of the dynamics in the PV plant operations, opening the possibility for a different performance indexing. In detail, all the plots distinguish three patterns possibly linked to the operations of the inverter block:

- A to B (from 00:00 of 20 May to 06:30 of 31 August), typical of the standard operation, features a degree centrality in the range 3 to 2, which is nearly proportional to the diameter of the network which experience a variation in the range 3 to 1.5.
- C to D (from 16:30 of 31 August to 18:00 of 15 September), represents the fault phase with the inverter shutdown. This cluster is characterized by low values of degree centrality (i.e., below 1) and low values of the network diameter (1.5 to 0.2).
- B to C (from 06:30 to 16:30 of 31 August), finally includes the sub-set of operations that transition the inverter to fault. This cluster contains the series of points connecting the previous groups. While the network diameter ranges between 2.5 and 1.5, the degree of centralities gradually decreases from 2 to about 1 for inverter 1 block A voltages and active power. This circumstance permits us to infer that this cluster, possibly, isolates the evolution of inverter operating conditions from the precursor (6:30) to the automatic block shutdown (13:30) including also the preliminary phase of fault evolution with non-zero AC/DC voltages (16:30). As such, it demonstrates the possibility of identifying at 7 h long pre-fault event which anticipates the evolution of inverter operating conditions on 31 August.

In this case, the results of the agglomerative clustering coincide with those emerging from the scatterplots and confirm the existence of a pre-fault behavior following a nearly proportional decrease in global/local graph topology measures.

5. Conclusions

This paper proposes a sensor fusion approach based on the use of graph modeling techniques to investigate the connectivity among several key parameters of the PV plant.

The results show that the study of the properties of the graphs through the application of Complex Network Analysis techniques is able to reveal a wealth of hidden information.

As reported in Korn et al. [61] the global behavior of the system is more than the sum of its parts, so these properties can be thought of as behavior deriving from the interaction between the components that can't be identified through their simple functional decomposition.

In particular, the visual analysis of the single topological metrics of the functional graph focused on the inverter block where the fault occurs (see Figure 4), reveals an evident variation in terms of correlation between the monitored variables, mainly associated with an anomalous deviation of the degree centrality of the solar irradiance with respect to the centrality of the key block parameters in the fault conditions.

However, by combining the multiple information deriving from the different network measurements (i.e., Degree Centrality and Network Diameter) with EDA techniques, it is possible to clearly distinguish not only the standard operation from the fault conditions, but also to isolate specific pre-fault conditions with an advance time with respect to the event of about 7 h. It is important to note that these conditions only emerge after applying CNA and are not observable with EDA techniques based on simple sensor signals. This latter type of analysis, in fact, is only able to discriminate standard operating conditions from fault conditions.

Thus, the results of this study show interesting potential in the evaluation of useful Key Performance Indicators and Control Charts based on topographic metrics of graphs for early fault detection in the PV plant.

Author Contributions: Conceptualization, A.C. and F.B.; Methodology, F.B.; Writing—Python Code, L.C.; Data Curation and Validation, F.L.; Formal Analysis and Interpretation, F.B. and L.C.; Writing—Original Draft Preparation, all authors; Writing—Review & Editing, all authors; Supervision, A.C.

Funding: This research received no external funding.

Conflicts of Interest: The authors declare no conflict of interest.

References

1. Lacal Arantegui, R.; Jäger-Waldau, A. Photovoltaics and wind status in the European Union after the Paris Agreement. *Renew. Sustain. Energy Rev.* **2018**, *81*, 2460–2471. [CrossRef]
2. Antonelli, M.; Desideri, U. The doping effect of Italian feed-in tariffs on the PV market. *Energy Policy* **2014**, *67*, 583–594. [CrossRef]
3. Moreno-Garcia, I.M.; Palacios-Garcia, E.J.; Pallares-Lopez, V.; Santiago, I.; Gonzalez-Redondo, M.J.; Varo-Martinez, M.; Real-Calvo, R.J. Real-Time Monitoring System for a Utility-Scale Photovoltaic Power Plant. *Sensors* **2016**, *16*, 770. [CrossRef] [PubMed]
4. Mellit, A.; Tina, G.M.; Kalogirou, S.A. Fault detection and diagnosis methods for photovoltaic systems: A review. *Renew. Sustain. Energy Rev.* **2018**, *91*, 1–17. [CrossRef]
5. Richter, M.; Tjengdrawira, C.; Vedde, J.; Green, M.; Frearson, L.; Herteleer, B.; Jahn, U.; Herz, M.; Kontges, M.; Stridh, B.; et al. *Technical Assumptions Used in PV Financial Models Review of Current Practices and Recommendations*; Report IEA-PVPS T13-08:2017; IEA: Paris, France, 2017.
6. Houssein, A.; Heraud, N.; Souleiman, I.; Pellet, G. Monitoring and fault diagnosis of photovoltaic panels. In Proceedings of the IEEE International Energy Conference and Exhibition, Manama, Bahrain, 18–22 December 2010; pp. 389–394.
7. Spagnuolo, G.; Xiao, W.; Cecati, C. Monitoring, Diagnosis, Prognosis, and Techniques for Increasing the Lifetime/Reliability of Photovoltaic Systems. *IEEE Trans. Ind. Electron.* **2015**, *62*, 7226–7227. [CrossRef]
8. Muhammad, N.; Zakaria, N.Z.; Shaari, S.; Omar, A.M. Fault detection approach in photovoltaic system using Mathematical method diagnosis. *J. Fundam. Appl. Sci.* **2018**, *10*, 270–281.
9. Davarifar, M.; Rabhi, A.; Elhajjaji, A.; Dahmane, M. Real-time Model base Fault Diagnosis of Photovoltaic Panels Using Statistical Signal Processing. In Proceedings of the International Conference on Renewable Energy Research and Applications, Madrid, Spain, 20–23 October 2013; pp. 599–604.
10. Hassan Ali, M.; Rabhi, A.; Elhajjaji, A.; Tina, G.M. Real Time Fault Detection in Photovoltaic Systems. *Energy Procedia* **2017**, *111*, 914–923.
11. Chen, Y.H.; Liang, R.; Tian, Y.; Wang, F. A novel fault diagnosis method of PV based-on power loss and I-V characteristics. *IOP Conf. Ser. Earth Environ. Sci.* **2016**, *40*, 012022. [CrossRef]
12. Garoudja, E.; Harrou, F.; Sun, Y.; Kara, K.; Chouder, A.; Silvestre, S. Statistical fault detection in photovoltaic systems. *Sol. Energy* **2017**, *150*, 485–499. [CrossRef]
13. Chiacchio, F.; Famoso, F.; D'Urso, D.; Brusca, S.; Aizpurua, J.I.; Cedola, L. Dynamic Performance Evaluation of Photovoltaic Power Plant by Stochastic Hybrid Fault Tree Automaton Model. *Energies* **2018**, *11*, 306. [CrossRef]
14. Dhimish, M.; Holmes, V.; Mehrdadi, B.; Dales, M. Multi-layer photovoltaic fault detection algorithm. *High Volt.* **2017**, *2*, 244–252. [CrossRef]
15. Bonsignore, L.; Davarifar, M.; Rabhi, A.; Tina, G.M.; Elhajjaji, A. Neuro-Fuzzy fault detection method for photovoltaic systems. *Energy Procedia* **2014**, *62*, 431–441. [CrossRef]
16. Ventura, C.; Tina, G.M. Development of models for on-line diagnostic and energy assessment analysis of PV power plants: The study case of 1 MW Sicilian PV plant. *Energy Procedia* **2015**, *83*, 248–257. [CrossRef]
17. Abdulmawjood, K.; Refaat, S.S.; Morsi, W.G. Detection and prediction of faults in photovoltaic arrays: A review. In Proceedings of the IEEE 12th International Conference on Compatibility, Power Electronics and Power Engineering, Doha, Qatar, 10–12 April 2018; pp. 1–8.
18. Chokor, A.; El Asmar, M.; Lokanath, S.V. A Review of Photovoltaic DC Systems Prognostics and Health Management: Challenges and Opportunities. In Proceedings of the Annual Conference of the Prognostics and Health Management Society, Denver, CO, USA, 3–6 October 2016; pp. 43–54.
19. Triki-Lahiani, A.; Bennani-Ben Abdelghani, A.; Slama-Belkhodja, I. Fault detection and monitoring systems for photovoltaic installations: A review. *Renew. Sustain. Energy Rev.* **2018**, *82*, 2680–2692. [CrossRef]

20. Gubbi, J.; Buyya, R.; Marusic, S.; Palaniswami, M. Internet of Things (IoT): A vision, architectural elements, and future directions. *Future Gener. Comput. Syst.* **2013**, *29*, 1645–1660. [CrossRef]

21. Macaluso, I.; Galiotto, C.; Marchetti, N.; Doyle, L. A complex systems science perspective on wireless networks. *J. Syst. Sci. Complex.* **2016**, *29*, 1034–1056. [CrossRef]

22. Corsini, A.; Bonacina, F.; Feudo, S.; Marchegiani, A.; Venturini, P. Internal Combustion Engine sensor network analysis using graph modeling. *Energy Procedia* **2017**, *126*, 907–914. [CrossRef]

23. Zhou, K.; Fu, C.; Yang, S. Big data driven smart energy management: From big data to big insights. *Renew. Sustain. Energy Rev.* **2016**, *56*, 215–225. [CrossRef]

24. Yu, N.; Shah, S.; Johnson, R.; Sherick, R.; Hong, M.; Loparo, K. Big Data Analytics in Power Distribution Systems. In Proceedings of the IEEE Power & Energy Society Innovative Smart Grid Technologies Conference, Washington, DC, USA, 18–20 February 2015; pp. 1–5.

25. Daliento, S.; Chouder, A.; Guerriero, P.; Massi Pavan, A.; Mellit, A.; Moeini, R.; Tricoli, P. Monitoring, Diagnosis, and Power Forecasting for Photovoltaic Fields: A Review. *Int. J. Photoenergy* **2017**, *2017*, 1356851. [CrossRef]

26. Riley, D.; Johnson, J. Photovoltaic Prognostics and Heath Management using Learning Algorithms. In Proceedings of the 38th IEEE Photovoltaic Specialists Conference, Austin, TX, USA, 3–8 June 2012; pp. 1535–1539.

27. Chine, W.; Mellit, A.; Bouhedir, R. FPGA-Based Implementation of an Intelligent Fault Diagnosis Method for Photovoltaic Arrays. In *Artificial Intelligence in Renewable Energetic Systems*; Springer: Cham, Switzerland, 2018; pp. 245–252.

28. De Benedetti, M.; Leonardi, F.; Messina, F.; Santoro, C.; Vasilakos, A. Anomaly detection and predictive maintenance for photovoltaic systems. *Neurocomputing* **2018**, *310*, 59–68. [CrossRef]

29. Jiang, L.L.; Maskell, D.L. Automatic Fault Detection and Diagnosis for Photovoltaic Systems using Combined Artificial Neural Network and Analytical Based Methods. In Proceedings of the IEEE International Joint Conference on Neural Networks, Killarney, Ireland, 12–17 July 2015; pp. 1–8.

30. Mohamed, A.H.; Nassar, A.M. New Algorithm for Fault Diagnosis of Photovoltaic Energy Systems. *Int. J. Comput. Appl.* **2015**, *114*, 26–31. [CrossRef]

31. Waltz, E.; Llinas, J. *Multisensor Data Fusion*; Artech House: Boston, MA, USA, 1990; Volume 685.

32. Newman, M.E.J. The Structure and Function of Complex Network. *Siam Rev.* **2003**, *45*, 167–256. [CrossRef]

33. Kauffman, S.A. *The Origins of Order: Self-Organization and Selection in Evolution*; Oxford University Press: New York, NY, USA, 1993.

34. Asbjørnsen, O.A. *Systems Engineering Principles and Practice*; SKARPODD Co.: Arnold, MD, USA, 1992.

35. Pestov, I.; Verga, S. Dynamical networks as a tool for system analysis and exploration. In Proceedings of the IEEE Symposium on Computational Intelligence for Security and Defense Applications, Ottawa, ON, Canada, 8–10 July 2009; pp. 1–8.

36. Carbone, A.; Jensen, M.; Sato, A.H. Challenges in data science: A complex systems perspective. *Chaos Solitons Fractals* **2016**, *90*, 1–7. [CrossRef]

37. Python Version 3.5.6, 2018. Available online: http://www.python.org (accessed on 15 October 2018).

38. Pedregosa, F.; Varoquaux, G.; Gramfort, A.; Michel, V.; Thirion, B.; Grisel, O.; Blondel, M.; Prettenhofer, P.; Weiss, R.; Dubourg, V.; et al. Scikit-learn: Machine learning in Python. *J. Mach. Learn. Res.* **2011**, *12*, 2825–2830.

39. Hagberg, A.A.; Swart, P.J.; Schult, D.A. Exploring Network Structure, Dynamics, and Function using NetworkX. In Proceedings of the 7th Python in Science Conference, Pasadena, CA, USA, 19–24 August 2008; pp. 11–16.

40. Stögbauer, H.; Grassberger, P.; Kraskov, A. Estimating Mutual Information. *Phys. Rev. E* **2004**, *69*, 066138.

41. Kraskov, A. Synchronization and Interdependence Measures and their Applications to the Electroencephalogram of Epilepsy Patients and Clustering of Data. Ph.D. Thesis, John von Neumann Institute for Computing (NIC) Series, Jülich, Germany, 2004.

42. Mason, S.P.; Barabási, A.L.; Oltvai, Z.N.; Jeong, H. Lethality and centrality in protein networks. *Nature* **2001**, *411*, 2–41.

43. Kern, A.D.; Hahn, M.W. Comparative genomics of centrality and essentiality in three eukaryotic protein-interaction networks. *Mol. Biol. Evol.* **2005**, *22*, 803–806.

44. Wibral, M.; Lindner, M.; Pipa, G.; Vicente, R. Transfer entropy—a model-free measure of effective connectivity for the neurosciences. *J. Comput. Neurosci.* **2011**, *30*, 45–67.
45. Quiroga, R.Q.; Bhattacharya, J.; Pereda, E. Nonlinear multivariate analysis of neurophysiological signals. *Prog. Neurobiol.* **2005**, *77*, 1–37.
46. Guye, M.; Bettus, G.; Bartolomei, F.; Cozzone, P.J. Graph theoretical analysis of structural and functional connectivity MRI in normal and pathological brain networks. *Magn. Reson. Mater. Phys. Biol. Med.* **2010**, *23*, 409. [CrossRef]
47. Freeman, L.C. Centrality in social networks: Conceptual clarification. *Soc. Netw.* **1979**, *1*, 39–215. [CrossRef]
48. Sànchez-Hernàndez, G.; Casabayò, M.; Agell, N.; Puigbò, J.Y. Influencer detection approaches in social networks: A current state-of-the-art. *Front. Artif. Intell. Appl.* **2014**, *269*, 261–264.
49. Hacid, H.; Favre, C.; Zighed, D.A.; Guille, A. Information diffusion in online social networks: A survey. *Acm Sigmod Rec.* **2013**, *42*, 17–28.
50. Granger, C.W.J. Testing for causality: A personal viewpoint. *J. Econ. Dyn. Control* **1980**, *2*, 329–352. [CrossRef]
51. Scardoni, G.; Laudanna, C. Centralities Based Analysis of Complex Networks. *New Front. Graph Theory* **2012**, 323–348. [CrossRef]
52. Das, K.C.; Maden, A.D.G.; Cangul, I.N.; Cevik, A.S. On Average Eccentricity of Graphs. *Proc. Natl. Acad. Sci. India Sect. A Phys. Sci.* **2017**, *87*, 23–30. [CrossRef]
53. Koschützki, D.; Schreiber, F. Centrality Analysis Methods for Biological Networks and Their Application to Gene Regulatory Networks. *Gene Regul. Syst. Biol.* **2008**, *2*, 193–201. [CrossRef]
54. Tamassia, R.; Tollis, I.G.; Di Battista, G.; Eades, P. *Graph Drawing: Algorithms for the Visualization of Graphs*; Prentice Hall: Upper Saddle River, NJ, USA, 1998.
55. Tukey, J.W. The Future of Data Analysis. *Ann. Math. Stat.* **1962**, *33*, 1–67. [CrossRef]
56. Yu, C.H. Exploratory data analysis in the context of data mining and resampling. *Int. J. Psychol. Res.* **2010**, *3*, 9–22. [CrossRef]
57. Seltman, H.J. *Experimental Design and Analysis*; Carnegie Mellon University: Pittsburgh, PA, USA, 2012.
58. Ho, T.K.; Basu, M.; Law, M.H.C. *Data Complexity in Pattern Recognition*; Springer Science & Business Media: Berlin, Germany, 2006.
59. Camacho, J.; Pérez-Villegas, A.; Rodríguez-Gómez, R.A.; Jiménez-Mañas, E. Multivariate Exploratory Data Analysis (MEDA) Toolbox for Matlab. *Chemom. Intell. Lab. Syst.* **2015**, *143*, 49–57. [CrossRef]
60. Komorowski, M.; Marshall, D.C.; Salciccioli, J.D.; Crutain, Y. *Exploratory Data Analysis. Secondary Analysis of Electronic Health Records*; Springer: Cham, Switzerland, 2016; pp. 185–203.
61. Korn, B.; Dohler, H. A System is More Than the Sum of Its Parts—Conclusion of DLR'S Enhanced Vision Project ADVISE-PRO. In Proceedings of the 25th AIAA/IEEE Digital Avionics Systems Conference, Portland, OR, USA, 15–19 October 2006; pp. 1–8.

Article

Deterioration Diagnosis of Solar Module Using Thermal and Visible Image Processing

Heon Jeong [1], Goo-Rak Kwon [2] and Sang-Woong Lee [3,*]

[1] Department of Fire Service Administration, Chodang University, Mu-An 58530, Korea; hjeong@cdu.ac.kr
[2] Department of Information and Communication Engineering, Chosun University, Gwangju 61452, Korea; grkwon@chosun.ac.kr
[3] Department of Software, Gachon University, Seongnam 13120, Korea
* Correspondence: slee@gachon.ac.kr

Received: 1 May 2020; Accepted: 27 May 2020; Published: 3 June 2020

Abstract: Several factors cause the output degradation of the photovoltaic (PV) module. The main affecting elements are the higher PV module temperature, the shaded cell, the shortened or conducting bypass diodes, and the soiled and degraded PV array. In this paper, we introduce an image processing technique that automatically identifies the module generating the hot spots in the solar module. In order to extract feature points, we used the maximally stable extremal regions (MSER) method, which derives the area of interest by using the inrange function, using the blue color of the PV module. We propose an effective matching method for feature points and a homography translation technique. The temperature data derivation method and the normal/abnormal decision method are described in order to enhance the performance. The effectiveness of the proposed system was evaluated through experiments. Finally, a thermal image analysis of approximately 240 modules was confirmed to be 97% consistent with the visual evaluation in the experimental results.

Keywords: thermal image; photovoltaic module; hot spot; image processing; deterioration

---

1. Introduction

Photovoltaic (PV) modules are made up of several solar cells. When an abnormality occurs in such a cell, the cell operates like an electrical resistor. As a result, the short circuit current decreases because of the power consumption and the series circuit characteristic. This phenomenon leads to an increase in temperature and accelerates the damage of the PV cell. Eventually, this abnormal process limits the short circuit current of the faulty PV module, reducing the power generation efficiency, performance and life of the PV module [1]. When the current capacity is reduced due to shadows produced by leaves, clouds, and dust or an abnormality occurs such that one cell is completely obscured or the cell is aged, a reverse bias is applied to the PV cell. This causes a so-called hot spot [2]. Some studies have found the modules that may cause failures by the thermal analysis of PV modules using an infrared thermal camera to prevent this loss [3–8].

Suárez-Domínguez's PV module thermal image analysis study [5] showed that the mean temperature of the PV module was 21 degrees. However, the temperatures at the two hot spots were found to be 26 degrees and 28 degrees. E. Kaplani's study [6], confirmed that the temperature of the hot spot area exceeded 100 °C in summer. Several other studies have also suggested that thermal imaging cameras are reliable, economical, and easy methods of inspecting hot spots, and suggest that periodic inspection can lead to optimal plant operation [6,7]. However, reflection and emissivity must be considered when inspecting a PV module using a thermal imaging camera. When PV modules are inspected from the front, a thermal imaging camera sees the heat distribution on the glass surface but only indirectly sees the heat distribution in the underlying cells. Glass reflections are specular, which means that the surrounding objects, with different temperatures, can be seen clearly in the thermal image. In the worst case, this results in misinterpretations (false "hot spots") and measurement errors. In order to avoid the reflection of the thermal imaging camera and the operator in the glass, it should not be positioned perpendicularly to the module being inspected. However, emissivity is at its highest when the camera is perpendicular and decreases as the angle increases [8].

Some solar power plants are considering installing fixed dual cameras. However, high-resolution thermal cameras are still very expensive. In order to install a large number of these expensive cameras, a large initial investment cost is required. In addition, it is necessary to establish and maintain an information processing system for multiple cameras. In recent years, in order to inspect the PV modules installed in a wide area or in a difficult accessing place, a thermal image of the PV module using a drone is taken and a thermal temperature analysis is performed [9,10]. However, these studies rely on visually inspecting the temperature distribution and finding faulty modules while viewing infrared thermal images on the monitor. Thus, a deterioration diagnosis program of a photovoltaic module is required to manage a module installed in various solar power plants in a wide area.

Thermal cameras generally measure the wavelength emitted by an object to extract temperature information. However, it is difficult to distinguish objects according to temperature conditions. The thermal image alone provides the temperature value of a specific object. Nevertheless, object boundaries cannot be distinguished by the thermal image alone in order to automatically extract the region and determine the abnormality through the temperature in the region [11]. Therefore, research has been carried out to realize object classification and object temperature distribution through a visible camera and a thermal camera. These studies express the temperature image of a specific region by performing a matching process between a visible image and a thermal image [12–14]. The image registration process is particularly essential in order to obtain necessary information by using multiple thermal and visible cameras. The matching process performs a geometric alignment process on the image so that data can be compared or integrated from different images. In these studies, the feature points must be extracted well for matching, and the matching relation of feature points is important. Davis [12] proposed a method of extracting background-subtraction and contour and synthesizing thermal image and visible images to extract the feature points. Conaire [11] evaluated the use of feature points by applying various methods such as four feature points, color and edge histograms, and weighted models.

Another study suggests a method of finding a plane homography that aligns the thermal spectrum with the visible spectrum. Using the obtained homography, the thermal image is image-warped to match the image [14]. In order to obtain such a precise homography, a number of correspondence points are defined in the image manually and perform a homography search process is performed that minimizes the least squared error [15]. However, the lack of correlation between the brightness intensity of the degradation image and the intensity of the visible image makes it difficult to generate an automatic correspondence point due to the brightness. Thus, a passive correspondence point designation or a specific landmark search method is required [16].

In this paper, we introduce an image processing method that can automatically identify the module generating hot spots in the solar module. In order to overcome the difficulties of generating the automatic correspondence points proposed in the previous studies, the proposed PV module analysis system adopts the image processing method that uses the characteristic elements of the PV module. Additionally, the proposed method extracts temperature information automatically and determines whether there is an abnormality through the rectified module image. The proposed method consists of four major image processing methods using the unique features of the PV module. The first method uses the shape and color characteristics of the PV module. The second exploits the fact that the PV modules are installed in multiple clusters. The third method involves the extraction of characteristic points from the visible image and deterioration. Finally, the fourth method derives a homography transformation equation for matching between minutiae points. To extract the feature points, we use the maximally stable extremal regions (MSER) algorithm, which computes the depth change value with the neighboring pixels and generates a tree up to the neighboring pixel level value [17]. Next, the proposed method derives the area of interest by using the inrange function using the blue color of the PV module. We propose an effective matching method for feature points and a homography conversion technique using the random sample consensus (RANSAC) algorithm [18]. The module area is rectified to increase the ease of the temperature distribution inspection of the module area after the matching of the thermal image and the visible image is completed.

The temperature information is extracted from the rectified thermal image and the failure prediction module is determined by calculating the abnormal area of high temperature area. Through this image processing method, it is possible to automatically identify the module generating the hot spot of the photovoltaic module, and it can help identify the cause of hot spots by displaying the thermal image and visible image generated by the hot spot. If there is contamination, shadows, or breakage, etc., in the visible image in the high temperature area, it is easy to identify the cause. However, if there is no abnormal point, it can be judged that the performance degradation is due to the characteristic change. In order to distinguish between the normal and abnormal modules, accurate region extraction of the photovoltaic module must proceed. If the photovoltaic module area is well extracted from the thermal image, the abnormal temperature distribution is not found as shown in Figure 1a, but it is judged as normal. If the module area is incorrectly extracted as shown in Figure 1b, an abnormal temperature distribution is found and it is judged to be an abnormal module.

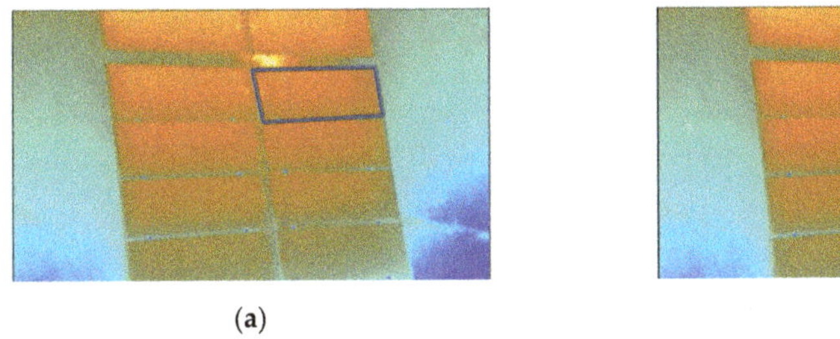

(a)   (b)

Figure 1. Examples of photovoltaic module area extraction and decision: (**a**) Correct extraction; (**b**) Incorrect extraction.

In Section 2, we describe an algorithm for detecting the occurrence of hot spots by analyzing the temperature distribution of a solar module. Furthermore, we describe the proposed visible image and thermal image registration, rectification process, temperature data derivation method, and normal/abnormal decision method. In Section 3, image analysis is performed on various solar modules and the effectiveness of the proposed algorithm is examined. In Section 4, we discuss the results and draw conclusions.

2. Deterioration Diagnosis Method of PV Module Using Thermal Image and Visible Image Processing

2.1. Proposed Method

The flow chart of the proposed method is shown in Figure 2.
The proposed algorithm can be summarized in the following five steps:

- Image acquisition.
- Solar module area search and block segmentation.
- Extraction of visible image and thermal image feature points and corresponding point matching.
- Matching and rectification based on corresponding points.
- Determining the abnormal module.

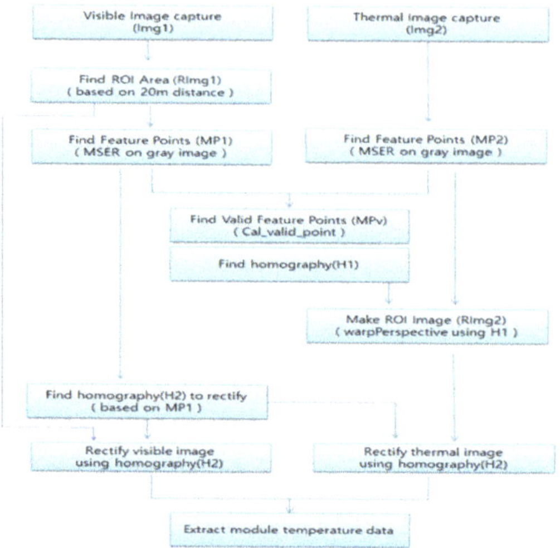

Figure 2. Flowchart of the proposed visible and thermal image matching and rectification algorithm.

Figures 3 and 4 show the algorithm proposed in this paper which includes the process of feature points extraction, feature points matching, segmentation, rectification and abnormal module judgment. The detailed description of the proposed method consists of following functional steps.

- A visible image (Img1) and a thermal image (Img2) are acquired by simultaneously photographing the PV module with a visible camera and a thermal camera.
- The region of interest corresponding to the region of Img1 is RImg1.
- Extract the PV module area from Rimg1 through the inrange function. The binary image represented by white in the module area portion is RMImg1.
- The feature points ($MP1$: mP11 to mP1n) corresponding to the vertexes of the box area are derived by the MSER algorithm in Rimg1.
- The feature points ($MP2$: mP21 to mP2n) corresponding to the vertexes of the box area are derived by the MSER algorithm in Img2.
- The effective matching point set ($M1,M2$) is derived from $MP1$ and $MP2$ through a decision criterion function.
- The homogrphy ($H1$) to convert $M2$ to $M1$ is found.
- Img2 is projected on Rimg1 to obtain a matching image (Rrimg).
- The temperature distribution calculation and module recognition process for each module consists of the following steps.
- In RMImg1, MB1, ... , MBn are obtained by segmenting by block using the findcontuors function.
- The homography Hb1 for rectifying the region of MB1, and the homography Hbn for rectifying the region of MBn are obtained. This homography is applied to obtain each rectified image RB1, ... , RBn.
- The rectified image RBImg is used to determine whether there is an abnormality through the temperature distribution and the abnormality determination equation for each module area.

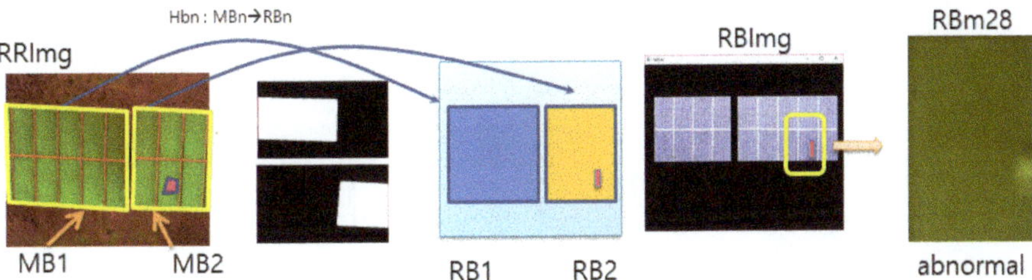

Figure 3. Feature points extraction and matching images creation through corresponding points.

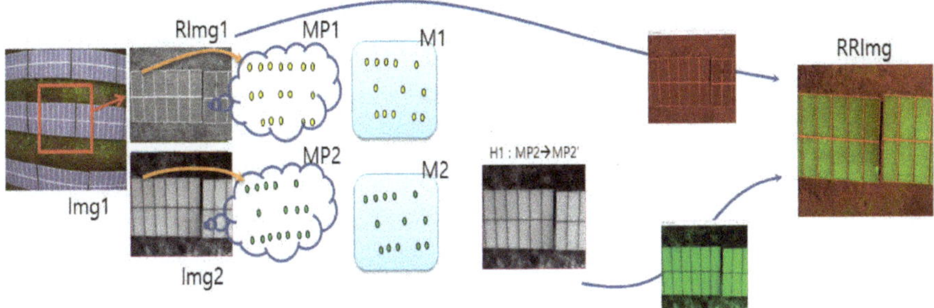

Figure 4. Segmentation, rectification, abnormal module judgment.

2.2. Image Acquisition

To acquire the image, we constructed a stereo camera with a visible image camera (resolution: 1280×720) and an infrared camera (640×512) as shown in Figure 5. Camera calibration was performed to derive a common image area with different angles of view between the visible image and the thermal image. Through the camera calibration process, the lens distortion correction and the main shooting distance were taken at positions of 8 to 18 m, and a strategic area of the thermal image was set in the visible image area.

Figure 5. Stereo camera configuration of visible and infrared camera.

2.3. PV Module Area Search and Block Segmentation

2.3.1. Extraction of Region of Interest through Color Inrange

We implemented an algorithm that derives the area of the photovoltaic module by exploiting the fact that the main color is in the blue region due to the characteristics of the PV module. The color image was transformed into the HSV image using the color inrange function. The hue value was 80~140, the saturation value was 65~255, and the value was 65~255. An example of the image processing using HSV conversion and inrange function is shown in Figure 6.

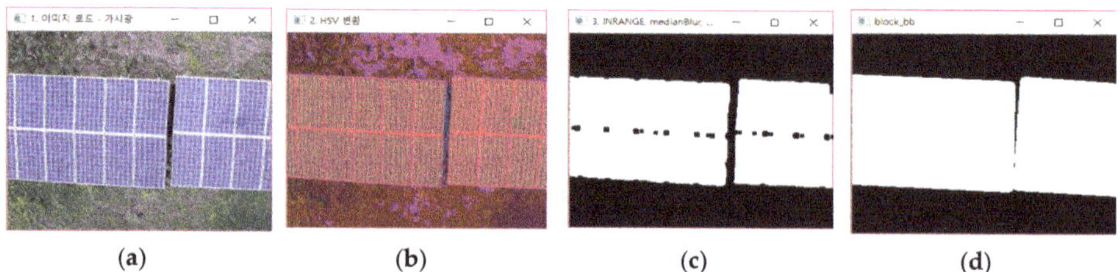

Figure 6. Image processing through HSV image and inrange function: (**a**) visible image; (**b**) HSV image; (**c**) inrange function result; (**d**) mask image (RMImg1).

2.3.2. Block Segmentation through Find Contours

Since PV modules have a cluster array structure for serial-parallel connection and the installation type may be changed in units of cluster blocks, the image was segmented in units of blocks. An example of the block segmentation by the findcontours function is shown in Figure 7.

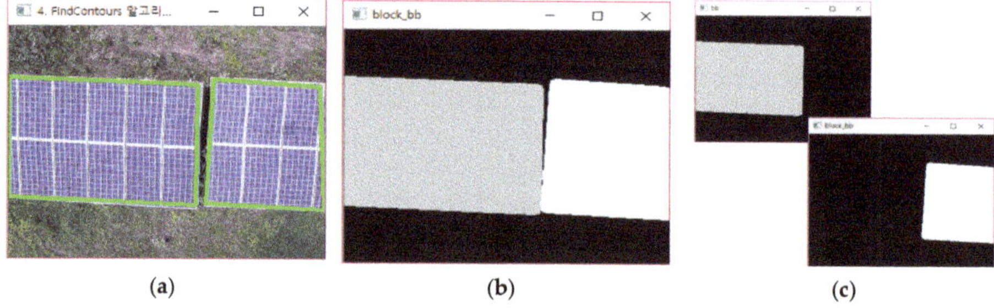

Figure 7. Block segmentation through findcontours: (**a**) find contours; (**b**) mask image; (**c**) sub-block mask image.

2.4. Module Feature Point Detection Using MSER Algorithm

This step converted the visible image into a gray image and extracted the module area using the MSER algorithm. Additionally, the thermal area image was extracted using the MSER algorithm. As an example, the results of using MSER without considering module size are shown in Figure 8.

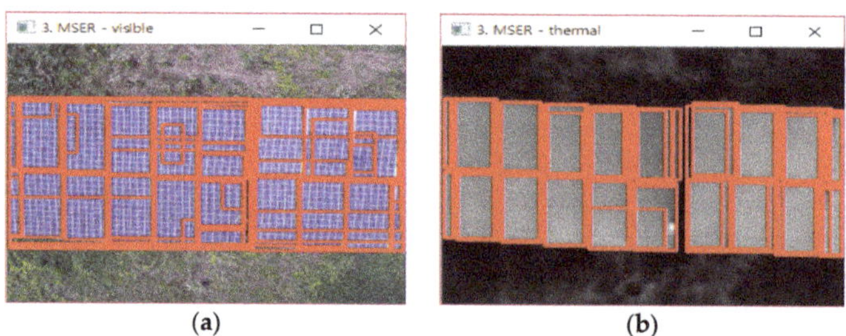

Figure 8. MSER algorithm without consideration of module size: (**a**) result of visible image gray MSER algorithm; (**b**) result of thermal image MSER algorithm.

The MSER rectangular areas of the visible image as shown in Figure 8a were sorted in ascending order, and are used as the predicted area value of one PV module based on the intermediate area value. In Figure 9, we show a graph of rectangular area vs. rank by area. MSER rectangular areas within 20%, the appropriate value according to various experiments, are derived by comparing these with the predicted area values. The calculation was performed in following steps.

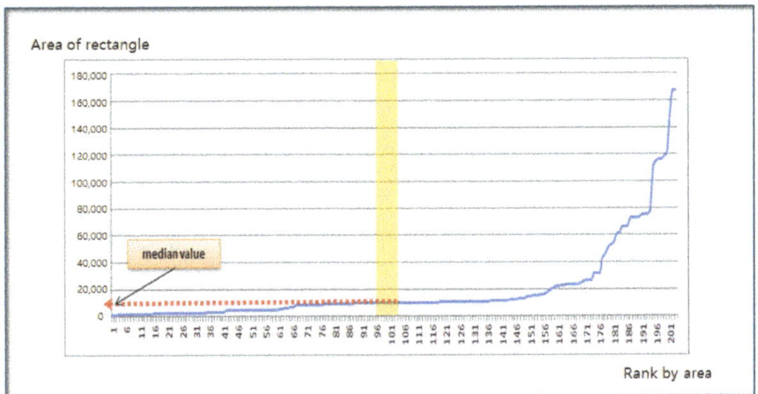

Figure 9. Choosing PV module area size by median value in a visible image.

- Arrange rectangular area elements by area.
- Find the median predicted area of the sorted result.
- Among the median predicted areas, the rectangular regions are filtered with those of a prediction area of ±20%.

In Figure 10, we show predicted areas by using the median value. The number of green rectangles in Figure 10 is reduced, compared to Figure 8, as a result of applying the MSER algorithm considering the module size and extracting the feature points. The yellow dots are used as feature points.

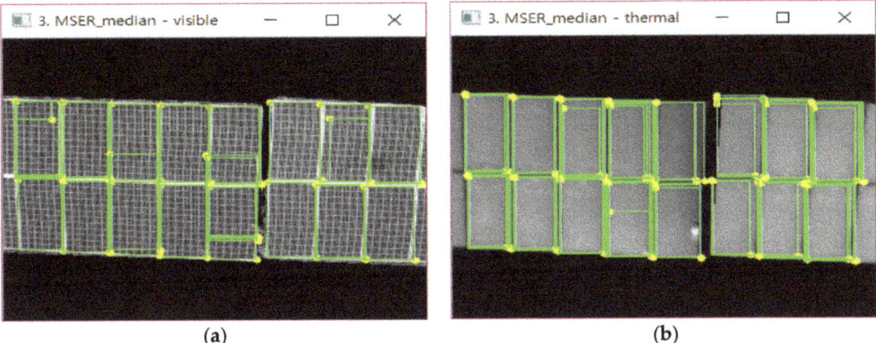

Figure 10. MSER algorithm and feature point detection result considering module size: (**a**) result of visible image; (**b**) result of thermal image.

2.5. Valid Matching Point, Homography and Registration

2.5.1. Valid Matching Point

An unnecessary matching point of a rectangular area is estimated as shown in Figure 10a,b. In order to remove the unnecessary matching points, the validity between the points of two images is determined, which is expressed as:

$$x1_i \in MP1, \quad x2_j \in MP2, \quad i : 1 \sim MP1_{size}, \quad j : 1 \sim MP2_{size} \tag{1}$$

$$r = \sqrt{\left(x1_i - x2_j\right)^2 + \left(y1_i - y2_j\right)^2}$$
$$if\ (r < th_r)\ then \quad M1_k = MP1_i, \quad M2_k = MP2_j \tag{2}$$

The set of feature points of the visible image and degradation images are called $MP1$ and $MP2$, respectively as shown in Figure 3. The expressions $(x1_i, y1_i)$ and $(x2_j, y2_j)$ respectively represent the x value and the y value of the i-th feature point of the $MP1$ and the j-th element of the $MP2$, Equation (1) performs an iterative search while changing to the introduction number. The validity evaluation step computes the distance value (r) between the two points. If the estimated distance is smaller than the boundary value (th_r), these two points are determined as an element of the effective feature point set $(M1, M2)$.

2.5.2. Homography Derivation through Valid Feature Points, Registration

A set $MP1$ of points $(x1_i)$ existing in the projection plane space corresponding to the set $MP2$ of points $(x2_i)$ existing in the two-dimensional projection plane space can be projected, and a projective transformation between the two images can be expressed as a relation function. Projectivity is defined as an invertible mapping h, which is a linear transformation through a non-singular 3×3 matrix [19].

$$\begin{pmatrix} x'_1 \\ x'_2 \\ x'_3 \end{pmatrix} = \begin{bmatrix} h_{11} & h_{12} & h_{13} \\ h_{21} & h_{22} & h_{23} \\ h_{31} & h_{32} & h_{33} \end{bmatrix} \begin{pmatrix} x_1 \\ x_2 \\ x_3 \end{pmatrix} \tag{3}$$

$$X' = HX \tag{4}$$

The homography converting the thermal image feature point $M2$ into the visible image feature point $M1$ was obtained as follows.

$$H1 : M2 \rightarrow M1 \tag{5}$$

Homography ($H1$) in the case of Figure 7 is as follows.

$$H1 = \begin{pmatrix} 1.011490762 & -0.0011793 & -0.17881936 \\ -0.0074779 & 0.95735177 & 5.3210135 \\ -0.0000029 & -0.0000903 & 1 \end{pmatrix} \tag{6}$$

Using the obtained homography $H1$, the thermal image is projected onto the visible image. Figure 11 shows a composite image of a red image with a visible image converted to green. The disparity between the images is shown by a simple synthesis before the conversion in Figure 11a,c. However, it can be confirmed that the images are well matched without any difference value in the composite image using the homography.

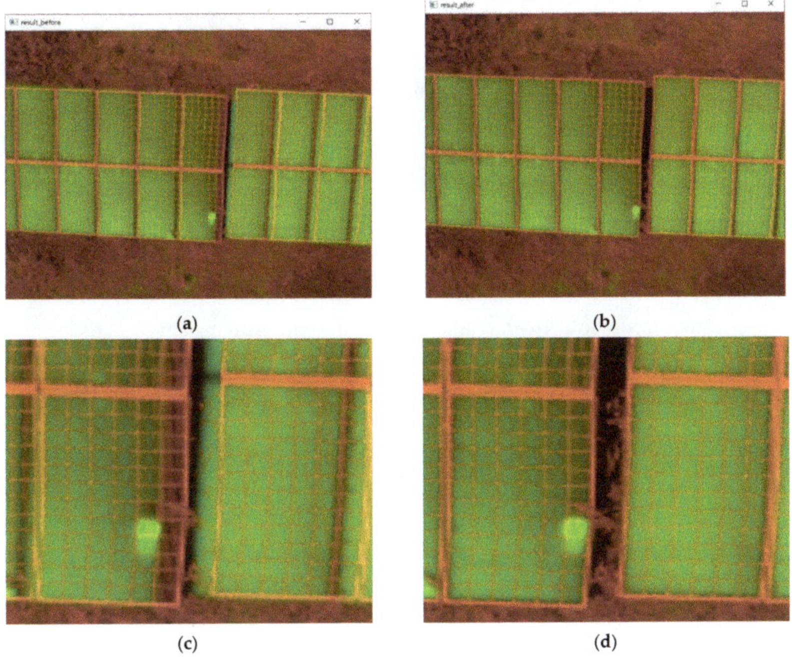

Figure 11. Registration images of visible and thermal using homography: (**a**) registration image without homography; (**b**) registration image with homography; (**c**) enlarged registration image of (**a**); (**d**) enlarge registration image of (**b**).

2.6. Rectification, Thermal Data Extraction, and Determination of Results

The module region was rectified to increase the ease of the temperature distribution inspection of the module area, once the registration of the thermal image and the visible image was completed. The faulty module was determined by calculating the high temperature area. Through this image processing procedure, the module generating hot spot in the PV module was automatically identified. This helped to identify the cause of the hot spot by displaying the thermal image and visible image generated by the hot spot. Visible images can be used to distinguish between contaminants, shadows, breakages, microcracks, or cell deterioration.

2.6.1. Homography Derivation and Rectification

In the blocks obtained through the segmentation, the entire area of the PV module was obtained, and the homography was derived using the edges of the valid feature points needing to be rectified.

$$HBn : MBn \rightarrow RBn, \; n : 1 \sim MB_{size} \tag{7}$$

$$RBimg = \sum_{n=0}^{blocksize} RBn \tag{8}$$

HBn uses a homography matrix for rectifying the sub-block *MBn* into the rectified block *RBn*, and obtains the rectified image *Rbimg* by summing the rectified images of sub-blocks. An example of obtaining the rectified images from sub-blocks is shown in Figure 12.

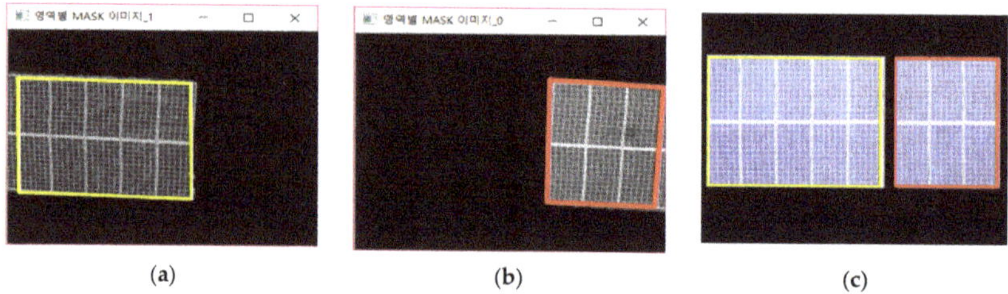

Figure 12. Process of obtaining a rectified image: (**a**) sub-block *MB1*; (**b**) sub-block *MB2*; (**c**) rectified block *RB1, RB2*.

2.6.2. Extracting Temperature Information and Determining the Abnormality

In order to determine whether an error had occurred in the PV module, the average temperature value (T_avg), the maximum temperature value (T_max), and the minimum temperature value, (T_min) information in the module area were extracted. Next, the high threshold temperature (Th_high) and the low threshold temperature (Th_low) were determined in the PV module area. If the temperature value was less than Th_high and less than Th_low, the count value (a_count) would be increased. If the value a_count exceeded 0.2% of the module area value (area_module), then the module would be considered as an abnormal module. Figure 13 shows an example of extracting temperature information and determining the abnormality. Figure 13a shows the temperature distribution for the abnormal module among several modules, and Figure 13b shows the results of the normal module (left) and the abnormal module (right) obtained through the proposed equation.

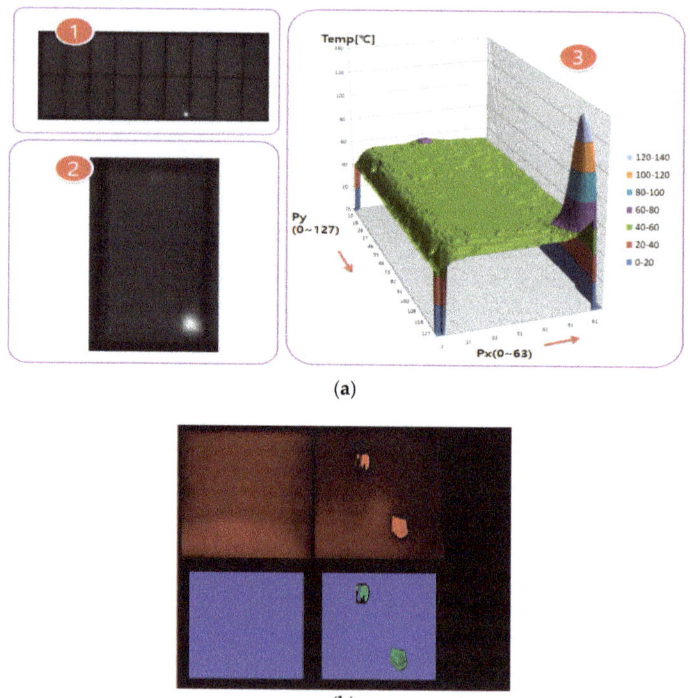

(a)

(b)

Figure 13. Extracting temperature information and determining the abnormality: (**a**) extracting temperature information; (**b**) normal module and abnormal module judgment.

$$TH\_high = T\_avg + T\_max \times 0.2$$
$$TH\_low = T\_avg + T\_min \times 0.2$$
$$if\ ((T\_val > Th\_high)\ or\ (T\_val < Th\_low))\quad a\_count\ ++;$$
$$if\ (a\_count > area\_module \times 0.002)\quad abnomal\ PV\ module$$

(9)

3. Experiment

To evaluate the validity of the proposed algorithm, 40 visible and thermal images were simultaneously captured using a drone. The shooting altitude was set to about 10 m, and the acquired image is shown in Figure 14.

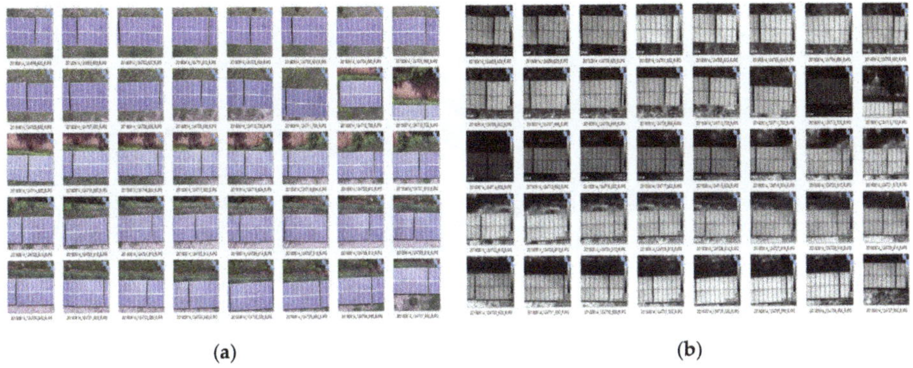

(a)

(b)

Figure 14. Samples for experiment: (**a**) visible image samples; (**b**) thermal image samples.

In order to evaluate correspondence consistency, ground-true images were manually generated for seven images as shown in Figure 15.

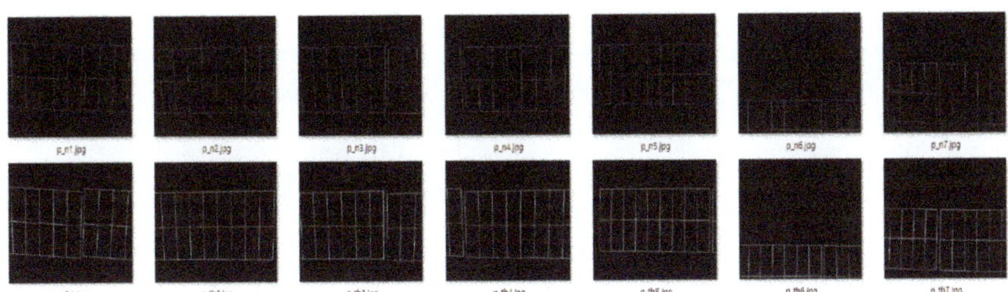

Figure 15. Ground-true images for evaluating matching consistency.

For each image, feature points were derived, and the homography for matching was derived, projected, and finally, matched. To obtain the validity, the homography was applied to the ground-true image to calculate the match rate. The results showed an average of 97% agreement.

The mean area value of the module area obtained by the MSER algorithm was 10,336 pixels, and the standard deviation of the PV area was 215 pixels. The average area accuracy of the module area was about 98%. Finally, the proposed method determined whether the module was normal or faulty, and estimated the faulty module by calculating the high temperature area. The thermal image analysis of approximately 240 modules was confirmed to be 97% consistent with the visual evaluation. The module with the error was confirmed to be faulty due to the reflection of sunlight.

Figure 16 shows the experimental results for finding the abnormal module. Based on the GPS information, the analysis results are shown on Google Maps in Figure 16a. A green mark indicates a normal module and a red mark indicates an abnormal module. When an user presses a red mark on a screen, the user can see the detailed analysis such as the corresponding visible image (Figure 16b), thermal image (Figure 16c), merge image before registration (Figure 16d), merge image after registration (Figure 16e), normalized image (Figure 16f), and the abnormal region derivation result by the determination algorithm (Figure 16g) are expressed. As a result of pressing the mark manually, it was confirmed that both normal and abnormal modules can be distinguished well.

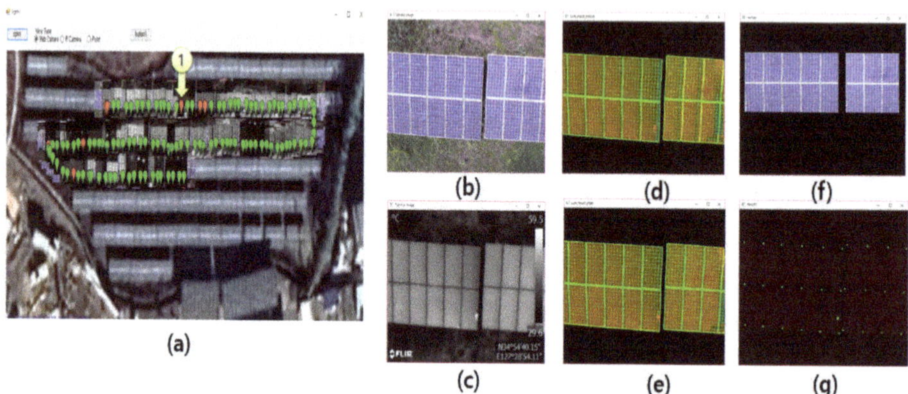

Figure 16. Abnormal module detection system: (**a**) main result screen; (**b**) visible image; (**c**) thermal image; (**d**) merge image before registration; (**e**) merge image after registration; (**f**) normalized image; (**g**) abnormal region derivation result.

4. Conclusions

In this paper, we introduced an image processing technique that can automatically identify the modules generating the hot spots in solar modules. The proposed PV module analysis system adopted the image processing method that uses the characteristic elements of the PV module in order to overcome the difficulties of generating the automatic correspondence point presented in the previous studies. Additionally, the proposed method determined an abnormality of the PV module. The image processing method using the characteristic elements of the PV module extracts the feature point derivation from the visible image and the degradation image by using the shape characteristic of the color module, the rectangular shape of the PV module, and the installation of a plurality of clusters, and a homography transform equation is derived from the matching points. To extract the feature points, we used the MSER algorithm, the inrange function, an effective proposed matching method and a homography conversion technique using the RANSAC algorithm.

Experimental results show that the match between the thermal image and the visible image was 97%. The accuracy of applying the MSER algorithm and the average area derivation algorithm was confirmed to be 98%. Similarly, through the rectification of the module area, the ease of the inspection of the temperature distribution of the module area was enhanced, and the failure prediction module was determined by calculating the abnormal area of high temperature. The thermal image analysis of approximately 240 modules was confirmed to be 97% consistent with the visual and visual evaluation. Through the application of these algorithms, it is possible to automatically identify the module generating the hot spots in the solar module, and to identify the cause of the hot spots by displaying the thermal image and visible image generated by the hot spot.

Author Contributions: Conceptualization, H.J. and G.-R.K.; methodology, H.J.; software, H.J.; validation, H.J., G.-R.K. and S.-W.L.; formal analysis, G.-R.K.; investigation, H.J.; resources, S.-W.L.; data curation, G.-R.K.; writing—original draft preparation, G.-R.K.; writing—review and editing, H.J.; visualization, G.-R.K.; supervision, S.-W.L.; project administration, H.J.; funding acquisition, H.J. All authors have read and agreed to the published version of the manuscript.

Funding: This work was supported by Institute for Information & communications Technology Promotion (IITP) grant funded by the Korea government (MSIP) (NO. 2017-0-01712) and National Research Foundation of Korea (NRF) grant funded by the Korea government (MSIT) (2019R1F1A1062075).

Conflicts of Interest: The authors declare that there are no conflicts of interest regarding the publication of this paper.

References

1. Alonso-Garcia, C.; Ruiz, J. Analysis and modelling the reverse characteristic of photovoltaic cells. *Sol. Energy Mater. Sol. Cells* **2006**, *90*, 1105–1120. [CrossRef]

2. Simon, M.; Meyer, E.L. Detection and analysis of hot-spot formation in solar cells. *Sol. Energy Mater. Sol. Cells* **2010**, *94*, 106–113. [CrossRef]

3. Fares, Z.; Becherif, M.; Emziane, M.; Aboubou, A.; Azzem, S.M. Infrared Thermography Study of the Temperature Effect on the Performance of Photovoltaic Cells and Panels. In *Advances in Theory and Practice of Computational Mechanics*; Springer Science and Business Media: Berlin/Heidelberg, Germany, 2013; Volume 22, pp. 875–886.

4. Hu, Y.; Cao, W.; Ma, J.; Finney, S.J.; Li, D.D.-U. Identifying PV Module Mismatch Faults by a Thermography-Based Temperature Distribution Analysis. *IEEE Trans. Device Mater. Reliab.* **2014**, *14*, 951–960. [CrossRef]

5. Suárez-Domínguez, F.J.; Prendes-Gero, M.-B.; Martín-Rodríguez, A.; Higuera-Garrido, A. IR thermography applies to the detection of solar panel. *Rev. De La Constr.* **2015**, *14*, 9–14. [CrossRef]

6. Kaplani, E. Detection of Degradation Effects in Field-Aged c-Si Solar Cells through IR Thermography and Digital Image Processing. *Int. J. Photoenergy* **2012**, *2012*, 1–11. [CrossRef]

7. Muñoz-García, M.; Muñoz-García, M.; Vela, N.; Chenlo, F. Early degradation of silicon PV modules and guaranty conditions. *Sol. Energy* **2011**, *85*, 2264–2274. [CrossRef]

8. Guide to the Use of Thermal Imaging Cameras in the Construction Industry and When Working with Renewable Energy Sources. Available online: http://www.flirmedia.com/MMC/THG/Brochures/T820325/T820325_EN.pdf (accessed on 2 April 2020).

9. Quater, P.B.; Grimaccia, F.; Leva, S.; Mussetta, M.; Aghaei, M. Light Unmanned Aerial Vehicles (UAVs) for Cooperative Inspection of PV Plants. *IEEE J. Photovoltaics* **2014**, *4*, 1107–1113. [CrossRef]

10. Addabbo, P.; Angrisano, A.; Bernardi, M.L.; Gagliarde, G.; Mennella, A.; Nisi, M.; Ullo, S.L. A UAV Infrared Measurement Approach for Defect Detection in Photovoltaic Plants. In Proceedings of the 2017 IEEE International Workshop on Metrology for AeroSpace (MetroAeroSpace), Padua, Italy, 21–23 June 2017; pp. 345–350.

11. Conaire, C.; O'Connor, N.; Cooke, E.; Smeaton, A. Comparison of Fusion Methods for Thermo-Visual Surveillance Tracking. In Proceedings of the 2006 9th International Conference on Information Fusion, Florence, Italy, 24 April 2006; pp. 1–7.

12. Davis, J.W.; Sharma, V. Background-subtraction using contour-based fusion of thermal and visible imagery. *Comput. Vis. Image Underst.* **2007**, *106*, 162–182. [CrossRef]

13. Istenic, R.; Heric, D.; Ribarič, S.; Zazula, D. Thermal and Visual Image Registration in Hough Parameter Space. In Proceedings of the 2007 14th International Workshop on Systems, Signals and Image Processing and 6th EURASIP Conference Focused on Speech and Image Processing, Multimedia Communications and Services, Maribor, Slovenia, 27–30 June 2007; pp. 106–109.

14. Mantel, C.; Spataru, S.; Parikh, H.; Sera, D.; Benatto, G.A.D.R.; Riedel, N.; Thorsteinsson, S.; Poulsen, P.B.; Forchhammer, S. Correcting for Perspective Distortion in Electroluminescence Images of Photovoltaic Panels. In Proceedings of the 2018 IEEE 7th World Conference on Photovoltaic Energy Conversion (WCPEC) (A Joint Conference of 45th IEEE PVSC, 28th PVSEC & 34th EU PVSEC), Waikoloa Village, HI, USA, 10–15 June 2018; pp. 0433–0437.

15. Zhao, J.; Cheung, S.-C.S. Human Segmentation by Fusing Visible-light and Thermal Imaginary. In Proceedings of the 2009 IEEE 12th International Conference on Computer Vision Workshops, ICCV Workshops, Kyoto, Japan, 27 September–4 October 2009; pp. 1185–1192.
16. Torabi, A.; Masse, G.; Bilodeau, G.-A. Feedback Scheme for Thermal-visible Video Registration, Sensor Fusion, and People Tracking. In Proceedings of the 2010 IEEE Computer Society Conference on Computer Vision and Pattern Recognition—Workshops, San Francisco, California, 13 June–18 June 2010; pp. 15–22.
17. Matas, J.; Chum, O.; Urban, M.; Pajdla, T. Robust wide-baseline stereo from maximally stable extremal regions. *Image Vis. Comput.* **2004**, *22*, 761–767. [CrossRef]
18. Chum, O.; Matas, J. Optimal Randomized RANSAC. *IEEE Trans. Pattern Anal. Mach. Intell.* **2008**, *30*, 1472–1482. [CrossRef] [PubMed]
19. Dubrofsky, E. Homography Estimation. Ph.D Thesis, University of British Columbia(Vancouver), Kelowna, BC, Canada, March 2009.

Article

Clustering-Based Self-Imputation of Unlabeled Fault Data in a Fleet of Photovoltaic Generation Systems

Sunme Park , Soyeong Park, Myungsun Kim and Euiseok Hwang *

School of Mechanical Engineering, Gwangju Institute of Science and Technology, 123 Cheomdangwagi-ro, Buk-gu, Gwangju 61005, Korea; pishp00200@gist.ac.kr (S.P.); soyeongp@gist.ac.kr (S.P.); rlaaudtjs@gist.ac.kr (M.K.)
* Correspondence: euiseokh@gist.ac.kr; Tel.: +82-62-715-3223

Received: 30 November 2019; Accepted: 3 February 2020; Published: 7 February 2020

Abstract: This work proposes a fault detection and imputation scheme for a fleet of small-scale photovoltaic (PV) systems, where the captured data includes unlabeled faults. On-site meteorological information, such as solar irradiance, is helpful for monitoring PV systems. However, collecting this type of weather data at every station is not feasible for a fleet owing to the limitation of installation costs. In this study, to monitor a PV fleet efficiently, neighboring PV generation profiles were utilized for fault detection and imputation, as well as solar irradiance. For fault detection from unlabeled raw PV data, K-means clustering was employed to detect abnormal patterns based on customized input features, which were extracted from the fleet PVs and weather data. When a profile was determined to have an abnormal pattern, imputation for the corresponding data was implemented using the subset of neighboring PV data clustered as normal. For evaluation, the effectiveness of neighboring PV information was investigated using the actual rooftop PV power generation data measured at several locations in the Gwangju Institute of Science and Technology (GIST) campus. The results indicate that neighboring PV profiles improve the fault detection capability and the imputation accuracy. For fault detection, clustering-based schemes provided error rates of 0.0126 and 0.0223, respectively, with and without neighboring PV data, whereas the conventional prediction-based approach showed an error rate of 0.0753. For imputation, estimation accuracy was significantly improved by leveraging the labels of fault detection in the proposed scheme, as much as 18.32% reduction in normalized root mean square error (NRMSE) compared with the conventional scheme without fault consideration.

Keywords: PV fleet; clustering-based PV fault detection; unsupervised learning; self-imputation

---

1. Introduction

A photovoltaic (PV) power plant is one of the most renewable, sustainable, and eco-friendly setups for converting solar energy, which is the most abundant and freely available energy source, into electrical energy [1]. The contribution of solar energy to the total global energy supply has rapidly increased in recent decades with PV installation capacity growing to more than 500 GW by the end of 2018 [2,3]. However, PV systems are exposed to harsh working conditions owing to uncertain outdoor environments and their complex structure. In [4], it was reported that the annual losses in PV generation reached 18.9% under zero or shading faults. Improving the reliability of renewable energy generation by fault detection and diagnosis (FDD) and correcting faulty data is essential for maintaining the efficiency of PV generation [5]. In addition, reliable information is required to be applied for various power applications, e.g., energy scheduling [6] and energy forecasting [7,8], to guarantee safe and stable grid systems.

For utility planners and operators, it is essential to examine the power output variability [9] to aggregate the fleet of PV systems, which is defined as the number of individual PV systems spread out over a geographical area [10]. Several studies presented station-pair correlation analyses by introducing virtual networks. A correlation was observed between short-term irradiance variability as a function of diverse distance and time scale [11]. Similarly, the maximum output variability of a fleet of PV plants was estimated by using the clearness index [12]. A variability model was built in [13] to integrate a large amount of generated solar power into power systems. To integrate the fleet as a distributed power source, it should be managed by an intelligent monitoring system that can correct abnormal data via real-time fault diagnosis and power generation forecasts.

PV faults occur because of various reasons at different locations, such as a module, string, or any other spot related to the PV systems. Visual and thermal methods employed in PV fault detection can detect superficial problems, such as browning, soiling or snow, discoloration, delamination, and hot spots, using auxiliary measurements [14]. However, this requires expensive and complicated equipment [15]. In recent years, many studies have employed methods using electrical variables via a data-driven approach. The electrical signal approaches are mainly referred to as maximum power point tracking (MPPT) with I–V characteristic analysis and power loss analysis. They are usually utilized to distinguish an open circuit, short circuit, degradation or aging, and shading faults that may typically occur on the DC side of a PV array [16–18].

For data-driven methods, automatic fault detection approaches can be categorized into conventional modeling-based methods and methods that utilize intelligent machine learning [19]. For the former case, the model can be built with respect to the physical attributes from the PV module specification for simulation settings to compare the desired output with the measured output [20]. Conventional statistical detection methods have been primarily presented in previous studies [16,21–25]. The exponentially weighted moving average has been used to identify DC side faults by comparing the one-diode model and estimated MPPs [21,22]. Lower and upper limits were set when the ratio of the measured to the modeled AC power exceeds 3-sigma [22]. In [23], outlier detection rules were proposed in their statistical details: The 3-sigma rule, Hampel identifier, and Boxplot rule using a PV string current. A symbolic aggregate approximation (SAX) scheme was used to convert the voltage profile, prior to performing clustering and anomaly detection [24].

With the advent of artificial intelligence (AI), which can be applied to various domains, particularly suited to the nonlinear behavior of PV systems, numerous studies have exploited AI-based monitoring systems [26]. The common artificial neural network (ANN) is widely used either to predict PV generation behavior or as a fault detection module based on several electrical parameters [15,27–31]. In comparison with a conventional back-propagation network, a probabilistic neural network (PNN) uses a probability density function as the activation function; thus, it is less sensitive to noisy and erroneous samples [32–34]. Fault detection by a support vector machine (SVM) has been used in several studies because it has the ability to separate objects by finding an optimal hyperplane that maximizes the margin in both binary and multiclass problems [35–37]. The decision tree (DT) builds repetitive decision rules within if/else instructions, which is intuitive. The model can be implemented conveniently with a large dataset [35,38]. The random forest (RF) has been applied to improve multiclass classification accuracy and to generalize performance [19]. Fuzzy classifications based on a fuzzy inference system (FIS) were developed by constructing logic rules [29,39]. A kernel extreme learning machine was investigated owing to its fast learning speed and good generalization [40]. Particle swarm optimization-back-propagation (PSO-BP) has been shown to improve the convergence and prediction accuracy of fault diagnosis systems [41].

Most of the data-driven approaches involve a supervised learning-based fault detection system that assigns a label for binary PV states as either normal or abnormal or as multi-class for corresponding fault types in advance. The detection or classification model learns the complex and unrevealed relation between input attributes and predefined labels in the training phase, and then the model is tested to determine whether it can distinguish PV states properly for new inputs. However, these processes require human effort to manually assign labels, and it is not easy to visualize the trained model. Graph-based semi-supervised learning (GBSSL) was proposed to detect line-to-line and open circuit faults using a few labeled data [42]. In [20], five types of faults were classified based on a single diode model with five input vectors associated with IV characteristics, solar irradiance, and temperature. Gaussian-fuzzy C-means was conducted using the distribution of each cluster and faults were diagnosed through PNN based on previous cluster center information [33]. A fuzzy membership algorithm based on degrees of fault data and cluster centers has been proposed [43]. Density peak-based clustering has also been proposed [44]. The 3-sigma rule was applied to determine each cluster center using the normalized voltage and current at the MPPs. Similarly, the PV local outlier factor (PVLOF) was computed from the current of the PV array to identify the degree of faults [45]. A single diode model-based prediction was implemented, enabling the generation of the residual, which was applied to the one-class SVM by quantifying the dissimilarity between the normal and faulty features [46].

In this study, we propose a framework of two stages of self-fault detection and self-imputation in a fleet of PV systems using neighboring PV power generation units based on correlation analysis. Because insolation data is not available with sufficient geographical resolution, especially for a small-scale PV system [12,37], neighboring PV generation data in the same fleet can be used jointly with distanced weather data. Since daily PV generation captures generally include unidentified erroneous samples, faulty data candidates were first labeled in the proposed scheme by an unsupervised manner with several extracted features. K-means clustering was employed to find out fault data point in the daily PV power outputs obtained from all the sites in the fleet. When the profile was considered as an abnormal pattern, restoration was accomplished by the following imputation step. Imputation schemes were implemented by autoregressive (AR) and multiple regression models with optional neighboring PV data of normal candidates obtained from the previous clustering step. For evaluation, several types of fault patterns observed in actual PV power profiles were simultaneously injected into a single or multiple sites, and proposed schemes were tested without injection information.

The remainder of this paper is organized as follows: Section 2 describes the PV fleet power output relationship between distance and the correlation with actual data measured on campus. Section 3 proposes an efficient fault detection and imputation methods for use with a PV fleet. The simulation setup, including the injected fault pattern, is provided in Section 4. Section 5 details the detection and imputation results, and Section 6 concludes the paper.

2. PV Fleet Power Data Analysis

2.1. Materials

In this study, actual PV fleet data were utilized for analysis and simulation. Hourly power generation data were measured in rooftop solar installations from 13 buildings at the Gwangju Institute of Science and Technology (GIST) campus located in Gwangju, South Korea, as shown in Figure 1, with installation details given in Table 1. In this densely distributed PV fleet, only the facility management building (site 3) collects environmental information, such as ambient temperature, module temperature, slope solar irradiance, and horizontal solar irradiance. The data were collected in 2019 for approximately nine months. To investigate the influence of solar irradiance with respect to the distance between the weather station and PV installation, local environmental data retrieved from the nearest weather station provided by the Korea Meteorological Administration (KMA) website were used [47]. The weather data including local solar irradiance were recorded every hour, the same as for the PV data. We selected 159 days of PV power data from 13 sites without missing values for an overlapping period. The meteorological data were acquired at the site weather station (SWS). In addition, a local weather station (LWS), which was approximately 7 km away from the campus, was subsequently employed for the same time period.

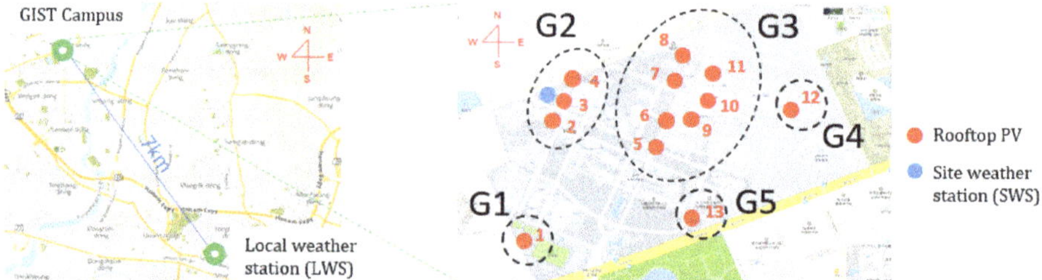

Figure 1. Weather stations and photovoltaic (PV) fleet on the Gwangju Institute of Science and Technology (GIST) campus.

Table 1. Installation information for rooftop PVs in the GIST fleet.

Site Number	Location	Capacity (kW)	Date of Installation
1	Soccer field	158.4	April 2011
2	Student union bldg.1	46.1	April 2011
3	Facility maintenance bldg.	115.9	January2009
4	Central storage	32.6	April 2011
5	Samsung environment science & research bldg.	54.4	March 2019
6	Dasan bldg.	51.8	February 2011
7	Central library	25.0	March 2019
8	LG library	21.8	March 2019
9	Central research facilities	122.9	March 2019
10	Renewable energy research bldg.	46.1	March 2019
11	GIST college bldg.C	21.0	December 2014
12	GIST college dormitory A	70.0	April 2012
13	Laboratory animal resource center	70.0	February 2017

2.2. Cross-Correlation Analysis

Because previous studies have demonstrated that the correlation coefficient decreases when the distance between the weather station and the PV system increases, we investigated the cross-correlation as a function of distance. The cross-correlation is computed using the Pearson's correlation coefficient as follows:

$$\rho_{\mathbf{x,y}} = \frac{\mathrm{cov}(\mathbf{x,y})}{\sigma_{\mathbf{x}}\sigma_{\mathbf{y}}} \tag{1}$$

The covariance of two time-series data $\mathbf{x}$ and $\mathbf{y}$, $\mathrm{cov}(\mathbf{x,y})$ is normalized by the product of their standard deviation $\sigma_{\mathbf{x}}$ and $\sigma_{\mathbf{y}}$. The correlation between each PV site with solar irradiance, which was obtained at SWS and LWS, was analyzed. As expected, the SWS data showed a stronger positive correlation than the LWS data for every site on the campus, as given in Table 2. The correlation analysis was conducted between PV generation data for each site in the same manner. Figure 2 shows the result of the correlation analysis, which produced five groups according to the degree of correlation. Apart from other sites, site 1 belonged to the independent group G1. Sites 2–4 were grouped into group 2 (G2) due to high correlation because of the relatively short distance from SWS. Group 3 (G3) comprised sites 5–11, which had a medium distance from SWS but were clustered together. Groups 4 (G4) and 5 (G5) comprised site 12 and 13, respectively, where one site was far from SWS and other sites. Based on the cross-correlation analysis, we determined that the use of neighbor PV power data can provide additional information that can help perform fault detection and imputation in the fleet of PV systems.

Table 2. Cross-correlation of the PV power output between the local weather station (LWS) and the site weather station (SWS).

Site Number	1	2	3	4	5	6	7	8	9	10	11	12	13
LWS	0.93	0.93	0.93	0.90	0.93	0.93	0.94	0.93	0.92	0.93	0.94	0.85	0.91
SWS	0.96	0.98	0.99	0.97	0.97	0.97	0.98	0.98	0.95	0.97	0.97	0.89	0.93

		1	2	3	4	5	6	7	8	9	10	11	12	13
G1	1	1.00	0.96	0.97	0.95	0.96	0.95	0.96	0.96	0.94	0.96	0.95	0.89	0.94
	2	0.96	1.00	0.99	0.98	0.97	0.96	0.97	0.97	0.95	0.97	0.96	0.90	0.93
G2	3	0.97	0.99	1.00	0.97	0.98	0.97	0.98	0.98	0.95	0.97	0.97	0.89	0.93
	4	0.95	0.98	0.97	1.00	0.95	0.94	0.96	0.96	0.93	0.95	0.94	0.90	0.91
	5	0.96	0.97	0.98	0.95	1.00	0.99	0.98	0.98	0.97	0.99	0.98	0.90	0.94
	6	0.95	0.96	0.97	0.94	0.99	1.00	0.98	0.98	0.97	0.98	0.98	0.89	0.93
G3	7	0.96	0.97	0.98	0.96	0.98	0.98	1.00	0.99	0.97	0.99	0.99	0.90	0.94
	8	0.96	0.97	0.98	0.96	0.98	0.98	0.99	1.00	0.97	0.99	0.99	0.90	0.94
	9	0.94	0.95	0.95	0.93	0.97	0.97	0.97	0.97	1.00	0.97	0.97	0.90	0.91
	10	0.96	0.97	0.97	0.95	0.99	0.98	0.99	0.99	0.97	1.00	0.99	0.91	0.95
	11	0.95	0.96	0.97	0.94	0.98	0.98	0.99	0.99	0.97	0.99	1.00	0.91	0.93
G4	12	0.89	0.90	0.89	0.90	0.90	0.89	0.90	0.90	0.90	0.91	0.91	1.00	0.87
G5	13	0.94	0.93	0.93	0.91	0.94	0.93	0.94	0.94	0.91	0.95	0.93	0.87	1.00

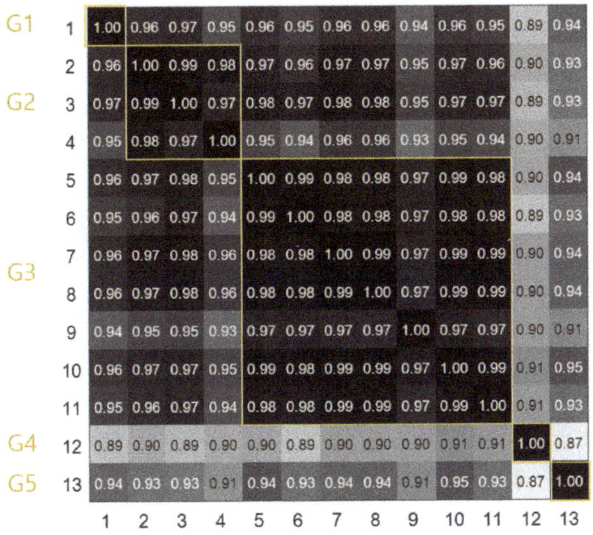

Figure 2. Cross-correlation of PV power output for the fleet.

3. PV Fleet Fault Detection and Imputation

The overall scheme of the proposed system is shown in Figure 3. SWS data and 13 power output data sets from the PV fleet were provided, from which features were extracted to distinguish the daily binary condition of normal or faulty. At the clustering stage, daily states for each PV site were estimated, which is reflected by the final step of imputation.

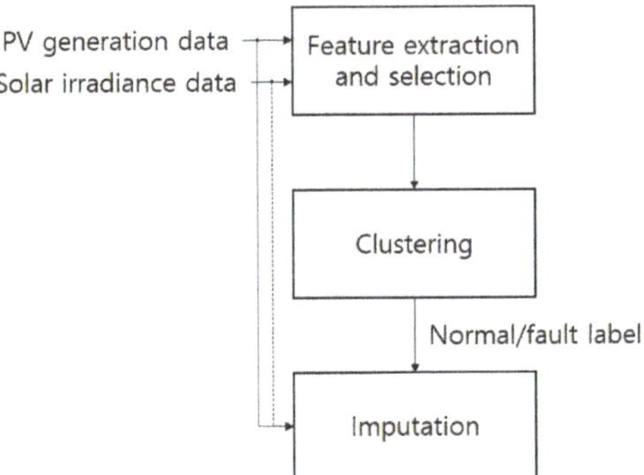

Figure 3. Overall architecture of the proposed system.

3.1. Operation Time

Operation time was considered to extract meaningful properties from the PV pattern accordance with environmental factor. The daylight hours or insolation duration can be obtained by mathematical modeling based on geographical components, such as the latitude of the PV site, using the following equation:

$$G_0 = G_{sc}\left(1 + 0.33\cos\left(\frac{360d}{365}\right)\right)(\cos\delta\,\cos\phi\,\cos\omega + \sin\delta\,\sin\phi) \tag{2}$$

where G_{sc} is the solar constant, which is 1368 [W/m^2], d is the day number in the year, δ is the solar declination, ϕ is the latitude of the site, and ω is the solar hour angle. During an hour after sunrise and an hour before sunset, the small amount of solar irradiance has a tidal effect on PV operation. We defined the operation time, T, by excluding these time periods from the insolation duration [37].

3.2. Feature Extraction

Several features were extracted from the given dataset (e.g., the PV fleet power data $\mathbf{x}$, and the solar irradiance observation data $\mathbf{y}$) to distinguish the fault patterns from normal data at the candidate PV site n. In this study, $\mathbf{y}_{1:N} = \mathbf{y}_3$ as there is one SWS at site 3. We distinguished F_s and F_f, which represent the feature set derived from daily solar irradiance and daily neighboring PVs, respectively. Seven features were introduced in this study for two types of datasets, which generated 14 features. The extracted features of the candidate PV can be written as $F^{(n)} = [F_s^{(n)}, F_f^{(n)}]$.

3.2.1. Total Coefficient of Determination

The coefficient of determination used in this study refers to how well a certain profile can explain the PV power generation profile for each PV site. For simulation comparison, the profile can be either solar irradiance data or neighboring PV generations, as follows:

$$\begin{aligned}
F_{s,1}^{(n)} &= 1 - R^2_{\mathbf{x}_n,\mathbf{y}_n}\\
F_{f,1}^{(n)} &= 1 - \frac{1}{N-1}\sum_k R^2_{\mathbf{x}_n,\mathbf{x}_k}
\end{aligned} \tag{3}$$

where R^2 denotes the coefficients of determination of the regression depending on the other independent variable.

3.2.2. Normalized Profile Distance

The distance between normalized profiles is related to the closeness of the pattern shape. Daily profiles were converted within the range of [0,1] through min–max normalization, and the distance for each time step was computed, as follows:

$$\begin{aligned}
F_{s,2}^{(n)} &= \frac{1}{T_p}\sum_{t\in T}|x_{norm\,n,t} - y_{norm\,n,t}|\\
F_{f,2}^{(n)} &= \frac{1}{N-1}\cdot\frac{1}{T_p}\sum_k\sum_{t\in T}|x_{norm\,n,t} - x_{norm\,k,t}|
\end{aligned} \tag{4}$$

where $x_{norm\,n,t} = \dfrac{x_{n,t}}{\max_{t\in T}\{x_{n,t}\}-\min_{t\in T}\{x_{n,t}\}}$ and $y_{norm\,n,t} = \dfrac{y_{n,t}}{\max_{t\in T}\{y_{n,t}\}-\min_{t\in T}\{y_{n,t}\}}$.

3.2.3. Degree of Consistency

To investigate the fluctuation consistency, the degree of consistency between two profiles was proposed in [48]. Regardless of the heterogeneous dataset, e.g., comparing solar irradiance and

PV generations or different capacities of the PV installations, it can reflect the variation tendency. During the operation time period, f_i, f_o, and f_z count whether the variation tendency is identical, opposite, or zero, respectively. Then, they are converted into F_3, F_4, and F_5 to be computed as a ratio.

$$F_{s,3}^{(n)} = 1 - \frac{1}{T_p} \sum_{t \in T} f_i(x_{n,t}, y_{n,t}) \quad \text{where} \quad f_i(x_{n,t}, y_{n,t}) = 1 \quad \text{if} \quad (x_{n,t} - x_{n,t-1})(y_{n,t} - y_{n,t-1}) > 0$$

$$F_{f,3}^{(n)} = 1 - \frac{1}{N-1} \cdot \frac{1}{T_p} \sum_{k} \sum_{t \in T} f_i(x_{n,t}, x_{k,t}) \quad \text{where} \quad f_i(x_{n,t}, x_{k,t}) = 1 \quad \text{if} \quad (x_{n,t} - x_{n,t-1})(x_{k,t} - x_{k,t-1}) > 0 \tag{5}$$

$$F_{s,4}^{(n)} = \frac{1}{T_p} \sum_{t \in T} f_o(x_{n,t}, y_{n,t}) \quad \text{where} \quad f_o(x_{n,t}, y_{n,t}) = 1 \quad \text{if} \quad (x_{n,t} - x_{n,t-1})(y_{n,t} - y_{n,t-1}) < 0$$

$$F_{f,4}^{(n)} = \frac{1}{N-1} \cdot \frac{1}{T_p} \sum_{k} \sum_{t \in T} f_o(x_{n,t}, x_{k,t}) \quad \text{where} \quad f_o(x_{n,t}, x_{k,t}) = 1 \quad \text{if} \quad (x_{n,t} - x_{n,t-1})(x_{k,t} - x_{k,t-1}) < 0 \tag{6}$$

$$F_{s,5}^{(n)} = \frac{1}{T_p} \sum_{t \in T} f_z(x_{n,t}, y_{n,t}) \quad \text{where} \quad f_z(x_{n,t}, y_{n,t}) = 1 \quad \text{if} \quad (x_{n,t} - x_{n,t-1})(y_{n,t} - y_{n,t-1}) = 0$$

$$F_{f,5}^{(n)} = \frac{1}{N-1} \cdot \frac{1}{T_p} \sum_{k} \sum_{t \in T} f_z(x_{n,t}, x_{k,t}) \quad \text{where} \quad f_z(x_{n,t}, x_{k,t}) = 1 \quad \text{if} \quad (x_{n,t} - x_{n,t-1})(x_{k,t} - x_{k,t-1}) = 0 \tag{7}$$

3.2.4. Relative Error Percentile of the Maximum Value

By comparing the relative maximum property of each profile, the relative error percentiles of the maximum values were determined. Two different attributes, i.e., the first order difference (F_6) and standard value (F_7), were used:

$$F_{s,6}^{(n)} = \frac{x_{n\,dmax} - y_{n\,dmax}}{y_{n\,dmax}}$$

$$F_{f,6}^{(n)} = \frac{1}{N-1} \sum_{k} \frac{x_{n\,dmax} - x_{k\,dmax}}{x_{k\,dmax}} \tag{8}$$

where $x_{n\,dmax} = \max_{t \in T} \left\{ \frac{|x_{n,t} - x_{n,t-1}|}{|x_{n,T} - x_{n,T-1}|} \right\}$ and $y_{n\,dmax} = \max_{t \in T} \left\{ \frac{|y_{n,t} - y_{n,t-1}|}{|y_{n,T} - y_{n,T-1}|} \right\}$.

$$F_{s,7}^{(n)} = \frac{x_{n\,smax} - y_{n\,smax}}{y_{n\,smax}}$$

$$F_{f,7}^{(n)} = \frac{1}{N-1} \sum_{k} \frac{x_{n\,smax} - x_{k\,smax}}{x_{k\,smax}} \tag{9}$$

where $x_{n\,smax} = \max_{t \in T} \left\{ \frac{x_{n,t} - \bar{x}_{n,T}}{\sigma_{x_{n,T}}} \right\}$ and $y_{n\,smax} = \max_{t \in T} \left\{ \frac{y_{n,t} - \bar{y}_{n,T}}{\sigma_{y_{n,T}}} \right\}$.

3.3. Clustering

Because the condition of the PV system is not labeled in the real environment, unsupervised learning was applied in this study. K-means clustering is the simplest method and requires low computation based on calculation of the Euclidean distance. The objective function of K-means clustering, J, is given by Equation (10). Determining the number of clusters and importing the initial center have a crucial impact on the clustering accuracy. In our case, we set the number of clusters as two to distinguish between normal and fault patterns, and they were used in the imputation part for filtering the PV site candidates of the explanatory variables. Initial centroids were set to zero vectors and one vectors multiplied by 0.5 for the normal and fault condition, respectively. Optimal features were selected by exhaustive search for the objective function as follows:

$$J = \sum_{j=1}^{s} \sum_{n=1}^{N} ||Fs^{(n)} - c_j||^2 \tag{10}$$

where s is the number of clusters ($s = 2$, binary condition for the fault or normal state), c_j is the center of cluster j, and $Fs^{(n)}$ are the selected features of site n. Feature selection results are shown in Section 5.

3.4. Imputation

Once the PV profile was labeled as a fault pattern, the fault data were restored using the regression method. In [49], the time series analysis of various meteorological data related to renewable energy systems was conducted using a statistical method. We built a linear regression model based on the linear minimum mean squared error (LMMSE) to estimate the regression parameter, as follows:

$$P = \beta X + \epsilon \tag{11}$$

$$\beta = (X^T X)^{-1} X^T P \tag{12}$$

where P is the expected imputation profile at the candidate site, β is the regression coefficient, X is the set of explanatory datasets, and ϵ is the random error.

In this study, several cases were studied to compare the explanatory dataset for the regression model and to investigate the impact of the labeling of neighboring PV profiles at the previous detection stage, as demonstrated in Table 3. In Case 1, we assumed that only PV historical power output data were available without any other measurements. In this case, we built an AR model for a single PV site to focus on its periodic power generation behavior. Case 2 exclusively considers local irradiance data. For our dataset, solar irradiance data provided at site 3 were used identically to construct the univariate simple regression model. Case 3 allows a neighbor PV profile that has unknown operation states in each PV system, and all of them are inputted into the multivariate regression model. Case 4 conducted a multiple regression model that was similar to Case 3 but the explanatory data were refined with the normal pattern classified at the previous detection stage. Case 5 merged Cases 2 and 4 that utilizes both local irradiance data and normal PV fleet data. Furthermore, as several previous works have confirmed that kNN is an efficient method for missing data imputation, we adopted kNN for fault data imputation as Case 6.

Table 3. Imputation case study.

Case	Description
Case 1	Autoregressive (AR) model that uses PV generation data from a single location
Case 2	Simple regression model that uses SWS solar irradiance data
Case 3	Multiple regression that uses all PV generation data without labeling
Case 4	Multiple regression model that uses normally labeled PV generation data
Case 5	Multiple regression model that uses normally labeled PV generation data and SWS solar irradiance data
Case 6	kNN

4. Simulations

4.1. Fault Pattern Injection

To evaluate the performance, we primarily sorted 159 days of PV generation patterns for all sites for the normal condition simultaneously. The overall normal dataset comprised $159 \times 13 = 2067$ PV power profiles. To characterize the fault patterns, the actual fault patterns detected from different PV plants were investigated. Six fault patterns were extracted, which were composed of whole zero, part zero, whole shift, part shift, constant padding, and spike, as shown in Figure 4. The faults were injected arbitrarily into a single or multiple power profiles of PV sites, thus retaining target labels to describe the performances of fault detection and imputation.

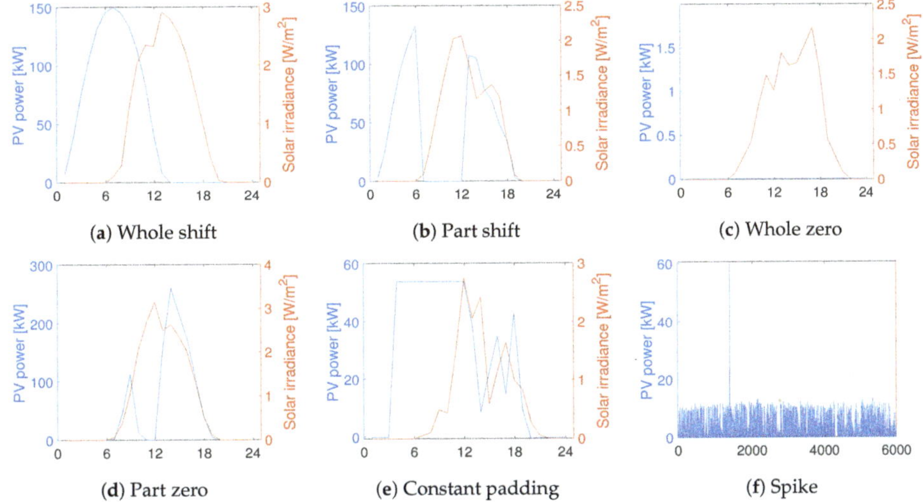

Figure 4. Fault pattern for the actual observation.

4.2. Evaluation Metric

4.2.1. Fault Detection Performance Metric

A confusion matrix is widely used to evaluate classification performance, as depicted in Table 4. It matches the output label and target label for testing samples one by one; thus, target output label sets are sorted as true positive (*TP*), false negative (*FN*), false positive (*FP*), and true negative (*TN*). Based on these attributes, accuracy is derived as the ratio of correctly classified samples among all samples. The error rate is the opposite of accuracy, which indicates the error proportion for all samples. Precision is the ratio of *TPs* for all positives that are categorized in the test phase. Lastly, recall indicates the ratio of *TPs* out of all pre-labeled positives, as follows:

Table 4. Confusion matrix.

		Output		
		Positive	Negative	Total
Target	Positive	*TP*	*FN*	*P*
	Negative	*FP*	*TN*	*N*
	Total	*P'*	*N'*	

$$Accuracy = \frac{TP + TN}{P + N}$$
$$Errorrate = \frac{FP + FN}{P + N}$$
$$Precision = \frac{TP}{TP + FP} = \frac{TP}{P'}$$
$$Recall = \frac{TP}{TP + FN} = \frac{TP}{P} \tag{13}$$

4.2.2. Imputation Performance Metric

To evaluate the accuracy of imputation performance, normalized root mean square error (NRMSE) is adopted since the PV capacities in the fleet are different. In this study, RMSE was normalized by mean as described in Equations (14) and (15) Low NRMSE implies valuable imputation performance where abnormal pattern is restored similar to the real profile. The daily abnormal pattern for a corresponding site is then substituted by the reconstructed normal pattern.

$$RMSE = \sqrt{\frac{\sum_{t \in T} (x_t - \hat{x}_t)^2}{T_p}} \ [kW] \tag{14}$$

$$NRMSE = \frac{RMSE}{\overline{\mathbf{x}}_{t \in T}} \tag{15}$$

where $\overline{\mathbf{x}}_{t \in T}$ is the mean of $\mathbf{x}$ during operation hours.

5. Results and Discussion

5.1. Feature Selection

The feature characterized in Section 3.2 can be assembled in various combinations. To determine the optimal parameter combination, an exhaustive search was followed by division into two scenarios: In one, combinations from F_s were selected and the other involved F_f. Tables 5 and 6 list the detection performance for all datasets that included 2067 feature sets and labels with and without neighboring PV data, respectively. Accuracy, error rate, precision, and recall for all the combinations were surveyed. We selected high accuracy with a similar portion of false alarm and missing fault to increase detection accuracy and prevent inclination to one side. Features 1, 2, and 5 were selected from F_s in the first scenario. However, the combination of features 1, 5, and 7 from F_s and 1, 2, 3, and 5 from F_f were selected for the second scenario to detect the fault pattern.

Table 5. Feature selection from F_s.

Feature Combination	Accuracy	Error Rate	Precision	Recall
$F_{s,1}, F_{s,2}, F_{s,3}, F_{s,4}, F_{s,5}, F_{s,6}, F_{s,7}$	0.9768	0.0232	0.9791	0.9964
$F_{s,1}, F_{s,2}, F_{s,3}, F_{s,5}, F_{s,6}, F_{s,7}$	0.9768	0.0232	0.9791	0.9964
$F_{s,1}, F_{s,2}, F_{s,4}, F_{s,5}, F_{s,7}$	0.9773	0.0227	0.9901	0.9855
$F_{s,1}, F_{s,2}, F_{s,3}, F_{s,5}$	0.9777	0.0223	0.9821	0.9943
$F_{s,1}, F_{s,2}, F_{s,5}$	0.9787	0.0213	0.9861	0.9912
$F_{s,1}, F_{s,5}$	0.9773	0.0227	0.9860	0.9896
$F_{s,2}$	0.9681	0.0319	0.9683	0.9984

Table 6. Feature selection from the combination of F_s and F_f.

Feature Combination	Accuracy	Error Rate	Precision	Recall
$F_{s,1}, F_{s,2}, F_{s,3}, F_{s,4}, F_{s,5}, F_{s,6}, F_{s,7}, F_{f,1}, F_{f,2}, F_{f,3}, F_{f,4}, F_{f,5}, F_{f,7}$	0.9836	0.0164	0.9861	0.9964
$F_{s,1}, F_{s,2}, F_{s,3}, F_{s,4}, F_{s,5}, F_{s,7}, F_{f,1}, F_{f,2}, F_{f,3}, F_{f,4}, F_{f,5}, F_{f,7}$	0.9840	0.0160	0.9907	0.9922
$F_{s,1}, F_{s,2}, F_{s,3}, F_{s,4}, F_{s,5}, F_{s,7}, F_{f,1}, F_{f,2}, F_{f,3}, F_{f,4}, F_{f,5}$	0.9860	0.0140	0.9877	0.9974
$F_{s,1}, F_{s,2}, F_{s,3}, F_{s,4}, F_{s,5}, F_{s,7}, F_{f,1}, F_{f,2}, F_{f,3}, F_{f,5}$	0.9874	0.0126	0.9892	0.9974
$F_{s,1}, F_{s,2}, F_{s,3}, F_{s,5}, F_{s,7}, F_{f,1}, F_{f,2}, F_{f,3}, F_{f,5}$	0.9884	0.0116	0.9897	0.9979
$F_{s,1}, F_{s,2}, F_{s,3}, F_{s,5}, F_{s,7}, F_{f,1}, F_{f,3}, F_{f,5}$	0.9869	0.0131	0.9887	0.9974
$F_{s,1}, F_{s,5}, F_{s,7}, F_{f,1}, F_{f,2}, F_{f,3}, F_{f,5}$	0.9874	0.0126	0.9933	0.9933
$F_{s,1}, F_{s,3}, F_{f,5}, F_{f,7}, F_{f,2}, F_{f,5}$	0.9874	0.0126	0.9897	0.9969
$F_{s,1}, F_{s,3}, F_{s,5}, F_{s,7}, F_{f,5}$	0.9860	0.0140	0.9882	0.9969
$F_{s,5}, F_{s,7}, F_{f,1}, F_{f,5}$	0.9869	0.0131	0.9912	0.9948
$F_{s,2}, F_{f,1}, F_{f,5}$	0.9855	0.0145	0.9852	0.9995
$F_{s,5}, F_{f,1}$	0.9850	0.0150	0.9847	0.9995
$F_{f,2}$	0.9700	0.0300	0.9712	0.9974

5.2. Analysis of Fault Detection Results

To validate the proposed clustering-based method compared with previous approaches, SolarClique [50] was implemented, and the obtained results were evaluated. To give a brief information of SolarClique, it detects anomalies on the basis of prediction by employing solar generation data from geographically nearby sites. Each candidate site is predicted for 100 iterations and implemented by random forest with bootstrapping. Then, the local factor is excluded from prediction error and used to detect anomalies. The data are categorized into fault when the prediction error is outside the threshold, i.e., 4-sigma.

By performing numerical evaluation, the overall results of fault detection are indicated in Table 7. The fault identification by the proposed method with and without PV fleet data show accuracies of 0.9874 and 0.9777, respectively, which are more accurate than accuracy of 0.9247 obtained by the SolarClique technique. In the same context, the detection error rates derived by clustering with and without PVs are 0.0126 and 0.0223 which are smaller than 0.0753 of error rate obtained by SolarClique. Precision and recall for the proposed method are also improved, compared with those of SolarClique. This is because the prediction-based approach is sensitive to random fluctuations that depend on subtle unknown changes; the detection depends only on a threshold applied to the univariate component. As a result, clustering-based fault detection using neighboring PVs shows the best detection performance.

Table 7. Fault detection comparison with SolarClique.

	Accuracy	Error Rate	Precision	Recall
SolarClique	0.9247	0.0753	0.9481	0.9707
Proposed (w/o PV)	0.9777	0.0223	0.9821	0.9943
Proposed (w/ PV)	0.9874	0.0126	0.9933	0.9933

Fault detection performances for each site are provided in Table 8. Fault detection tested for all sites in a campus fleet showed a high accuracy rate that exceeded 98%, except for site 12. For interpretation, normal profiles and injected profiles that are distinguished as a fault are depicted in Figure 5. The green-shaded profile shown in Figure 5c indicates the missing fault that the clustering machine identified from an abnormal condition as a normal status. In contrast, the yellow-shaded area in Figure 5b represents false positives or false alarms with wrong identification despite the normal condition. However, the normal condition is not guaranteed in this case in that the false alarm profile seemed to spike, which showed a disparate profile with irradiance compared with Figure 5a. Hence, these false positives should be reassigned as true negatives to provide a more accurate detection rate.

Table 8. Fault detection of individual sites.

Site Number	Accuracy	Error Rate	Precision	Recall
1	0.9874	0.0126	0.9932	0.9932
2	0.9874	0.0126	0.9867	1.0000
3	0.9874	0.0126	0.9868	1.0000
4	0.9874	0.0126	0.9932	0.9932
5	0.9937	0.0063	0.9933	1.0000
6	0.9937	0.0063	1.0000	0.9934
7	0.9937	0.0063	0.9933	1.0000
8	0.9937	0.0063	0.9933	1.0000
9	0.9937	0.0063	0.9932	1.0000
10	0.9937	0.0063	0.9933	1.0000
11	0.9937	0.0063	1.0000	0.9932
12	0.9434	0.0566	0.9929	0.9456
13	0.9874	0.0126	0.9933	0.9933
Overall	0.9874	0.0126	0.9933	0.9933

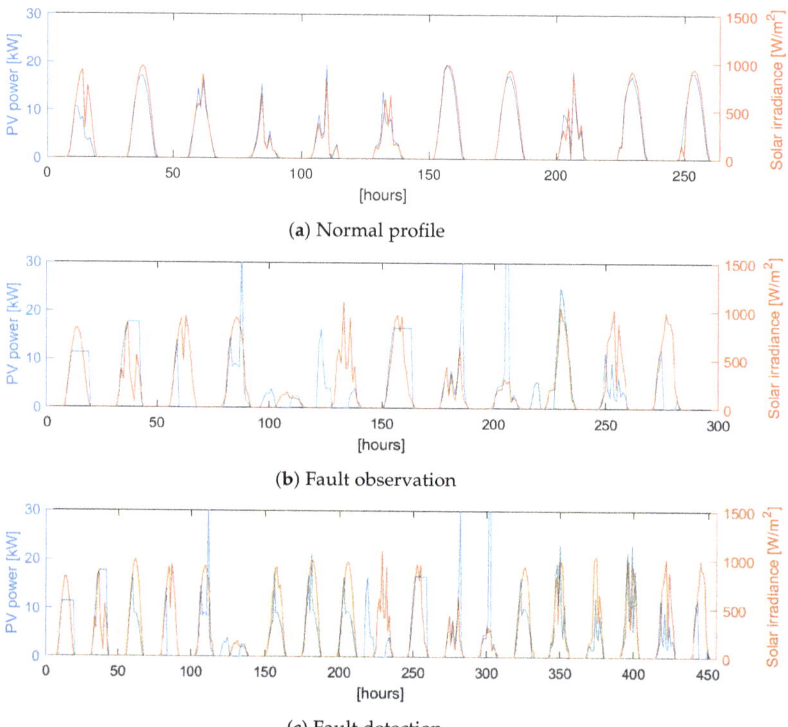

(a) Normal profile

(b) Fault observation

(c) Fault detection

Figure 5. Profiles of PV power and solar irradiance for (**a**) normal, (**b**) fault injected, and (**c**) fault detected days.The green-shaded profile in (**b**) denotes a missing fault and the yellow-shaded profiles in (**c**) correspond to false alarms occurring at site 12.

5.3. Analysis of Fault Imputation Result

The imputation accuracy computed with the NRMSE metric is presented in Table 9. As generally expected, most of the sites in the PV fleet show improved imputation performance when using each other self-labeled PV data, corresponding to Cases 4 or 5 as their overall NRMSEs are the lowest at 0.2359 and 0.2367, respectively. Comparing to Case 2 (only SWS data utilized), where overall imputation accuracy is 0.2925, the improvement is shown to be 19.35% and 19.08%, respectively. Likewise compared to Case 3 (PV without label) where NRMSE is 0.2888, Cases 4 and 5 have been improved by 18.32% and 18.04%, respectively. In addition, as depicted in Figure 6, Cases 4 and 5 are shown to have slightly tighter deviations in comparison with the other cases, which means more stable. In the following we present the details for each imputation case.

Case 1 supposed that only historical PV data from a certain site is applicable, which means that information is restricted when considering weather variations. Therefore, an AR model was constructed in this case that focused only on the periodicity of the operation time without reflecting climatic properties. The imputation result showed insufficient performance compared to the other case studies for all sites in the fleet. Case 2 was a general case that applied the nearest meteorological data. The weather stations were lacking in the fleet, which resulted in representative station data being used after all. We concluded that when the PV site was close to the weather station, Case 2 showed better performance than when using surrounding PV data. For example, the imputation results from sites 3 and 4 that belonged to G2 were the best in this case. Case 3 used additional data obtained from neighboring PV systems beyond the labels of the system conditions. Because abnormal patterns were not removed in the previous stage, they were fed into the imputation phase as provided. This case was feasible only when the assumption that faults do not occur was satisfied. Case 4 was used to examine the effectiveness of a neighbor PV generation profile with labeled status. Because normal profiles were selected as candidates of explanatory data for the regression model, they showed reliable imputation results. In particular, PV sites from G4 showed improved results to those used in Case 2. This was because, even though the groups were slightly away from the weather station, they were crowded collectively, which provided more relevant information on the PV power output than distant weather data. As mentioned earlier, Case 5 merged Cases 2 and 4. It is observed that only sites in close distance to SWS have effects on combining SWS data to neighbor PVs. Otherwise, it did not appear to have a significant effect in faraway PVs despite of belonging to PV fleet. This demonstrates that even neighboring PVs in the same fleet have subtle differences in solar irradiance. The kNN in Case 6 shows low performance in PV fleet fault imputation. Since it depends on similar historical patterns of solar irradiance, it may not reflect random fluctuation especially in overcasting day which results in low imputation accuracy.

Table 9. Imputation normalized root mean square error (NRMSE).

Site Number	1	2	3	4	5	6	7	8	9	10	11	12	13	Overall
Case 1	0.88	0.74	0.95	1.12	0.63	0.82	0.82	0.85	0.90	1.37	0.89	0.92	0.98	0.9156
Case 2	0.29	0.38	0.12	0.30	0.18	0.29	0.14	0.18	0.38	0.24	0.21	0.55	0.48	0.2925
Case 3	0.33	0.27	0.23	0.58	0.10	0.38	0.14	0.19	0.33	0.22	0.19	0.34	0.49	0.2888
Case 4	0.28	0.35	0.14	0.33	0.08	0.16	0.08	0.13	0.29	0.16	0.17	0.36	0.52	0.2359
Case 5	0.28	0.36	0.13	0.32	0.08	0.17	0.08	0.13	0.30	0.18	0.17	0.35	0.53	0.2367
Case 6	0.45	0.62	0.49	0.63	0.34	0.46	0.36	0.35	0.62	0.52	0.43	0.54	0.60	0.4930

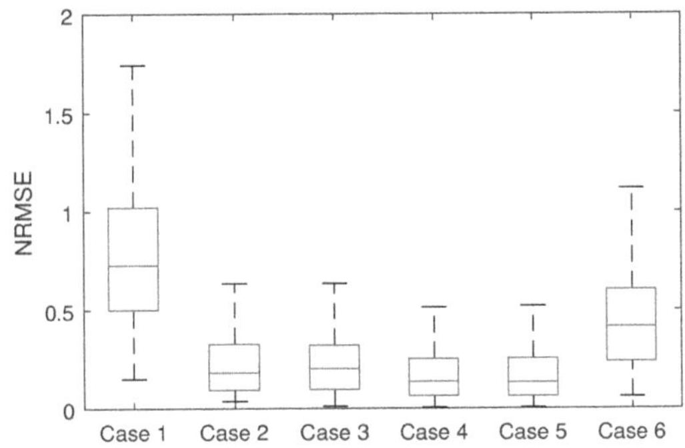

Figure 6. Boxplot of imputation NRMSE for each imputation case.

6. Conclusions

This paper presents a framework for PV fault detection and imputation method on PV fleets without the manual annotation of the state of the PV systems. We supplement the meteorological data measured at LWS, which had a relatively low value for the cross-correlation with PV generation, using the neighboring PV fleet data. Several features were derived to be used as input for K-means clustering to label normal or abnormal patterns. PV fleet on the campus and solar irradiance data measured at one of the PV sites in the fleet were utilized to extract the features for fault pattern detection. We arbitrarily injected a fault pattern based on actual observations, and the detection accuracy was evaluated using a confusion matrix. The detection error rate was compared for three cases: Using SolarClique (a conventional prediction-based detection method), a clustering-based method without PV fleet data, and a clustering-based method with PV fleet data which is the proposed method. The error rates for these three cases were 0.0753, 0.0223, and 0.0126, which means that the proposed clustering-based detection using neighboring PV fleet data can effectively detect the faults.

Data imputation was conducted for the distinguished abnormal patterns. Five cases of regression-based imputation and kNN were evaluated by NRMSE. In general, Cases 4 and 5, which utilized neighboring self-labeled PV data, showed better imputation performance than imputation without nearby sites or without labeled PV data by reducing NRMSE over 19% and 18%, respectively. In addition, according to earlier grouping information based on cross-correlation analysis, G3, which were close SWS, generally showed better performance when only solar irradiance data were used. However, the imputation result for G3 sites and neighboring PV profiles provided more relevant information than weather data obtained at SWS that was relatively far away. In summary, neighboring PV data are effective in improving fault detection and imputation accuracy in a dense PV fleet.

Author Contributions: All authors contributed to this work by collaboration. S.P. (Sunme Park) developed the idea and methodologies and is the first author in this manuscript. S.P. (Soyeong Park) reviewed previous literature and conducted data analysis for implementation, M.K. conducted detection performance comparisons and E.H. led and supervised the research and is the corresponding author. All authors have read and agreed to the published version of the manuscript.

Funding: This research was funded by the ETRI R&D program (19ZK1140) and the government of Korea and by the Korea Institute of Energy Technology Evaluation and Planning (KETEP).

Acknowledgments: This work was supported by the ETRI R&D program (19ZK1140), funded by the government of Korea and by the Korea Institute of Energy Technology Evaluation and Planning (KETEP) and the Ministry of Trade, Industry and Energy (MOTIE) of the Republic of Korea (No. 20171210200810).

Conflicts of Interest: The authors declare no conflict of interest.

Abbreviations

The following abbreviations are used in this manuscript:

F_s	Feature associated with candidate PV and solar irradiance
F_f	Feature associated with candidate PV and neighboring PVs
N	Total number of PVs in PV fleet
n	ID number of candidate PV
k	ID number of neighboring PVs i.e., $k \in \{1, 2, \cdots, n-1, n+1, \cdots, N\}$
t	Hourly time index
T	Set of PV operation time indexes
T_p	Duration of T
$\mathbf{x}$	Daily profile of PV
$\mathbf{y}$	Daily profile of solar irradiance

References

1. Mellit, A.; Tina, G.M.; Kalogirou, S.A. Fault detection and diagnosis methods for photovoltaic systems: A review. *Renew. Sustain. Energy Rev.* **2018**, *91*, 1–17. [CrossRef]

2. Pvps, I. Strategic PV Analysis and Outreach. 2019. Available online: http://www.iea-pvps.org/fileadmin/dam/public/report/statistics/IEA-PVPS_T1_35_Snapshot2019-Report.pdf (accessed on 27 November 2019).

3. Murdock, H.E.; Gibb, D.; André, T.; Appavou, F.; Brown, A.; Epp, B.; Kondev, B.; McCrone, A.; Musolino, E.; Ranalder, L.; et al. *Renewables 2019 Global Status Report*; UNEP: Nairobi, Kenya, 2019.

4. Firth, S.K.; Lomas, K.J.; Rees, S.J. A simple model of PV system performance and its use in fault detection. *Solar Energy* **2010**, *84*, 624–635. [CrossRef]

5. Daliento, S.; Chouder, A.; Guerriero, P.; Pavan, A.M.; Mellit, A.; Moeini, R.; Tricoli, P. Monitoring, diagnosis, and power forecasting for photovoltaic fields: A review. *Int. J. Photoenergy* **2017**, *2017*. [CrossRef]

6. Yoon, S.; Hwang, E. Load Guided Signal-Based Two-Stage Charging Coordination of Plug-In Electric Vehicles for Smart Buildings. *IEEE Access* **2019**, *7*, 144548–144560. [CrossRef]

7. Park, K.; Yoon, S.; Hwang, E. Hybrid load forecasting for mixed-use complex based on the characteristic load decomposition by pilot signals. *IEEE Access* **2019**, *7*, 12297–12306. [CrossRef]

8. McCandless, T.C.; Haupt, S.E.; Young, G.S. The effects of imputing missing data on ensemble temperature forecasts. *J. Comput.* **2011**, *6*, 162–171. [CrossRef]

9. McCandless, T.; Haupt, S.; Young, G.S. A model tree approach to forecasting solar irradiance variability. *Solar Energy* **2015**, *120*, 514–524. [CrossRef]

10. Hoff, T.E.; Perez, R. Quantifying PV power output variability. *Solar Energy* **2010**, *84*, 1782–1793. [CrossRef]

11. Perez, R.; Kivalov, S.; Schlemmer, J.; Hemker Jr, K.; Hoff, T.E. Short-term irradiance variability: Preliminary estimation of station pair correlation as a function of distance. *Solar Energy* **2012**, *86*, 2170–2176. [CrossRef]

12. Hoff, T.E.; Perez, R. PV power output variability: Correlation coefficients. In *Research Report of Clean Power Research*; Clean Power Research: Napa, CA, USA, 2010.

13. Widén, J. A model of spatially integrated solar irradiance variability based on logarithmic station-pair correlations. *Solar Energy* **2015**, *122*, 1409–1424. [CrossRef]

14. Alsafasfeh, M.; Abdel-Qader, I.; Bazuin, B.; Alsafasfeh, Q.; Su, W. Unsupervised Fault Detection and Analysis for Large Photovoltaic Systems Using Drones and Machine Vision. *Energies* **2018**, *11*, 2252. [CrossRef]

15. Chine, W.; Mellit, A.; Lughi, V.; Malek, A.; Sulligoi, G.; Pavan, A.M. A novel fault diagnosis technique for photovoltaic systems based on artificial neural networks. *Renew. Energy* **2016**, *90*, 501–512. [CrossRef]

16. Garoudja, E.; Harrou, F.; Sun, Y.; Kara, K.; Chouder, A.; Silvestre, S. Statistical fault detection in photovoltaic systems. *Solar Energy* **2017**, *150*, 485–499. [CrossRef]

17. Espinoza Trejo, D.; Bárcenas, E.; Hernández Díez, J.; Bossio, G.; Espinosa Pérez, G. Open-and short-circuit fault identification for a boost dc/dc converter in PV MPPT systems. *Energies* **2018**, *11*, 616. [CrossRef]

18. Islam, H.; Mekhilef, S.; Shah, N.B.M.; Soon, T.K.; Seyedmahmousian, M.; Horan, B.; Stojcevski, A. Performance evaluation of maximum power point tracking approaches and photovoltaic systems. *Energies* **2018**, *11*, 365. [CrossRef]

19. Chen, Z.; Han, F.; Wu, L.; Yu, J.; Cheng, S.; Lin, P.; Chen, H. Random forest based intelligent fault diagnosis for PV arrays using array voltage and string currents. *Energy Convers. Manag.* **2018**, *178*, 250–264. [CrossRef]

20. Madeti, S.R.; Singh, S. Modeling of PV system based on experimental data for fault detection using kNN method. *Solar Energy* **2018**, *173*, 139–151. [CrossRef]

21. Harrou, F.; Sun, Y.; Taghezouit, B.; Saidi, A.; Hamlati, M.E. Reliable fault detection and diagnosis of photovoltaic systems based on statistical monitoring approaches. *Renew. Energy* **2018**, *116*, 22–37. [CrossRef]

22. Platon, R.; Martel, J.; Woodruff, N.; Chau, T.Y. Online fault detection in PV systems. *IEEE Trans. Sustain. Energy* **2015**, *6*, 1200–1207. [CrossRef]

23. Zhao, Y.; Lehman, B.; Ball, R.; Mosesian, J.; de Palma, J.F. Outlier detection rules for fault detection in solar photovoltaic arrays. In Proceedings of the 2013 Twenty-Eighth Annual IEEE Applied Power Electronics Conference and Exposition (APEC), Long Beach, CA, USA, 17–21 March, 2013; pp. 2913–2920.

24. Alam, M.; Muttaqi, K.M.; Sutanto, D. A SAX-based advanced computational tool for assessment of clustered rooftop solar PV impacts on LV and MV networks in smart grid. *IEEE Trans. Smart Grid* **2013**, *4*, 577–585. [CrossRef]

25. Mekki, H.; Mellit, A.; Salhi, H. Artificial neural network-based modelling and fault detection of partial shaded photovoltaic modules. *Simul. Model. Pract. Theory* **2016**, *67*, 1–13. [CrossRef]

26. Pérez-Ortiz, M.; Jiménez-Fernández, S.; Gutiérrez, P.A.; Alexandre, E.; Hervás-Martínez, C.; Salcedo-Sanz, S. A review of classification problems and algorithms in renewable energy applications. *Energies* **2016**, *9*, 607. [CrossRef]

27. Jones, C.B.; Stein, J.S.; Gonzalez, S.; King, B.H. Photovoltaic system fault detection and diagnostics using Laterally Primed Adaptive Resonance Theory neural network. In Proceedings of the 2015 IEEE 42nd Photovoltaic Specialist Conference (PVSC), New Orleans, LA, USA, 14–19 June 2015; pp. 1–6.

28. Jazayeri, K.; Jazayeri, M.; Uysal, S. Artificial neural network-based all-sky power estimation and fault detection in photovoltaic modules. *J. Photonics Energy* **2017**, *7*, 025501. [CrossRef]

29. Dhimish, M.; Holmes, V.; Mehrdadi, B.; Dales, M. Comparing Mamdani Sugeno fuzzy logic and RBF ANN network for PV fault detection. *Renew. Energy* **2018**, *117*, 257–274. [CrossRef]

30. Chouay, Y.; Ouassaid, M. An intelligent method for fault diagnosis in photovoltaic systems. In Proceedings of the 2017 International Conference on Electrical and Information Technologies (ICEIT), Rabat, Morocco, 15–18 November 2017; pp. 1–5.

31. Jiang, L.L.; Maskell, D.L. Automatic fault detection and diagnosis for photovoltaic systems using combined artificial neural network and analytical based methods. In Proceedings of the 2015 International Joint Conference on Neural Networks (IJCNN), Killarney, Ireland, 12–17 July 2015; pp. 1–8.

32. Garoudja, E.; Chouder, A.; Kara, K.; Silvestre, S. An enhanced machine learning based approach for failures detection and diagnosis of PV systems. *Energy Convers. Manag.* **2017**, *151*, 496–513. [CrossRef]

33. Zhu, H.; Lu, L.; Yao, J.; Dai, S.; Hu, Y. Fault diagnosis approach for photovoltaic arrays based on unsupervised sample clustering and probabilistic neural network model. *Solar Energy* **2018**, *176*, 395–405. [CrossRef]

34. Akram, M.N.; Lotfifard, S. Modeling and health monitoring of DC side of photovoltaic array. *IEEE Trans. Sustain. Energy* **2015**, *6*, 1245–1253. [CrossRef]

35. Wang, Z.; Balog, R.S. Arc fault and flash detection in photovoltaic systems using wavelet transform and support vector machines. In Proceedings of the 2016 IEEE 43rd Photovoltaic Specialists Conference (PVSC), Portland, OR, USA, 5–10 June 2016; pp. 3275–3280.

36. Yi, Z.; Etemadi, A.H. A novel detection algorithm for line-to-line faults in photovoltaic (PV) arrays based on support vector machine (SVM). In Proceedings of the 2016 IEEE Power and Energy Society General Meeting (PESGM), Boston, MA, USA, 17–21 July 2016; pp. 1–4.

37. Jufri, F.H.; Oh, S.; Jung, J. Development of Photovoltaic abnormal condition detection system using combined regression and Support Vector Machine. *Energy* **2019**, *176*, 457–467. [CrossRef]

38. Zhao, Y.; Yang, L.; Lehman, B.; de Palma, J.F.; Mosesian, J.; Lyons, R. Decision tree-based fault detection and classification in solar photovoltaic arrays. In Proceedings of the 2012 Twenty-Seventh Annual IEEE Applied Power Electronics Conference and Exposition (APEC), Orlando, FL, USA, 5–9 February 2012; pp. 93–99.

39. Belaout, A.; Krim, F.; Mellit, A.; Talbi, B.; Arabi, A. Multiclass adaptive neuro-fuzzy classifier and feature selection techniques for photovoltaic array fault detection and classification. *Renew. Energy* **2018**, *127*, 548–558. [CrossRef]

40. Chen, Z.; Wu, L.; Cheng, S.; Lin, P.; Wu, Y.; Lin, W. Intelligent fault diagnosis of photovoltaic arrays based on optimized kernel extreme learning machine and IV characteristics. *Appl. Energy* **2017**, *204*, 912–931. [CrossRef]

41. Liao, Z.; Wang, D.; Tang, L.; Ren, J.; Liu, Z. A heuristic diagnostic method for a PV system: Triple-layered particle swarm optimization–back-propagation neural network. *Energies* **2017**, *10*, 226. [CrossRef]

42. Zhao, Y.; Ball, R.; Mosesian, J.; de Palma, J.F.; Lehman, B. Graph-based semi-supervised learning for fault detection and classification in solar photovoltaic arrays. *IEEE Trans. Power Electron.* **2014**, *30*, 2848–2858. [CrossRef]

43. Zhao, Q.; Shao, S.; Lu, L.; Liu, X.; Zhu, H. A new PV array fault diagnosis method using fuzzy C-mean clustering and fuzzy membership algorithm. *Energies* **2018**, *11*, 238. [CrossRef]

44. Lin, P.; Lin, Y.; Chen, Z.; Wu, L.; Chen, L.; Cheng, S. A density peak-based clustering approach for fault diagnosis of photovoltaic arrays. *Int. J. Photoenergy* **2017**, *2017*. [CrossRef]

45. Ding, H.; Ding, K.; Zhang, J.; Wang, Y.; Gao, L.; Li, Y.; Chen, F.; Shao, Z.; Lai, W. Local outlier factor-based fault detection and evaluation of photovoltaic system. *Solar Energy* **2018**, *164*, 139–148. [CrossRef]

46. Harrou, F.; Dairi, A.; Taghezouit, B.; Sun, Y. An unsupervised monitoring procedure for detecting anomalies in photovoltaic systems using a one-class Support Vector Machine. *Solar Energy* **2019**, *179*, 48–58. [CrossRef]

47. Korea Weather Information (Korean). Available online: https://data.kma.go.kr/cmmn/main.do (accessed on 27 November 2019).

48. Wang, F.; Zhen, Z.; Mi, Z.; Sun, H.; Su, S.; Yang, G. Solar irradiance feature extraction and support vector machines based weather status pattern recognition model for short-term photovoltaic power forecasting. *Energy Build.* **2015**, *86*, 427–438. [CrossRef]

49. Shams, M.B.; Haji, S.; Salman, A.; Abdali, H.; Alsaffar, A. Time series analysis of Bahrain's first hybrid renewable energy system. *Energy* **2016**, *103*, 1–15. [CrossRef]

50. Iyengar, S.; Lee, S.; Sheldon, D.; Shenoy, P. Solarclique: Detecting anomalies in residential solar arrays. In Proceedings of the 1st ACM SIGCAS Conference on Computing and Sustainable Societies, San Jose, CA, USA, 20–22 June 2018; p. 38.

MDPI
St. Alban-Anlage 66
4052 Basel
Switzerland
Tel. +41 61 683 77 34
Fax +41 61 302 89 18
www.mdpi.com

Energies Editorial Office
E-mail: energies@mdpi.com
www.mdpi.com/journal/energies